Storey's Guide to
TRAINING
HORSES

Heather Smith Thomas

Storey Publishing

The mission of Storey Publishing is to serve our customers by publishing practical information that encourages personal independence in harmony with the environment.

Edited by Deborah Burns and Marie A. Salter
Copyedited by Doris Troy
Art direction and cover design by Meredith Maker
Text production by Kelley Nesbit
Cover photographs by Giles Prett; photo on facing page by Heather Smith Thomas
Illustrations by Jo Anna Rissanen, except those on page 5 by Brigita Fuhrmann, pages 8 and 280 by Jim Dyekman, and pages 470 and 471 by N.J. Wiley
Indexed by Susan Olason/Indexes & Knowledge Maps

Printed in the United States by Malloy Printing
10 9 8

Library of Congress Cataloging-in-Publication Data

Thomas, Heather Smith, 1944–
　Storey's guide to training horses / Heather Smith Thomas.
　　p. cm.
　ISBN-13: 978-1-58017-467-1 (pbk. : alk. paper)
　ISBN-10: 1-58017-467-1 (pbk. : alk. paper)
　ISBN-13: 978-1-58017-468-8 (hardcover : alk. paper)
　ISBN-10: 1-58017-468-X(hardcover : alk. paper)
　　　　1. Horses—Training. I. Title: Guide to training horses. II. Title.
SF287 .T46 2003
636.1'0835—dc21

TO MY DAUGHTER, ANDREA

Your early love of horses and riding was a joy and a help to me as we rode on the range and worked our cattle. As soon as you were big enough to get on a horse by yourself, you began helping train our ranch horses.

I wanted someday to dedicate one of my books to you — if I ever wrote one on horse training — because you were my training partner with so many young horses. Then in July 2000, you nearly lost your life in a terrible fire. I feared that our wonderful days of riding and training together were over. But you hung on and eventually fought your way back to physical fitness. It's a long, hard road for a person recovering from severe burns, and some things are never the same again, but now, three years later, you are riding and helping me train horses again.

I am very glad to have this opportunity to dedicate a book to you, and so thankful that you are here and able to read it. I am also grateful for your help, in looking over the manuscript, just as I have been thankful for all your help, during the past twenty-two years, in training our horses.

CONTENTS

PREFACE

Before you begin to train a horse, familiarize yourself with the basics of good horsemanship and be aware of general safety practices. You should also be adept at handling horses and be a good rider. The first two chapters of this book are an overview of the fundamentals of good horsemanship and safety and a discussion of horse psychology — essentials a trainer must know to properly handle and understand an equine pupil.

We then cover training the horse, starting when he is a baby, step-by-step through his growing years. You will learn the things he should be taught as he grows and as you start his education as a riding or driving horse. Finally, we will discuss further schooling and retraining a spoiled horse.

You can use this book as a basic training manual for starting any horse — the child's riding or driving pony; horses for Western pleasure or trail riding; horses for English riding, jumping, or any other horse sport or competition. You'll also get advice on how to handle and correct an older horse with bad habits.

It's always nice to start with a young, unspoiled horse, but sometimes you acquire a horse that's already spoiled in some way or another and you must figure out ways to correct his bad habits. It's also important that you do not spoil a young horse as you train him, or create more bad manners or problems in an older horse. This book can help you avoid common mistakes that novice horse owners often make.

Many methods of handling and training horses are discussed, and some may seem contradictory. There are many ways to train a horse or to deal with a certain problem; some work well for some horses but not for others, and sometimes you must resort to something completely different.

As trainers, we don't always have ideal conditions or the ideal horse. You don't always have the opportunity to imprint a newborn foal or a chance to trailer-train a young horse before you have to haul him somewhere. You may acquire a yearling or two-year-old that had no early training and have to start "kindergarten" lessons with a horse that is bigger and stronger than you.

To help you, this book will cover the basics and alternatives, and you may occasionally have to be creative on your own. The important thing is to be in tune with your pupil, constantly evaluating what is best for him at that particular phase of his training, in that particular lesson. If you always put the horse first — choosing methods that will work best for him and working with him at his own pace — you will do a good job. Your ultimate goal is to see how well you can train your horse, not how far you can progress within a certain time frame. Be patient, and progress at a speed that's right for him.

Horses are often our best teachers. We shouldn't force a horse to conform to our favorite method but instead should strive to accommodate his needs, adapting our training programs to whatever it takes to gain his trust and respect.

Some advice and some discussion of methods will be repeated in various chapters, as they apply to different phases of training or working with a horse under different circumstances. The goal of this book is to help you with all of these phases or circumstances to create a positive and willing partnership.

Author's note: For simplicity, in this text the horse is always referred to as *he*, unless otherwise noted.

INTRODUCTION

The training principles that follow will help you provide your horse with a solid foundation for lifelong learning. Consider them touchstones to be relied on and returned to each time you work with your horse.

The benefits of a positive approach to training cannot be overstated. How you teach the basics makes a big difference in how solidly the horse learns them. If a training relationship is built on trust rather than trauma, the horse will accept a lesson more quickly and remember it better. If he is compelled to do something through pain and punishment, he'll remember the pain instead of the lesson. Always favor a positive approach.

Practice Makes Perfect

The horse learns through repetition. He may make the correct response accidentally when first learning, but by having the stimulus (pressure) removed when he gives the correct response, and through repetition, he soon learns what he is supposed to do. He becomes comfortable with it — giving the correct response more quickly the next time he receives that same stimulus. Good habits are learned through repetition. You can refine your cues, to a slight pull on the rope, for example, and a slight press with your leg when riding, because he has learned what is being asked. If you are consistent in your requests, repetition reinforces good habits.

Develop Correct Patterns of Behavior

A horse reacts to your actions. When you make the right thing easy and the wrong thing more difficult, he will choose the proper response. If his reaction reduces the pressure he feels, he will in the future keep doing whatever it was he did that relieved the pressure. If, however, pressure is not reduced or is increased, his reaction is either to try to get away from it or to fight it. Always try to help him make the correct response.

Work Progressively

There is much a horse must learn before he can perform advanced maneuvers. Start with the basics and build a foundation for the next steps. If you skip some of the early steps, the foundation won't be as solid, nor will the end result. The shaky foundation will eventually reveal itself and your horse will develop a problem.

Building on what he already knows gives your horse a sense of security. He is at ease and confident with things you ask. He knows what to expect from you because you have been consistent in your requests, and he knows that when he gives the correct response, he will be rewarded. He is willing to try new things because he has confidence from earlier lessons that were accomplished without trauma or confrontation. If he accepts a new step, and finds the new response equally nonconfrontational, he will do it willingly from then on. Work step-by-step.

Master Each Detail

Take one thing at a time; you'll be less apt to confuse your horse or to alarm him. A horse becomes confused or upset when you go too quickly and proceed before he is ready. If this happens, drop back to something your horse already knows, so you can both feel good about the lesson. If you try to do something difficult before he has mastered earlier steps, you may create problems that are more time-consuming to correct. Some days you're better off not to try any new steps at all but just to concentrate on things he already knows. To insist on a new step when the horse is not ready may set back your training several steps. Always wait for the right time to ask for something new.

Quit before He Gets Bored

If you can sense your pupil's mood and always stop before the horse does, you'll keep him fresh and willing in his lessons rather than sour and resentful. Be alert to any signs of overtraining, whether mental or physical. This will usually show up as resistance or reluctance. If the horse starts resenting lessons, back off and do something easier or do it more slowly.

Don't Overdo Lessons

This advice is repeated elsewhere in this book but it is always important. Take your time. Occasionally, you must be content with a small amount of progress or even just holding ground. Your horse is the best judge of how long or intense a lesson should be. There are no timetables for training or for how long any specific lesson should last. A lesson is always good if you end it on a positive note.

Let's get started!

1

BASIC SAFETY PRACTICES

A horse and his human handler can forge a great partnership if they understand each other. Much of this understanding comes from the horse being handled enough to become relaxed and comfortable with the human and handled consistently so he knows what to expect. On his part, the human tries to know and understand the horse and to be attuned to the horse's body language. If you understand a horse, you are able to anticipate his reactions and will be better prepared for what he might do next.

Safety First

One key to working safely with a horse is good training; that is, you want to get him used to what is expected of him so that he will react in predictable ways. Another key is preparedness and attentiveness. A good horse handler is always tuned in to the horse and aware of what his reactions might be to any given situation.

The horse is a large, strong animal, and if he becomes upset or frightened or moves suddenly, you may be injured if you're in the wrong place at the wrong time. You can prevent most problems with common sense; make it a practice to handle yourself and your horse in such a way that there's less chance of unexpected or serious trouble. If you always have safety in mind, there will be less chance of getting caught in a dangerous situation.

BODY LANGUAGE

Horses communicate their feelings and intentions, and you can tell what a horse is thinking by watching his body language. Ears forward means alert interest; ears flat back signals a threat that could be followed by a bite or kick; ears to the side means boredom or sleepiness. Tenseness or relaxation of the body can also be a clue to a horse's mood. Tail swishing means irritation and sometimes anger — a prelude to a kick.

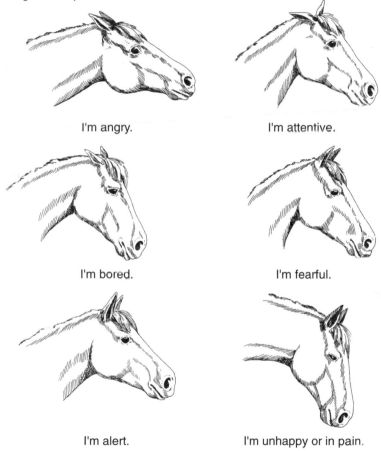

I'm angry.   I'm attentive.

I'm bored.   I'm fearful.

I'm alert.   I'm unhappy or in pain.

These common positions of the equine ear reveal much about a horse's state of mind. Remember, though, that each horse is unique, and the meaning of these ear positions may vary from horse to horse and from circumstance to circumstance.

Attitude Is Important

A good horseman has a trusting, respectful rapport with his horse and is never careless. *Never* take any horse for granted. Even the most dependable horse may move suddenly if startled and can hurt you unintentionally if you happen to be in the way. *Always* have proactive, safety-conscious work habits, even when training a horse you know and trust. This is part of good horsemanship, and it makes for fewer stepped-on toes, bumped heads, and other, more serious, mishaps. Anticipate which way a horse will move next and be prepared to move with him.

An important factor in minimizing accidents is a good working manner, handling a horse with quiet confidence. A gentle but firm manner transmits "good vibes" to the horse, making him less apt to try to test you if he's an aggressive individual and less apt to be insecure, afraid, or flighty if he's timid.

When you are angry or afraid, a horse will sense that. He won't be able to relax and trust you. A nervous horse, uneasy about your handling, is more likely to become unmanageable and give you problems than is a horse that feels secure about what you are doing. A large part of getting along with a horse and avoiding trouble that might lead to an accident or an injury is your attitude and feelings as you handle him.

Know Your Horse

You control a horse through your mind and your body. Controlling the way a horse thinks comes with familiarity, mutual understanding, and the horse knowing what you want and being conditioned to obey. He knows, from previous work, that *Whoa* means stop and stand still, that he must respect restraint by the halter, and that he must behave when you pick up a foot. This is all part of the relationship you develop as you work with him.

The horse is stronger than you are, but through training and your confident attitude he accepts your dominance. If he is momentarily frightened or upset, however, he may forget his manners and become difficult to work with. You must be able to calm and restrain him. If

SAFETY TIP

You may want to wear gloves to shield your hands from rope burns when working with horses. The gloves should be close-fitting and flexible so they don't inhibit your finger dexterity.

he respects you, he'll be more apt to listen when you want him to stand still and behave even under difficult conditions. Use proper leverage and contact with him to best advantage, to keep him under control and to keep from being kicked, bumped, or seriously injured if he becomes alarmed. (See Safe Horse Holding and Safe Leading on pages 16 and 19, respectively.)

Safety Precautions

To minimize accidents and injuries, always wear proper attire when working with horses and use safe equipment.

Proper Attire

When working with a horse, wear protective footwear, not soft shoes or sandals. Sturdy shoes and boots, preferably leather, help protect against the weight and scraping of a horse that bumps or steps on your toes; a good leather boot may also lessen the impact and damage because a horse's foot may slide off it.

> **SAFETY TIP**
>
> If your stirrups have rubber pads, don't wear rubber-soled boots or shoes; they will stick. Alternatively, remove the rubber pads from the stirrups. You want your feet to be able to move and kick out of the stirrups if necessary.

English riding boot

Western boot

Lace-up riding shoe

Sturdy boots and shoes are the safest footwear when you work around and ride horses.

Choose soles with good traction. Slippery soles may cause you to lose your footing and fall down in front of a horse or slide underneath him.

When riding, wear boots or high-top riding shoes that have heels. These will keep your ankles from being rubbed raw on the stirrup leathers and a foot from sliding clear through a stirrup. People have been dragged to death when a foot slipped through a stirrup before they fell off. The high, slanted heel on old-style cowboy boots made it nearly impossible for the foot to slip through a stirrup bow. If a foot does get hung up, a cowboy boot or riding boot will generally pull off; your foot will come out of a boot caught in the stirrup more easily than out of a lace-up boot or shoe.

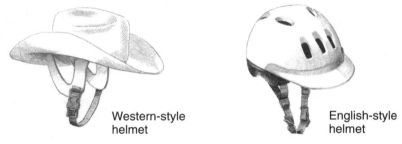

Western-style helmet

English-style helmet

Helmets should be sized appropriately and always worn with the chin strap snug and securely fastened.

Helmets are important when working with horses. A helmet can save your life and prevent serious head injury if you strike your head when falling from a horse or if a horse knocks you down or kicks you. Just as many accidents occur when handling horses from the ground as when riding them. Your head is the most vulnerable part of your body; a broken arm and cracked ribs will mend, but a broken skull or a brain injury can be fatal or leave you permanently impaired. The most common cause of death in horse accidents is head injury.

Tests required for riding helmets to meet safety standards are stricter than for any other sports helmet. When a helmet is approved for horse-back riding by the American Society for Testing and Materials (ASTM) or the Safety Equipment Institute (SEI), it has undergone rigorous testing and meets those standards. Helmets should fit securely, with a snug chin strap, for good protection. Today there are Western-style helmets (beneath a Western hat) as well as English.

Protective vests that incorporate the same kind of shock-absorbing foam used in helmets are sometimes worn by riders when jumping or working with an unreliable horse. A vest may reduce risk for serious injury if you are thrown from a horse or if a horse falls with you.

Safe Equipment

Use a strong halter that is of proper size for the horse. A halter too small and tight will be uncomfortable; a halter too large may pull off if the horse sets back. A too-large halter also risks the horse catching on something or getting a foot caught if he scratches his head with a hind foot.

A snap-on lead rope or strap may be fine for leading, if the snap is strong and well attached, but it may be unsafe for tying. For each horse, have an appropriately sized halter with lead rope securely tied to it — something that will never break or pull loose. When tying, use a rope attached to the halter, not a leather lead strap, and *never* tie with bridle reins.

A bridle should fit well, cause no discomfort, and keep the bit in the proper place for good control. Bridles and reins should be made of strong material and kept in good repair. A broken rein or headstall puts the rider at risk if it breaks during a ride.

Tie a rope, not a lead rope with a snap, securely to the halter. Any simple knot that becomes tighter when pulled will do, or you can attach the rope to the halter permanently by braiding it back into itself.

The saddle should fit both you and your horse. A saddle that doesn't fit the horse will cause him discomfort and he won't perform as he should — he may even try to buck you off. A saddle that doesn't fit you will be harder to ride properly; when you are not in harmony with your horse, you are more at risk for accidents. The saddle should be well built of quality materials. A cheap saddle is often made from inferior materials that won't hold up — a tree might break, stitching may come apart, the leather pulls apart, for example. An inexpensive saddle is not a bargain if it causes an accident.

Girths and cinches should be strong, and they must be comfortable to the horse. Stirrups should be wide enough that you can easily kick your foot free — so it won't catch if you fall — but not so wide that a boot could easily go through one.

Many English saddles have breakaway stirrups. These are designed to pull loose from the saddle if pulled toward the rear of the horse, enabling a fallen rider to break free before he is dragged. Other designs help the foot come out of the stirrup during a fall. Some Western saddles have *tapaderos*, stirrup covers over the toes to protect them when riding through brush and also to keep a foot from going clear through the stirrup bow.

Check Your Equipment

An important part of horse safety is regular inspection of tack. You don't want an accident caused by a broken rein or girth. When cleaning tack, undo all buckles and thoroughly clean, oil, or soap underneath them. Leather becomes thin and weak when buckles are in the same spot for a long time. The off-side billet on a cinch or girth can wear too much

---

RATING ROPE

A nylon rope is stronger and more weather resistant than is cotton rope of the same diameter (a ½-inch nylon rope will hold as well as a ¾-inch cotton rope) but has a slicker surface and is likely to cause a friction burn if it slides through your hands. A small-diameter hard-twist rope such as a lariat is too stiff for easy handling and is also apt to cause a friction burn to you or your horse. The best kind of lead rope is a ⅝-inch or ¾-inch soft cotton rope, kept clean and dry so it won't rot or fray.

in one spot (the fold that rests against the buckle or metal rigging of the saddle) and will eventually break if it is not checked and replaced occasionally. Many riders fail to check it because they are always saddling and working with the horse on the left.

If you are using a Western saddle with double rigging, make sure the connector or hobble strap fastening the back girth to the front one is secure, otherwise the flank girth (rear cinch) can swing back too far and startle or irritate the horse, thus causing him to buck.

Safe Ground Handling

Good horsemanship is just as important while you are doing ground work as when riding, and this includes conscientious, consistent safety practices. Haltering and leading are the most basic elements in the horse-human relationship. How you control a horse in these fundamental steps sets the tone for the rest of your interaction.

HALTERS ARE ESSENTIAL

Always use a halter. In an emergency situation when you don't have one — when encountering a loose horse, perhaps — you might use a rope, belt, scarf, or hay twine held together under his neck, but none of these gives proper control and none provides a safe way to handle a horse. If you must use a rope or hay twine, you'll have more control in an emergency if you loop it around his nose as well.

Safe Catching

The term *safe catching* is misleading because it implies catching up with or snagging something in motion (a ball or fish, for example). You want to approach or meet and halter your horse, rather than having to "catch" him with a lariat or corner him in his pen. The goal is to halter him in a nonconfrontational manner, with him walking up to you or you walking up to him so you can put the halter on.

Make sure a lead rope or strap is attached to the halter; a halter by itself is not adequate for handling a horse because he can jerk away from you. Use a lead 8 to 12 feet long. A rope too short doesn't allow you to give any slack (without losing it) if he pulls back or bolts; a longer one may get in your way or entangle you.

When preparing to catch your horse, drape the unfastened halter over your arm or shoulder in a neat configuration, with the lead rope coiled or looped so it won't drag on the ground and trip you. You want the halter ready to slip easily onto the horse without having to unfasten or untangle it.

How to Safely Approach a Horse

To keep from startling a horse, always let him know you are coming. Speak to him as you approach. That way, if he's standing half asleep or busy eating, he'll be aware of your presence and won't jump or kick because you surprised him. And always speak to a horse before you touch him.

Never come up behind a horse unless there is no other way to approach — and you are sure he knows you are there and is comfortable with your approach. Horses have wide lateral vision but a blind spot directly behind the rump. It's best to approach the shoulder, if possible. With a strange horse, always approach on his left, as this is the side from which horses are most accustomed to being handled. Move slowly and speak softly; the horse may be startled by sudden movement or noise.

Approach the horse and touch him in preparation for haltering.

Safe Haltering

After you've touched the horse, hold the unfastened halter in your left hand and slip the rope around his neck (holding it together under his throatlatch with the right hand) so he knows he is "caught" and will stand still for you to put the halter on. A horse may try to avoid the halter, but the rope around his neck will keep him from taking off.

If the horse holds his head in a normal position (or lowers it for you, as a well-trained horse will do), it's easy to slip on the halter and fasten it while still holding the rope around his neck. Slip the noseband up over his muzzle with your left hand. Use your right hand to reach under the throatlatch to put the long strap or halter loop up over his neck behind his ears so you can grab it from the near side to fasten it.

A. *Slip the lead rope around the horse's neck but don't tie it. (See Safety Tip.)*

B. *With your right hand, reach under the throatlatch to put the long strap or halter loop up over the horse's neck and behind his ears, and then reach up and fasten it.*

If he raises his head, you still have control with the rope around his neck. After you slip the nose piece over his muzzle, you may have to reach with your right hand to place the long end of the crownpiece behind his ears, but it's generally long enough to flip over his neck, even with his head in the air, to where you can reach it with your left hand and then buckle or otherwise fasten it.

In the Stall

Most horses in a stall are willing to be haltered because they want to be taken out. When you enter a stall, be sure your horse knows you are there, especially if he is eating or napping. If he's timid, approach slowly and speak to him reassuringly, carefully cornering him if he is reluctant to have you near him. A horse that associates people with unpleasant things or hard work may try to avoid being caught. If any horse is resistant, be careful; there's not much room to get out of his way if he turns his hindquarters to you to kick.

The bold, spoiled horse needs a more assertive approach. Have a confident, firm manner so he knows he can't bully you.

The horse in a stall may be so eager to get out that he steps on you or pushes into you as you try to halter him. Leave his door closed until you have the halter on him. Make him stand quietly a moment, even after you open the door, so he learns he must wait for your signal to move. When you're ready to lead him out, open the door wide so there is no risk of his bumping the door or pushing you into it as he leaves.

Know your horse and work accordingly. The timid horse needs slow, careful cornering and a lot of patience. In all cases, position yourself where the horse will not be likely to step on you, push you into the wall, rush past you, or turn his hindquarters to you.

REMOVING THE HALTER

When taking off a halter, loop the lead rope around your horse's neck and hold it under the throatlatch while you take off the halter. Don't let the halter drop or dangle — you don't want either of you to step on it or get tangled in it. Hold the horse for a moment with the rope so he knows he cannot take off the instant you unhalter him, then release the rope and walk away.

In Pen or Pasture

The well-trained horse comes when called, or at the very least stands still for you to approach and halter him. If a horse is evasive about being caught in a pasture, see chapter 17 to learn how to train him to be caught easily.

Whether you bring a treat when you catch him will depend on the horse and his training. Some are trained to come when called or to stand as you approach, with no bribes; others expect a treat or a bite of grain. In a paddock with no grass, a handful of green grass may serve as a reward. Whatever your method, be consistent and expect consistency from the horse. If he expects a treat and is polite — doesn't charge over you or nip your hand to get it — this is the easiest compromise. You expect him to come, and he expects a small reward for coming.

If he tries to grab a bite of grain without being caught, don't take grain with you. It's better to teach him to allow himself to be haltered, knowing he will then receive his reward. This makes for better manners and is usually safer for you.

TREAT SAFETY

If you must use bribes to catch a horse, don't let him eat from your hand. Horses fed by hand tend to get pushy and nippy. Always feed grain in a bucket or some other container. Even treats like carrots and apples are best fed from a container instead of from your hand. If you keep treats in a pocket, an impatient horse may look for them, and a jealous horse may snatch at them if more than one horse is involved.

In a Group of Horses

Don't take grain or treats to catch a horse that is living with other horses; the group may come charging for the grain and fight over it, putting you at risk of being run over or kicked. When several horses live together, this complicates the job of catching one of them. You may have to catch the dominant one first so he will not hinder your attempts at catching the other one.

Lead the boss horse; the others will follow you to the barn or pen, where you can catch the one you want in a more controlled situation. If you must feed the group in the paddock or pasture in order to catch one, put out as many piles of hay or tubs of grain as there are horses, so they won't fight over the feed as you try to catch the one you want.

When working with a group of loose horses, to make sure none invades your personal space: carry a crop and use it. Horses in groups can be very dangerous as they chase one another or squabble, not paying attention to the person among them. Even if you're not the target of aggression, a subordinate herd member may crash into you while trying to avoid a kick or bite from a dominant individual.

Safe Release

Turning a horse loose in his pen or pasture after you've worked with him can be dangerous if he has poor manners. Even turning out a stabled horse for exercise can pose a danger if he is eager to be loose. Some horses want to take off bucking and kicking as soon as you free them; stay out of the way.

A horse that is inclined to be exuberant when turned loose requires a controlled situation. Lead him through the gate and make him stand for a moment. Before you take off the halter, loop the end of the rope around his neck; this way, you still have control after you slip off the halter. After the halter is removed, wait until he is calm and relaxed, then step back and walk away. He must learn that you are the one who does the leaving, not him.

If a horse is really rambunctious when released, first turn him around to face the gate (leave it unlatched so you can slip back through it). Make him stand still before taking off the halter, then hold the rope around his neck. When you finally slip off the rope, quietly back through the gate so that you are out of the pen or pasture completely before he turns around to take off. Now you are not in danger if he takes off bucking and kicking. Always make him wait until he is calm before you allow him his freedom, thus enforcing the rule that you are the one to leave first.

Safe Horse Holding

Sometimes you must hold a horse still for the farrier or veterinarian. The person holding the horse can determine, in large part, whether the procedure is difficult or easy, safe or dangerous. Keeping a horse calm, keeping him from moving at the wrong time, and preventing him from kicking and striking are all critical.

Use Proper Restraint

Use a properly fitted halter. Some horses may need a chain over the nose to restrain them during certain procedures, or even a twitch or Stableizer (see page 119, chapter 4). If you use a chain, make sure it doesn't cause pain or annoyance; the horse may become even more upset and unmanageable by fighting the restraint as well as whatever is being done to him. When using a chain, make sure you keep your fingers clear of it.

A horse's instinct is to run away from whatever he perceives as danger or to fight if he can't get away. Anything the horse is not used to may provoke a defensive response. The best way to prevent dangerous behavior is with training — that is, with consistent handling from the time the horse is young. With training, your horse will be relaxed about the things you do with him. He will trust and respect you instead of shifting into flight-or-fight mode whenever something upsets him.

A young horse may not have enough training to be at ease during procedures he is unaccustomed to, such as vaccination, foot trimming, clipping, deworming, or treatment of an injury, however. Even a dependable horse may hurt you if he moves at the wrong time and you're not expecting it. Be prepared for evasive or defensive actions, know how to prevent or minimize these actions, and be in a position where the horse cannot inflict harm if he does move.

Choose a safe place to hold the horse. Sometimes an open area is best, a place with no obstacles to bump into if you have to move around. If the horse must be quite still, however, hold him next to a stall wall or solid fence (never a wire fence). If you think he might rush backward, back him up against a fence or wall so he can't use this evasive tactic. When there's a solid barrier on one side and both the horse holder and the person working on the horse are on the other side, the horse will generally stand still.

When holding a horse, never stand in front of him. You don't want to be stepped on or bumped into. Instead, stand at his shoulder, where you'll be out of harm's way if he lunges forward or strikes. Or stand facing his shoulder, with one hand on his halter and the other on his neck or withers, with your feet positioned so you can move with him if he moves. Bracing yourself against the horse also enables you to move with him.

Holding the halter with your left hand, facing your horse, you can read his expression and intentions. With your right hand, rub his neck or withers to help calm and distract him. A soothing voice or a steady humming or a soft whistle while rhythmically rubbing his neck will help keep his mind off whatever else is being done to him, and he'll more readily tolerate the situation.

You can usually keep him from moving at the wrong time or from kicking. Raising his head or pulling it toward you makes it more difficult for him to kick at the person working on him. If the halter is loose, hold it more snugly under his chin, holding the loose part together, so your hand has contact with his jaw as well. The more contact you have with a horse, the better control you have, and the less likely it is that he will bump into you if he jumps around.

> ### SAFETY TIP
> Body contact is a crucial part of safe horse handling. You not only have better leverage for holding him and working on him, but you can also tell whether he is relaxed or tense. In this way, you will be better prepared for any sudden movements.

When holding a horse still, stand at his shoulder, with one hand on his halter and the other on him, so you can move with him if he moves.

Close contact with a horse is safer than being a distance away. This is especially true when working around the hindquarters, where you could get the full force of a kick.

When you are touching a horse and are confident, relaxed, and matter-of-fact in your actions, he is more relaxed and secure. He knows exactly where you are and what you are doing, and is less apt to jump or kick. Sudden contact, such as a spray of fly repellent, may startle or alarm him. But if you're touching him in that spot and reassuring him, you can usually accomplish the procedure without disturbing him. Safety involves knowing the horse as well as possible and being alert to potential reactions — automatically keeping yourself in a position that puts you at least risk. This is usually a position very close to him, touching him, and transmitting your confidence through that touch.

Safe Leading

When leading a horse, you'll have the most control over his movements and actions by walking beside his left shoulder, about a foot away, holding the halter rope 6 to 8 inches from the halter, or even holding on to the halter itself if more precise control is needed for a horse that is not yet well trained. When leading a horse with a bridle, hold on to both reins, a few inches from the bit, with a finger between the reins and the extra length of reins looped neatly in your other hand.

Never wrap the end of a lead around your hand. Keep the extra length in neat loops with your hand over them, not through them. Then you can drop a loop or two to give the horse slack or take it up again in a hurry. Don't hold the rope in a coil that might slip around your hand or arm if he bolts.

When leading with loop reins (rather than split reins), always take down the reins over the horse's head when you start to lead him. If the reins are still on his neck and he balks, backs up, or bolts, you won't have control and may make the problem worse, because one side of the reins will be pulling on his mouth.

SAFETY TIP

Always have a rope or lead shank attached to the halter. Your horse will be less able to pull away or dislocate your shoulder if he jumps, rears, or turns quickly. *Never* lead a horse with just a halter. If a rope is attached to the halter, you can always let out some slack; that way, he can't jerk you as hard as he could if you were just hanging on to the halter.

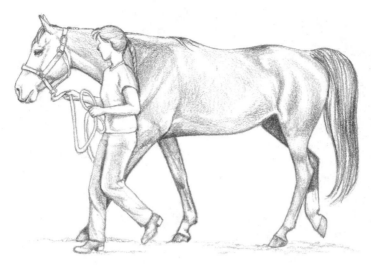

Walk beside your horse's shoulder, holding the halter rope 6 to 8 inches from the halter. Keep the leg closer to him in stride with his closest leg, which reduces the likelihood that your foot will be stepped on.

Don't walk in front of a horse you are leading. If he spooks and leaps forward, he could bump into you, jump on top of you, or step on your heels. You have no control of his movements when you are in front of him. Walk beside his left shoulder, moving along with him. Then you have control over his head and, thus, all of his actions. You can keep him at the speed you want and halt him when necessary. If he tries to go too fast or bolt, use body leverage to halt or slow him: Lean your elbow into his shoulder and pull his head around so he has to circle you and can't get away. When you have this kind of contact, you can be as strong as he is.

Hold the lead rope fairly close to the halter when leading a rambunctious horse. If there is too much distance between your hand and the halter, the horse can get up too much speed too quickly and can pull directly into the halter; you may not be able to stop him. To have control, you must be able to "take his head away," pulling it around toward you. You are better able to do this when you have a short hold on the rope and the ability to brace yourself against his shoulder.

When walking beside him or leading at a trot, keep your feet from being stepped on by

It your horse tries to go too fast, use body leverage to halt or slow him. Lean your elbow into his shoulder and pull his head around so he can't bolt.

moving your feet in cadence with his, at the same stride. Your leg closer to him moves in unison with his leg that is closest to you. The only time he can step on your foot is if yours is already on the ground while his is coming down. If you are walking in stride with him, your foot and his will be hitting the ground together; thus, there is little chance of him stepping on your foot.

Safe Tying

Knowing how to tie a horse properly is one of the fundamentals of good horsemanship. An improperly tied horse is potentially dangerous. If he pulls back when startled, he may break loose. If he flies back on the rope when you are trying to tie or untie him, you may get your hand or fingers caught in the rope, perhaps losing a finger or part of your hand. The horse may also crash into you if he sets back, then flies forward again.

Halters and ropes should be strong and in good repair. Web halters should be at least three-ply, with sturdy hardware that won't break or come apart. A rope halter without buckles or metal rings is best. A horse should be tied with something strong enough to hold him even if he pulls back with all his strength. If he breaks loose when pulling back,

USE A HALTER OR NONSLIP KNOT

Never tie a horse with a rope around his neck, unless you use a nonslip knot such as a bowline. If you use a knot that slips, the horse could tighten the rope around his neck and strangle. And remember this: *It's always safer to use a halter.* The halter distributes pressure more evenly; if the horse does set back, he won't be so apt to injure his neck.

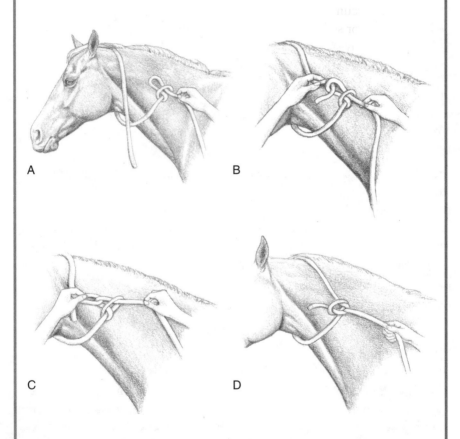

A

B

C

D

*To tie a bowline knot, pass the rope around the neck and put a simple loop in the standing part of the rope **(A)**. Insert the free end of the rope through the loop formed; double back the end and hold this loop with the thumb and forefinger of your left hand **(B)**. With your right hand, pull the standing part of the loop **(C)** using an upward motion that will make the loop of the first knot slip over the loop in the left hand to form a bowline **(D)**.*

he'll likely try it again and thus become dangerous to tie up. A horse that never breaks loose when pulling is more apt to accept restraint readily. Make sure the rope is securely fastened to the halter. Many snaps used on halter ropes will break if a horse pulls back hard; use a lead rope you can tie to the halter with a knot that will not come loose.

Some people feel that if their horses are well trained, they don't need to tie them properly. But careless tying — or using a flimsy halter or rope, using a too-light lead shank or snap, tying with bridle reins or to an insecure object — is dangerous. Even the most dependable horse may spook in unusual circumstances. Any number of things may cause a placid horse to jump or set back: A gust of wind could blow something into his face; a noise may startle him if he's dozing.

A horse that falls over backward when the rope breaks may seriously injure himself or suffer fatal brain damage if he hits the top of his head on the ground. If he pulls a board or pole from the fence and takes off with it swinging and hitting him, he may become injured or injure people or other horses. The trauma of any of these experiences will render the horse undependable about being tied up again.

Never tie to anything that might move or come loose. When tying to a pole fence, don't tie to a pole unless it is nailed securely to the other side of the posts and cannot be pulled off. Loop your rope around a solid post as well as to the pole. Never tie to a wire fence or netting; if the horse paws at the fence, he might get his foot caught. If tying to a horse trailer, be sure the trailer is attached to the pulling vehicle or securely blocked so the wheels can't move.

Don't tie to a metal fence panel or gate, even if it's solidly connected to a post. Tie to the post. If there is no post, take several turns around the top metal rail with the rope, rather than tying hard and fast. Then if the horse does pull back, the rope will slip a little rather than pulling the gate or panel too much. Don't leave him unattended with this type of "tie."

Do not tie a horse for bridling or unbridling (or even saddling, if he's still green) or get on him while he is tied. Don't have him tied for clipping, deworming, vaccinating, or any other procedure he may be uneasy about. Don't walk under his neck while he's tied; walk around him to change sides or untie him. Approach him from the side to work with him or to untie him. If you walk straight at his head and startle him, he may pull back. Always make sure he knows you are there, and approach in a calm and relaxed manner when he is tied.

Using Cross Ties

Restraining a horse in a stall or barn aisle for grooming and saddling, for example, is often done with cross ties, using ropes of proper length fastened to rings in the wall, with snaps on the free ends to attach to the horse's halter. Position the horse in the center of the aisle or stall with his head between the cross ties. Fasten one of the ties to his left side, snapping it to the cheek ring on the side of his halter, then fasten the one on his right.

For the safety of the horse, the space in which he is tied should be no more than about 15 feet wide. Position ties only slightly higher than the horse's head to allow some movement without being loose — the horse should be able to move his head at least 6 inches either way. The snaps on the rope ends should be easy to release even if there is pressure on them. Make sure the footing is nonslippery. It's best if there is a wall behind the horse so he will not be inclined to pull back.

> ### SAFETY TIP
>
> Some horses feel overly confined in cross ties and may become nervous or try to pull back. These horses do better if tied with just one line, which gives them more freedom of head movement.

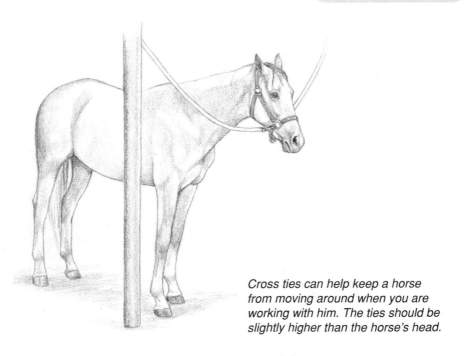

Cross ties can help keep a horse from moving around when you are working with him. The ties should be slightly higher than the horse's head.

How to Tie a Horse

Tie with enough slack for freedom of head and neck, but not enough for the horse to put his head down under the rope or lift a foot over the rope. If he reaches for grass or to scratch his face on his leg or rub on the fence, he could get into trouble. Tie short — about 16 inches — but not so short that he cannot move his head; this would make him feel threatened by the confinement.

Tie to something level with the horse's head or higher. It is best to tie a little higher than his poll. Then if he pulls back, he won't damage his neck muscles or hurt his neck.

When tying to a tree, make sure the rope can't slip down the trunk. Choose a tree with a branch or some irregularity in the trunk that will keep the rope at proper height. Make sure there are no sharp branches that could injure the horse's head or eyes. A stout horizontal limb above his head, well above your saddle so it can't catch on the branch, is safest. Loop the rope over the branch and tie with a quick-release knot. Never tie to a dead limb or tree; it may break off if the horse pulls back.

> **SAFETY TIP**
>
> When tying, choose a place with good footing. *Never* tie a horse where the ground is sloping or covered with rocks.

Properly tied, a horse has some freedom of head and neck but won't be able to get his foot over the rope. The rope should be tied level with his head or higher.

BE PREPARED

Always carry a sharp pocketknife when working with horses. You may find yourself in a situation where you must cut a horse loose — from a rope that's entangling or choking him, for example, from a harness strap, or from a blanket strap. *Never* tie a horse with something that can't be cut in an emergency.

Use a knot that will hold securely, yet be easy to untie quickly. Never tie with bridle reins; sooner or later you'll end up with broken reins and a horse with an injured mouth. On a ride, take a halter and rope if you plan to tie up, or keep a halter on under the bridle. Then you're prepared for any situation in which you might need to tie the horse.

Manger Tie

The manger tie is an effective knot that can be undone with a quick pull on the loose end, yet is very secure. It won't come loose if the horse pulls back. If a horse nibbles at the rope or tries to untie himself, you can put the loose end through the loop of the tie, making it impossible for him to undo the knot, especially if you put it around the side of the post so the loose end hangs down on the other side of the fence.

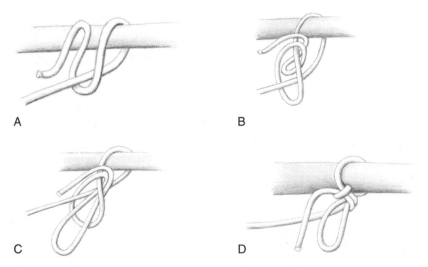

A   B

C   D

*To tie a manger tie, pass the end of the rope under and then over the top of the pole and loop the rope under itself (**A**). Double the end of the rope into a loop (**B**) and pass that loop through the first (**C**). Pull the loop through and tighten (**D**).*

Safe Grooming

Keep your grooming tools organized and out of the way of the horse. Stand at his shoulder, not in front of him, and keep in close contact. If you have a hand on him at all times, you won't startle him. Don't suddenly reach for his head; begin at his shoulder and work forward.

Standing close to the hindquarters is safer than at arm's length away; if the horse kicks, his leg will merely push you rather than hit you with full force. Don't stand directly behind a horse. When brushing the tail, bring it around to the side to work on it. Go easy on sensitive areas and ticklish spots.

Always be prepared to move with your horse if he moves or startles to avoid being stepped on. When working near his hindquarters, be aware that he could accidentally kick you as he kicks at flies. He may also swat you with his tail. Sometimes a horse that's upset may swat you across the face. This not only is uncomfortable — the hairs can sting if he swats hard — but also can cause serious injury if he lashes the tail hairs across an eye.

When working around his legs and feet, as when putting on hoof dressing, never kneel on the ground. You won't be able to move fast enough to get away if he happens to move in your direction. If you crouch, you always have your feet under you and can quickly pop up to the standing position or move fast to get out of his way.

BE NEAR OR FAR

When working with a horse, stay either very close for leverage or at least 12 feet away, out of kicking range. Let him know you are there and what you intend to do before you do it. Touch him with your hand before you touch him with a brush or comb. Run a hand down his leg before you pick up a foot. Keep a hand on him and talk to him so he knows where you are. Keep in contact so he won't be startled if he gets relaxed and sleepy.

Safe Bridling

Use a careful and conscientious routine every time you bridle and saddle. It's generally best to saddle a horse first, while he's still tied, if he ties well, and then to bridle him. When you are finished with your ride, take off the bridle first, tie the horse, then remove the saddle and groom.

When bridling, put the bridle on over the halter, so you still have control of the horse. If you are taking off the halter before bridling, untie the horse and temporarily buckle the halter around his neck to maintain control or hold the rope or bridle reins around his neck for restraint.

When bridling, don't clank the horse's teeth with the bit or cram his ears into the headstall. Slip in the bit — pressing thumb or finger against the interdental space if he won't open up — with your left hand while holding up the headstall with your right, ready to slip it over his ears as soon as the bit is in his mouth. Slip in the right ear first. Make sure the bridle is adjusted properly and fasten the throatlatch snugly but not tight.

A. *Hold the bit in your left hand and the headstall in your right.*

B. *With the bit resting in your hand, put a finger into the corner of the horse's mouth and press on his gum line, to encourage him to open his mouth. Use your thumb to guide the curb chain behind his chin. Slip the bit into his mouth and pull up the headstall.*

C. *Slip the right ear into the headstall, then the left.*

Bridle Fit

For the comfort of the horse and for optimum communication, the bridle should fit well — the cheek straps should be tight enough that the bit fits snugly but does not overly wrinkle the corners of the mouth — and the bit should be appropriate to his stage of training (see chapter 7 for more on bits and their uses). When you bridle your horse, make sure all straps, especially the curb strap or chain, are straight and smooth so they don't rub and cut into the skin. The throatlatch should be just loose enough for you to slip your hand between it and the horse's throat. Its purpose is to keep the horse from being able to rub off the bridle.

Unbridling

When removing the bridle, unfasten the throatlatch and slip off the headstall over the ears, letting the horse spit out the bit. Don't pull it out or let it clank against his teeth.

Be careful with loop reins. An inexperienced or nervous horse may injure his mouth if he gets a foot in his reins and panics when he puts up his head and is restrained by the reins.

Safe Saddling

Before saddling a horse, brush his back to clean off any dirt, mud, or debris that could end up under the saddle and cause discomfort or irritation. Place the saddle pad or blanket well forward of the proper position, then slide it back into place. Never slide it forward; this ruffles the hair the wrong way and can create a sore.

Put on the saddle gently; don't swing it up and flop it on the horse. Dropping it onto his back may startle him, causing him to jump forward or set back. If he jumps, he not only causes the saddle to end up in the wrong place on his back, but also may crash into you or set back on his tie rope and create a dangerous situation. The saddle may fall underneath him, spooking him even more.

Don't let the off stirrup bang against him as you put on the saddle. With a Western saddle, hook the off stirrup over the saddle horn to keep it from flopping down on his off side; with an English saddle, keep the stirrup irons run up close to the saddle until the saddle is in place. Position the cinch or girth over the saddle seat so it won't flop down, either. Once the saddle is situated, go around to the off side and lower the stirrup and girth, and make sure the girth is straight.

If you are using a Western saddle with a rear cinch, always fasten the front cinch first, so the saddle is secure before you fasten the rear one, in case the horse moves. Fasten a rear cinch snug but not tight. You don't want it to irritate the horse, but you also don't want so much slack in it that the horse could stick a foot through it while kicking at a fly.

When cinching up, do it slowly and gradually. If you cause him pain, your horse may set back or reach around to bite you. Cinching a horse too tight and too fast will make him resentful of being cinched in the future. He may then hold his breath to keep it from being too tight (the saddle will be loose when he lets out his breath). Check the girth again just before you mount, to make sure it's snug, and check it again after you've ridden the horse a few minutes. It should be tight enough that your saddle will not slip or turn but not so tight that you can't get a couple of fingers under it.

If you make bridling and saddling a pleasant experience for the horse, he'll be more apt to have good manners and stand quietly for this routine. You won't have to worry about him trying to pull back or moving around to avoid the saddle or bridle — and risk him stomping on your feet or pinning you against a fence or stall wall.

Saddle Fit

The saddle should fit both horse and rider. Make sure the stirrups are not too long or too short so you have the proper leg contact. You should be able to put weight in the stirrups and still get your heels down, and be able to stand in the stirrups and be up out of the saddle. The saddle seat should be of proper size: A seat too small is uncomfortable and won't distribute your weight evenly on the horse; a seat too large makes it more difficult for you to keep proper position and balance on the horse.

The saddle should be the proper width for the horse. Some horses have a narrow back and high withers; others have a wide back and low withers. A wide saddle on a narrow horse will sit too low, rubbing his withers and making them sore; a narrow saddle on a wide horse will perch precariously and may possibly pinch the horse.

Unsaddling

To take off the saddle, run the stirrups up on an English saddle and put the girth over the seat. On a Western saddle, put the off stirrup, cinch, and latigo over the seat so that when you pull the saddle off, the stirrup and cinch won't be dragged over the horse's back or on the ground. If the saddle has a breast collar, tie-down, or martingale, fasten these after you tighten the cinch or girth, when saddling and unfasten them first when unsaddling. You want everything unfastened from the horse before you loosen and unfasten the cinch or girth.

Don't let a horse drag his reins. If he isn't trained to stand still for unsaddling, take off his bridle first and tie him with a halter while you unsaddle and put away equipment. Never let him wander around with a bridle on. A horse dragging his reins may step on them and break them. An inexperienced horse should not be allowed to graze with a bit in his mouth. If he's not accustomed to eating with a bit, he may get grass wadded in and not be able to chew it. Horses have been known to choke when stuffing too much grass or hay into their mouths while wearing a bit.

> ### SAFETY TIP
>
> Unfasten the rear cinch first on a Western saddle with double rigging. You risk a serious injury if you unfasten the forward cinch first and the horse jumps or moves. The saddle would then be held only by the rear cinch and could slip backward, causing the cinch to drop around his flanks — a sensitive area on most horses.

Safe Mounting and Dismounting

Always check the girth or cinch before mounting; it should be snug enough that the saddle will not turn. Take hold of the mane and reins with your left hand so you still have control over the horse if he starts to move before you are in the saddle. It's better to hold on to the mane rather than the saddle; you are less apt to pull him off balance or pull the saddle toward you. This could happen on a round-backed horse if you use both hands on the saddle — pommel and cantle — to pull yourself up.

Facing the rear of the horse, turn the stirrup toward you with your right hand, place your foot in the stirrup, then use your hand to grasp the saddle as you spring up lightly. Swing your right leg over the horse, taking care to not bump or drag it across his rump, as this might startle him. Ease gently into the saddle; never plop down hard.

There are two proper ways to mount: standing facing the same direction as the horse and standing facing to the rear. The first works well on a horse that is well trained to stand; the second is better if there's a possibility he might take a step forward as you mount. If he moves, he moves into you and you are still able to swing on, and you are not left behind.

Mounting, facing the same direction as the horse.

Mounting, facing the rear of the horse.

Whichever way you mount, keep a hand on the reins at all times and keep your horse under control. Don't leave so much slack in the reins that you'd have to gather up a lot of rein before you could stop the horse. On the other hand, don't have the reins so tight that you are putting pressure on the bit.

You should be able to check your horse with just a slight pull on the reins if he starts to move while you are mounting. If he is over-eager, teach him to stand still while you mount. If you check him with the reins every time he starts to move, he'll soon realize it's easier to wait until you're ready.

SAFETY TIP

Don't mount inside a barn with a low roof or overhang. Mount in an open area away from fences, trees, vehicles, and other obstacles.

PROPER FINGER POSITION

When using a snaffle and holding a rein in each hand, have the rein between your last two fingers or outside the little finger. When using two reins in one hand such as with a Pelham or double bridle and four reins or in one hand with a curb bit, the little finger divides the reins. When using a snaffle and one hand, divide the reins either by your little finger or by all four fingers, depending on the horse's level of training and how widely separated you need the reins to be for your bit signals.

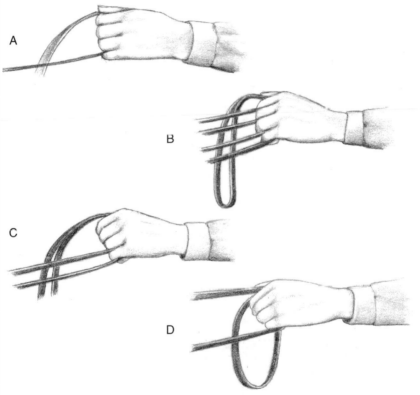

A

B

C

D

*Holding the reins. **A.** With a snaffle bridle (two hands on the reins), the rein goes between your little finger and the next. Some riders prefer the rein outside the little finger. **B.** When riding with one hand and four reins (double bridle or Pelham), the reins are divided by the fingers. **C.** Using two reins in one hand (both reins in one hand when using a snaffle or Western curb; two reins in each hand using a double bridle or Pelham), the two reins are divided by the little finger. **D.** When using one hand with a snaffle, the reins are often widely divided by all fingers, to give more flexibility for individual rein signals.*

When dismounting, take both feet from the stirrups, lean slightly forward, swing your right leg over the horse's back, and, in the same smooth motion, slide down. Don't leave your left foot in the stirrup; this will get you in trouble if the horse moves when you have one foot on the ground and one still in the stirrup. It's best to swing completely free and land lightly on your feet.

Take down the reins for leading the horse; don't leave them up on his neck. If you are riding on an English saddle, run up the stirrups after dismounting so they won't swing or bounce and bump you or the horse or catch on something.

Safe Foot Handling

When handling a foot, the more contact you have with the horse, the better. Not only are you more adequately braced against him for holding the leg, but you can also sense his mood and be able to anticipate any movement he might make. When picking up a foot, encourage the horse to shift his weight to the other three legs by leaning your hand against him, then running your hand down the leg and gently squeezing the back tendon at the fetlock joint.

To pick up a front foot, run a hand down the back of the horse's front leg, press your shoulder against his shoulder to encourage him to shift his weight off the foot, and squeeze gently at the fetlock joint if he doesn't pick up the foot for you.

PICKING UP A HIND FOOT

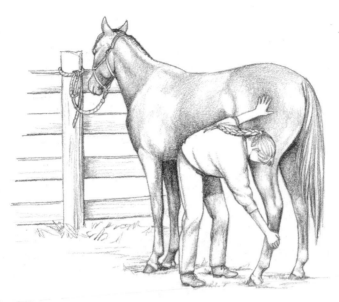

A. *To pick up a hind foot, lean against the horse as a signal for him to shift his weight, run a hand down his leg, and gently pinch the back tendon, if necessary, to encourage him to pick up his foot. Rest the hind foot against your thigh.*

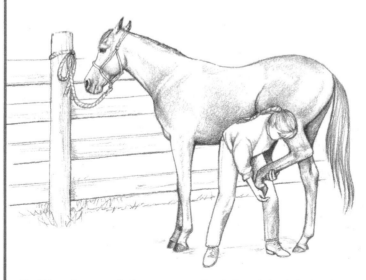

B. *If he tries to pull his foot away, firmly bend the fetlock joint, pulling the toe upward. With the joint flexed, it will be more difficult for him to kick at you or pull away his foot.*

When holding a front foot to clean or treat it, your upper arm should touch the horse's side. This gives you more leverage if you must lean against him and hold tightly to the foot to keep him from pulling it away. Keep your feet away from where he will put his foot back down, in case he puts it down faster than you anticipate. Take care never to be right in front of a leg as you bend over to work on the foot. The horse could unintentionally bump you just by picking up his foot, and his knee could hit your face.

When holding a hind foot, most of the horse's leg should be in contact with your leg. If the horse tries to jerk it away from you, bend the foot (toe upward) and it will be harder for him to pull it away or kick you.

Safety When Riding

The keys to a safe ride are being prepared and not asking more of your horse than he can give you. Be properly attired and have your horse properly prepared. Use well-fitting tack. Don't try to ride with just a halter or jump on him bareback unless he has been trained for this and is accustomed to being ridden this way. Most accidents happen when a rider tries something unplanned or foolish.

Safe Rein Handling

When riding, keep the reins short enough so you can control the horse but sufficiently loose that you don't bump him in the mouth. If the reins are too floppy, you won't be able to check him if he decides to go faster or suddenly shies. If you're using a snaffle bit, you'll have a rein in each hand. When riding Western, with a curb bit, you will have both reins in your left hand.

Safe Halting

To slow or stop the horse, never give a steady pull or sharp jerk. The best way is to pull and release, pull and release, then release all pressure the instant he responds. Keep communication with the bit as smooth and comfortable as possible. He should experience discomfort from the bit *only* when he misbehaves.

Exercise Good Judgment and Think Ahead

Consider the horse's training level and your abilities; avoid situations that might get you into trouble. Don't do fast work or play games on horseback unless you have prepared the horse with careful training. Always think first of the horse and how he might react to a new experience. Will a certain situation aid or hinder his training progress? Will it jeopardize your safety? If riding by yourself, let someone know where you are going and approximately when you will be back.

Consider the Weather

Some days are more challenging for working with horses, and not just because the weather is nasty and uncomfortable. Bad weather also creates hazardous conditions. Rain and snow may make footing unsafe and increase the risk for slipping and falling. Storms may make a horse more nervous and unpredictable, more unruly, and more likely to slip and fall because he is not under control.

Wind makes a horse nervous because everything is in motion. He is instinctively on the lookout for danger, prepared to leap and run from it. Motion, in his mind, could mean a predator creeping up behind him in the waving grass or poised to pounce on him from the wiggling bushes. When things are blowing about, he is keenly aware of every movement and becomes jumpy and anxious. Windy days are not good for training a horse; he will be paying little attention to you. An inexperienced horse may be quite upset, and you could lose ground in training. It's better to skip a lesson when the weather is bad; safety is always your first consideration.

> **SAFETY TIP**
>
> If a storm is brewing and there might be lightning, skip the ride or lesson. The storm will be frightening to the horse and lightning puts both you and the horse at risk.

No Dogs Allowed

Don't take along a dog when training a green horse, even if the dog is well behaved and the horse is used to the dog. A dog may create a hazardous situation by attracting or fighting other dogs, causing a problem with traffic if you're riding along a road, distracting you from keeping your attention focused on the horse, or running under your horse for protection from something that frightens him.

Footing and Terrain

Make your horse walk — do *not* allow him to trot or gallop — when you go through difficult terrain, creek crossings, bogs, or slippery or rocky footing. If you're reckless when traveling where your horse might have bad footing, he could pull a shoe, strain a tendon, even break a leg. He might fall down and break *your* leg. Insecure footing makes a horse anxious; he may rush or scramble when he should be going slowly and cautiously.

Be careful when galloping or making a sharp turn on grassy sod. Grass that is wet is slippery. If the horse hits bad footing unexpectedly, he may not have time to catch himself. Think ahead and adjust his speed when footing looks precarious.

Crossing Water. Give your horse plenty of time to get through a bog or deep water. If he panics, he may strain a leg. When going through a stream — or any obstacle he hasn't encountered before — he'll be less afraid to cross if he sees other horses crossing. As they cross and continue on the trail, he will not want to be left behind. Be prepared, however, in case he tries to jump the obstacle or stream or lunge through it in a hurry.

Steep Hillsides. Travel steep slopes at a walk. Don't trot or gallop up a steep hill unless it is absolutely necessary. If the horse is not in top athletic condition, galloping up a steep hill may pull muscles. Keep him at a controlled gait. Don't gallop downhill; a horse always has better balance, control, and traction at a trot rather than a gallop.

When descending a steep hill, stay on your horse, especially if the footing is bad. You are more apt to fall down than he (he has four legs and better traction). Keep him going straight down the hill and he will be able to keep his

SAFETY TIP

- When crossing shallow streams, do *not* get off and try to lead him if he balks. He may lunge into you or try to jump onto the same rocks you are using.

- When leading a horse up or down a hill, walk a little in front and to the side — never directly in front, in case he charges or falls. Going downhill, he may slide right into you. Be careful leading across a ditch or gully or up a steep bank. He may jump the gully and land on top of you or lunge up the bank. Don't be in the way.

- When snow is sticky and slippery, coat the bottom of your horse's feet with margarine to keep snow from packing and balling up in the hoofs. This will allow better traction.

feet, even if he slides. Don't try to go down a steep slippery hill crooked or sideways, or his feet may slip out from under him.

The only time you might be safer to get off your horse and lead him is if he is very green and does not yet know how to handle himself going downhill with someone on his back. If you think an inexperienced horse may panic going up or down a steep hill, get off and lead him.

Pavement. Never trot or gallop on pavement or on hard-packed ground. Fast gaits on hard surfaces can cause lameness from the concussion and jarring, damaging the inner structures of the feet and legs. The constant pounding can cause splints, cracked hoofs, even a broken bone or navicular disease.

If you must take your horse on pavement, walk. Pavement can be slippery even when dry for a horse shod with regular shoes. If the shoes provide traction with toe and heel calks or other devices for walking on rocky terrain, ice, or slippery grass, these increase the risk of damage to feet and legs when on pavement. There is no "give" to pavement, and the sudden stopping of the foot each time it is put down can create too much strain and concussion.

Safety When Riding in Groups

Keep your horse a safe distance from others to avoid accidents. Your horse may want to become chummy with another horse, but this often results in one of them getting bitten or kicked. A kicking horse may break a rider's leg. Never let your horse rub on another horse or bite or kick; always keep some distance.

MARES IN HEAT

Some mares exhibit temperamental behavior when in heat and become more difficult to handle. A few are hard to work with even on the ground — they become touchy while being groomed, for instance. More common, a mare in heat is a challenge when you are riding with other horses. She may squeal and kick or strike at a strange horse that gets too close. Be careful when riding in groups if you are on a mare in heat or if there is a mare in the group in heat. Keep a safe distance from other horses.

If your horse is inclined to kick when another horse comes up behind him, tie a red ribbon in his tail, a warning to other riders that he kicks. A green ribbon means he is inexperienced — also a signal for other riders to give him more room.

When following other horses, yours may become more difficult to control if he falls behind and begins fretting and prancing. This can be dangerous going through timber or up and down hills; he may charge through the trees — putting you at risk for a branch across the eyes — or rush down a hill paying no attention to footing. If your fellow riders know you are on a nervous or inexperienced horse, or if anyone else in the group is on a green horse, good manners will have the other riders wait for the green horse to catch up. Ride at the speed or gait that is safest and most comfortable for the least skilled rider or greenest horse in the group.

Arena Etiquette

Horses circling to the left have the right of way when several riders are in the same arena. Slower-moving horses should stay in from the rail and allow faster-traveling horses to go by them on the outside; the fast horses will not have to swerve away from the rail to get around the slow horses. Don't block another rider; always look behind you when stopping or turning.

Trail Etiquette

Don't hold up the line. If the group is crossing a stream and your horse wants to drink, move to one side. If you have to tighten your cinch, get off the trail to do so. If someone in your group must get off, wait for him, or for any lagging rider, or for the person shutting a gate, so he is not left behind with an upset horse. When riding in brushy country or through trees, be careful not to let branches snap back into the next rider's face.

Many problems can be avoided if all riders in a group keep a safe distance between horses. When traveling single file, keep at least one horse length between you and the next horse when walking (you should be able to easily see the horse's feet ahead of you; if you can't, you are too close), and a two-horse space when trotting. When riding side-by-side, keep at least 12 feet between horses. This will usually prevent horses from nipping or making faces at one another or kicking, which is often more dangerous for the riders than for the horses.

No Fast Gaits. Most horseback accidents occur when riders use poor judgment. It's wiser to travel at a walk or trot than a gallop when riding with other horsemen. Racing can be especially dangerous. A race for fun on a racetrack with good footing may be fine, however, if horses are well trained and controllable and you keep a safe space between them, but it is dangerous to race anywhere else.

A smooth-looking open field may have uneven ground, gopher holes, or hidden rocks that could cause injury if your horse stumbles or falls. Much can happen when a horse is going at top speed. Racing can make a horse nervous or bad mannered, wanting to race whenever there are other horsemen riding with you.

Don't Tie Horses Close Together

At home you may tie horses near one another if they get along well, but never tie strange horses close enough to kick one another. To be safe, always allow three body lengths between them. Never let a loose horse wander among tied horses; if you do, there will be horse fights and broken ropes or bridles.

Safety When Riding Along Roads with Traffic

A road can be a dangerous place to ride, but sometimes you have no choice. Often it's safest to ride on the left edge, facing the oncoming traffic, so you always see what's coming and no vehicle can rush up behind you. In some states, however, horses are considered vehicles rather than pedestrians and must travel on the right — in the same direction as traffic. Whichever side you ride on, don't take a horse on a busy road until he is far enough along in training to stay under control and has experience with vehicles. (Getting a horse accustomed to traffic is covered in chapter 8, which discusses early training of a driving horse learning to pull a cart, and in chapter 10, which discusses first training rides out in the open.)

In a group, travel single file along a road, even when there's little traffic. On a back road with no traffic, move into single file as soon as you see or hear a car coming, with all riders going to the same side of the road to get out of the way. If a noisy truck or frightening vehicle approaches, try to find a wide spot where you can get your horse clear off the road. As a last resort on a skittish horse that might whirl into the path of a vehicle, dismount and hold him until the vehicle is safely past.

How to Stop a Bolting Horse

To control a horse that gets up speed before you can stop him, quickly shorten your reins, brace a hand against his neck, grab the other rein close to the bit, and give a strong, quick pull to bring his head around by your knee, thus pulling him in a tight circle. If the terrain is such that you can't circle (slippery or rocky) or if you put yourself at risk by circling (a road with traffic, trail through trees, a group of horses), pull strongly up and back with one rein while bracing with the other hand.

Emergency Dismount

In an emergency dismount at any gait, first slip both feet out of the stirrups. Lean forward with your hands on the horse's neck or on the pommel of the saddle, swing a leg over his back, and slide or jump off, pushing up and away from the horse as you leave him, in order to land safely away from him. Land on the balls of your feet with your knees bent, moving in the same direction as the horse to avoid too much jarring. Remember to keep hold of the reins, so you will still have your horse!

For the emergency dismount, slip both feet out of the stirrups, lean forward, swing a leg over the horse's back, then slide or jump off. Land with your knees bent and move in the same direction as the horse to minimize the risk of injury.

2

HORSE SENSE AND TRAINING PSYCHOLOGY

Part of our fascination with horses stems from the vast potential of the horse-human relationship. There is almost no limit to what a sensitive rider and sensitive horse can accomplish when they work together. You can school a good horse to do many things — leap tall obstacles, run a barrel-racing course, perform difficult maneuvers in a dressage ring, outsmart and outrun the wildest cow, pull a wagon or cart. A good horse has great athletic ability and is swift. But even more important than speed and agility is his receptiveness to training — that is, his willingness to work with humans.

Horses are capable of learning many things. They have excellent memories, but they don't reason in the same way humans do. Often a rider or trainer makes the mistake of assuming they do. This leads to communication problems and frustration, defeating the rider's purpose and creating behavioral problems in the horse, especially if the horse is punished for something he does not understand to be "wrong." But with patience and "horse sense" (that is, understanding how a horse thinks), you can accomplish much — even with a "difficult" horse — by making the most of how horses learn.

A good horseman senses and measures his pupil's abilities, leading the horse step-by-step to greater levels of achievement and response without asking too much too soon or going beyond his ability. The horseman is constantly evaluating a horse's capacity to perform. A sensitive, understanding horseman is the best judge of a horse's mind and is able to get the most from that horse in terms of training and performance.

43

Understanding Horses

Horses evolved as grazing animals that stayed in groups for protection. For millions of years, they relied on their keen eyesight and hearing to warn of approaching predators and on their quick reflexes and speed to escape harm. As trainers, these are the basic characteristics we must work with and overcome, because horses instinctively view humans as predators.

We must first allay a horse's fears and suspicions, so he can trust us, and then gradually acquaint him with the many "unnatural" things we want him to accept. It is unnatural for a horse to be confined by rope or stall, for example; in the wild a horse's only guarantee of not being eaten was his ability to flee. It is also unnatural for him to carry a rider — in the wild, predators were the only things that leaped onto his back — and to submit to the will and direction of that rider, who is hindering his ability to flee from danger. A horse's herd instinct, or instinctive desire to be part of a group, helps us convince him to accept human dominance; it is our job to win his trust so that he transfers his allegiance from the herd "boss" to us, his new partners. Take an assertive leadership role appropriate for, and in tune with, a particular horse's temperament and ability. Then you can forge a willing partnership without having to resort to force, excessive punishment, or some gimmick to "make" the horse perform.

Horses Are Herd Animals

For successful training, you must understand how the horse thinks and learns and how he relates to other herd members and to you as "boss" or "leader." The collective strength of the herd allowed horses to survive in the wild, and horses maintain that herd instinct even when domesticated. Many horses are reluctant to leave their herd mates and become *barn sour* — that is, they refuse to leave home — or may frustrate riders because of their eagerness to return to their "herd." The horse feels safer as part of a group. Animals in a herd depend on one another for mutual protection: More eyes, ears, and noses increase the likelihood of detecting predators. Another advantage of herd life is collective experience; the older animals know the safest routes and where to find food and water, both of which benefit youngsters.

Horses are herd animals and feel safest when part of a group.

Social Order

To survive in a group, horses develop a social, or threat, order to determine who leads and who follows. This order is critical for survival in the wild, and is actually the reason that horses are easily domesticated and trainable.

The leader in a group of horses is generally an older mare, an aggressive individual who has established her role as boss. She determines when the herd flees, which direction it goes, and when it rests, grazes, and goes to water. She makes the decisions, drinks first, and eats the best feed. She leads the herd, and the stallion defers to her, bringing up the rear to keep any stragglers moving along. Because her dominance is clear, she rarely has to assert herself to keep her position; a mere threatening gesture with ears back, tail swishing, or a raised hind foot is enough to keep lower-ranking herd members in check. In any relationship with a horse, the horseman assumes the dominant position, taking on the role of lead mare in the herd.

And each horse does have a place in the herd. The second-ranking individual can threaten any horse but the lead mare. Each horse has individuals it can boss and those it is bossed by, except the horse that is bossed by everyone and threatens no one. Immature horses are submissive to all adults, but as they grow up, the more aggressive ones challenge the ranking and find their own place in the hierarchy.

ESTABLISHING DOMINANCE

A combination of bonding and appropriate body language gives you true dominance over the horse, placing you in the position of boss, leader, and teacher. This enables the horse to be at ease in the relationship; he won't aggressively disobey you through lack of respect, nor will he be unduly afraid of you. If you have this kind of control, you rarely have to resort to using physical strength or correction to have him do what you ask.

The Herd "Boss"

A horse is willing to transfer his allegiance to humans because at some point he has submitted to an older member of the herd or to a dominant herd boss and therefore can accept a subordinate position. Horses are very trainable and manageable if we understand this and assume the proper role, and never allow a horse to dominate the relationship. The trainer acquaints the horse with "threat" cues, the pressure of a leg, for example, and a voice command such as *Whoa*, reducing the need for physical reprimands, just as the lead mare uses a gesture to make a herd mate obey her. (See pages 55–59 for more on use of natural and artificial aids in training.)

Threatening gestures are a basic means of communication among horses. Here, a mare protects her foal.

Social ranking is part of herd life; horses react to one another according to the rules of dominance and submission. A horse must figure out how humans fit into that picture. If a horse succeeds in dominating people, he will do as he pleases, disregarding your commands. Horses need ground rules that they know they must obey when being handled by people.

Training Horses

Both dominant and submissive individuals can be trained successfully, particularly when training begins at an early age. The point at which we enter and influence a horse's life makes a difference, especially when handling dominant individuals. A captured wild horse is easier to train as a foal or yearling than as an old stallion or lead mare. Some independent and aggressive mature horses that aren't handled until late in life are difficult to fully domesticate. Submissive individuals train easily at any age.

As a general rule, the younger the horse is when training starts, the easier it is to gain his trust if he's timid and his respect if he tries to be dominant. Until about two years old, horses tend to be submissive to adults in the herd and to human handlers who assume a dominant role. The aggressive youngster, however, may challenge an unskilled trainer but will probably accept discipline from an older, dominant equine. Herd life is often the most effective way to impress upon the young, aggressive horse that he is *not* the boss.

Develop Good Communication

Strive for quiet confidence and consistency in your actions and methods. As leader, seek to achieve a level of benevolent dominance, which makes the horse feel more secure and trusting. If you are nervous, inconsistent, timid, or angry, the horse will be uneasy and resistant. When handling a horse and trying to communicate with him, your attitude and emotions will contribute significantly to your degree of success.

Working with a horse is easier when you are attuned to what he is doing and thinking. He should also be comfortable with the signals he picks up from you. Training is a two-way street: The horse must learn to respond to the rider and the rider must learn how to communicate with

---

BE AWARE

You are always sending signals to the horse, whether or not you realize it. Each time you handle him, you are training (or untraining) him, making things better or worse, depending on what you do. You are always teaching him something, good or bad. Everything you do with a horse — whether catching and haltering him or asking for an advanced maneuver while riding him — should be done with some thought as to how he will respond. Constantly scrutinize your actions and methods to determine whether you are helping or hindering your training goals.

---

that particular horse. As communication improves, the horse learns new things more quickly. Horses "learn how to learn" when handled with patience and consistency. They become easier to train after they know the basics, especially if they trust the rider and the rider always strives to make training a pleasant experience, with good communication.

The Importance of Body Language

Horses can read us well. They sense changes in physical tension — particularly in a mounted rider — which tell them whether a person is angry or frightened. Our body language reflects moods and feelings, no matter how we try to cover them up. Animals have an acute ability to read body language; this is the way they communicate among themselves.

To get a horse to trust you, first you must be relaxed, both physically and mentally. If you are tense and nervous, he becomes tense and nervous. He thinks that whatever is alarming you may be potentially dangerous to him, so he becomes agitated. When you are at ease, he reads this through your body language and tends to become more at ease himself.

"Bonding" with a Horse

When horse and rider understand each other, they develop a strong bond of mutual trust and respect. This level of communication comes by working together: Through proper training the horse becomes conditioned to obey you and knows precisely what you want and you are relaxed, clear, and consistent in how you deal with him. Sensing your quiet confidence, he looks to you as leader and respects you as the "boss horse" in his life.

You must be at ease in your relationship with the horse. Even if you go through the motions of being dominant in words and actions, if in your mind you are afraid of the horse or tense, he will sense it. This over-shadows your outward actions and makes him nervous, or even aggressive if he is an individual who wants to challenge your dominant role.

Temperament and Learning Ability

There is a direct relationship between the temperament of a horse — is he bold or timid, for example, or calm and easygoing or insecure and flighty? — and how easy or difficult he is to train. An emotional horse that is readily distracted or upset may take longer to train and be more of a challenge than a calm horse that can keep his attention focused on the lessons.

But ease of training is not always an accurate indicator of a horse's ultimate level of achievement. Intelligence and physical ability are often quite separate from temperament. The calm, unflappable animal may be a dullard or a genius. The flighty one may be a brainless fruitcake or a sensitive soul. The latter can become an outstanding performer once you know how to communicate with him and gain his trust. The task is to get to know and understand each horse's personality and abilities, then to develop a harmonious relationship through good communication.

When handling an inexperienced horse, try to work with him according to his temperament. Make sure a bold one doesn't dominate you, a lazy one doesn't cheat and avoid a lesson, and a timid one doesn't become more afraid. Make sure first lessons go well, always keeping in mind the horse's exceptional memory: Early experiences, good and bad, are never forgotten.

Work with the Horse's Natural Responses

Always work with the abilities and inclinations of your horse. This makes the job of training easier by minimizing confrontations and battles of will that you might not win. Even if a horse is reluctant or unable to do what you ask, it's far better to change your tactics than to belabor a point and lose ground. Flexibility is key. Continue the lesson smoothly and lead him to think the response he has given is what you wanted, thus maintaining the impression that you are still in control. Progress at a rate appropriate for the horse's abilities, using his natural responses to your advantage rather than fighting them.

Plan Ahead

Working with a young horse in a friendly environment — that is, a setting where everything is in your favor — helps early lessons go smoothly, and he'll have a good first impression about being caught, led, tied, loaded into a trailer, bridled, and saddled. If you plan to take a nervous horse for his first ride outside the pen, choose a nice day with no distractions. Taking him on his first outing on a windy day with everything in motion will make your training job harder; the horse will think everything is spooky out there. If he shies, bolts, or is startled into bucking, it can set back your training program considerably. Plan each lesson, and expose him to new experiences only as he becomes ready. Make sure the timing and the situation are right. Build good memories and good habits.

Motivating a Horse

In training a horse, we ask him to do a specific action in response to a signal. For a horse to learn this, there must be some kind of motivation that makes sense to him. The best motivator is reward, and the best reward for the horse is release of pressure. Your cue is pressure; his response is to perform the desired action to ease the pressure. Reward in the form of release of pressure is a better motivator than the application of more pressure.

ENCOURAGE AND PRAISE

A good horseman rarely has to punish a horse. Instead, he tries to encourage the horse, communicating his wishes in such a way that the horse knows exactly what is wanted. Be precise and consistent, leaving no room for misunderstanding. Praise the horse for a correct response or even an attempt at a correct response. Cultivate his respect and cooperation.

Pain and fear are *never* good motivators; they distract the horse from the desired response. If you continue pulling on the rein or kicking the horse, eventually he will stop responding. He becomes heavier on the bit and harder to stop, or quits paying attention to your legs and you have to kick him harder or use spurs to make him go. It's a vicious cycle. The more pain used, the more futile he feels it is to respond. He becomes resistant to your signals and less cooperative. By contrast, if he is rewarded for a proper response by release of pressure from bit or leg, his response will be quicker and better.

If you ask the horse to do something in a manner that makes him uncomfortable or afraid, he will resist. But if you structure the request in a way he can relate to, his response will be much better. If he thinks he is putting pressure on himself (as he moves into leg pressure or into the bit, into a fixed hand on the halter, or into the elbow that is situated to intercept the playful nip), he will yield more readily. In such a case, it becomes his own idea to give the correct response; you are not trying to force him to respond. Thus, the horse maintains a better attitude and is more receptive to your cues. He becomes motivated to give the correct response for his own comfort and is more open to learning. He has the proper motivation.

You must remove the pressure (the leg, for example, or the bit) when the horse starts to respond properly. Then he will quickly learn what is desired, for he thinks it's his own idea. Allow him to make choices and reward him immediately for making the correct ones, and your horse will learn quickly to make the correct response.

Cue Clearly

Your goal is to give signals to which the horse can react properly. You must be able to predict how he will react; thus, be consistent in your signals so his reactions will be predictable. If you give the proper amount of pressure and release it as soon as he responds, the horse should give the desired reaction.

If a horse doesn't respond at all or responds improperly, you may not be cuing him correctly. Before you consider punishment, examine your methods; you may need to improve your communication. Usually, an improper response is the fault of the trainer (unclear cues), not the horse. (See Aids on page 55 for more on how to achieve and maintain good communication.)

Reinforce through Repetition

A horse learns to respond a certain way to a cue or stimulus by repetition. He gets used to your handling him a certain way and knows what to expect and how to respond properly. He learns to tolerate things he knows won't hurt him, such as walking into a trailer, and you gradually build on earlier lessons as he makes proper responses and masters them. As you get to know each other, the relationship blossoms and communication expands to encompass a variety of cues and responses.

Repetition and refinement of cues are the basis for good training. Most horses try to please and, if you do not abuse them with punishment they don't understand, will keep learning any new lessons we try to teach. A good trainer teaches a horse a new movement and is satisfied with a few steps in the right direction or an attempt at the maneuver.

Complete understanding and refinement of the movement may actually take place between lessons, when the horse develops "muscle memory," which will make the exercise easier the next time. Because his body has already done a particular movement before, he recognizes it as familiar. You can often use lighter or more subtle aids during the next lesson for a particular movement, because you know he understands what you want and will try the movement again.

Reward and Punishment

Training is faster and easier if you carefully consider *beforehand* how you will reward and punish the horse. (The term *punishment* implies reprimand for doing wrong. In most instances, however, the term *correction* is more appropriate when training a horse, for he often does not under-

---

GET IN SYNC WITH THE HORSE

Books, videos, and articles describe or demonstrate ways to teach horses various lessons, and there are many good methods. Yet even the best may fail if you apply them without an understanding attitude: You must be in tune with your horse to know what will work best for him. Treat each horse as an individual. It is not so important *what* you do as *how* you do it. Make cues and requests reasonable, something your horse will feel comfortable with and able to perform successfully.

AVOID EXTREMES IN TRAINING

Two extremes in training methods — neither of which works well — are force and bribery. Because he will fight back, force may make an aggressive horse even more bold and resentful, and it typically makes a timid horse more fearful. Bribery quickly spoils a horse; he soon learns that he is in control. Giving the horse what he wants, whether it's food to keep him from pawing or letting him head back toward home to keep him from throwing a fit, constitutes bribery. Bribery can take many forms, but it always compromises your role as leader.

---

stand what he did wrong.) But first you must understand the particular horse's temperament and know how to ask him to do something in a way that will make sense to him.

Some horses require a lot more reward and less correction. If you have to correct a horse, make sure he understands what he is being corrected for by acting quickly, so he associates the error with the correction. And reward him in such a way that he'll want to repeat the proper response the next time he is asked. Often the only reward needed is the release of pressure, followed by praise and encouragement.

The sound of your voice is important. An approving voice can be a powerful reward or reinforcement of positive behavior. A disapproving voice is often sufficient correction; the horse knows when his human is displeased. Reward and correction via body language, voice, and attitude should suffice when you and your horse have good communication.

Too often a rider or trainer becomes angry or frustrated and corrects a horse for actions prompted by human error — incomplete or inadequate cues, for example, or expecting him to do things beyond his ability or training level. Some riders take it as a personal affront when a horse doesn't perform properly or misbehaves because of fright, confusion, or discomfort. It's all too easy to blame him for our mistakes. Many people put too much emphasis on correction and not enough on reward. Obedience is something we tend to take for granted, but the horse needs to be praised for everything he does well.

Reward

Horses tend to repeat rewarded behavior. When you ask a horse to respond to a cue, the reward may merely be a release of pressure (leg pressure, bit pressure). When he starts to respond properly, immediately release the pressure. The next time he is asked, he gives the desired response because he knows it will result in the release of pressure. He will repeat the responses that consistently give him comfort.

Timing is just as important for reward as it is for punishment; you must respond instantly with the reward, such as the release of pressure or giving praise, so the horse will associate his action with the reward.

Punishment

Should some kind of physical reprimand be necessary, make it instant and appropriate. You may need only a stronger seat and a firmer leg, or a strong half-halt with the bit, or a sharp kick or tap with the whip, depending on the horse and the situation. These things are covered in later chapters.

For some horses, vocal disapproval is enough. For others, especially the bolder or aggressive individual who may be testing you, a sharp swat may be needed. If he's challenging your dominance, as he might do in a herd situation (trying to be boss horse), you must remind him that you are still at the top of the social order.

Excessive correction makes a horse lose respect for the handler, replacing trust with blind fear or resentment. Continually picking at a horse can also confuse him and is counterproductive. If you continue punishing him, he will either quit trying or become afraid of everything you do.

When a horse does something wrong because he doesn't understand a cue, correction just makes the situation worse, destroying the trust necessary for good training. If a horse misunderstands a cue, often it helps to revisit something simpler that he *does* understand and can do, then gradually work back up to the request. It can be frustrating for you when a horse becomes insecure or is threatened by harmless things like shadows

---

BE CONSISTENT

Use of repetition, routine, and lots of small rewards to reinforce positive habits is important at every stage of training. The reward can be as simple as saying "Good boy!" in a pleasant voice, gentle stroking, or allowing the horse to relax after doing something strenuous or complex. Often the most effective reward is to end the lesson and let the horse go back to the stall or pasture.

or a rock beside the trail, but if you react to such behavior with anger and punishment, the punishment will justify his anxiety and his alarm reaction will intensify. You must help the horse overcome his anxiety rather than increase it. If a horse spooks at a bicyclist along the road or at something along the trail, he needs reassurance, not punishment.

Aids

The rider uses voice, attitude, and body language, as well as riding aids such as weight shift and leg and rein signals to "talk" to the horse, communicating his or her wishes. As horse and rider become partners, they become in sync with each another. The horse learns what is wanted and is able to respond more precisely to cues. There are also artificial aids you can use in training.

Hands

To have perfect communication, you must keep your hands in contact with the horse's mouth. Your hands are your communication link: They give subtle signals, collect him, allow him to extend, keep him at exactly the speed you want, maintain him perfectly under control. Your reins control the forehand (the front part of the horse), and your legs control the hindquarters.

If you are using one hand on the reins (a curb bit), check the horse with a flex of your fingers and wrist for slowing, stopping, or backing. Neck-rein him for turns, moving your hand forward and upward to press one of the reins against his neck to signal the turn.

If you are using both hands on the reins (a snaffle bit), you will have them fixed momentarily or squeezing and releasing to encourage the horse to slow down, halt, or back up. You will use one hand more actively when turning. To make a sharp turn, the active hand is opened slightly outward while increasing tension on the bit. This slows the horse as he turns; the passive hand gives as much as the active hand takes. In a wide turn, the active hand is carried outward and leads the horse around the turn while the passive hand gives.

You can also use your hands as a cue to move your horse sideways as he travels. To move him to the left, for example, carry the active hand (the right one, in this case) to the left without crossing the withers, which produces tension to the left and rear. This keeps the horse rela-tively straight as he moves forward and to the left.

Legs

Use your legs to indicate direction and to increase speed. Leg pres-sure causes the horse to use his hind legs more vigorously or to move his hindquarters to one side or the other. Leg pressure on one side of the horse makes him move over. Using your legs helps him make smoother turns and keeps him traveling straight when he doesn't want to.

Using both legs with equal pressure at the same time signals the horse to move forward. Sometimes you use just the inside leg when making a turn, to steady him and keep him from turning too sharply. Often you use one leg farther back behind the girth to push the hindquarters around a turn or to move them over. A good rider uses legs often, though with very subtle movement or pressure. Leg cues, in conjunction with subtle bit cues, are much more effective than relying on the bit alone.

Weight Shifts

Body weight — shifting forward, back, or to the side — indicates speed and direction. Shifting forward slightly changes the horse's center of gravity, causing him to move forward to restore it. Shifting back encourages him to slow or stop.

Voice

Your voice relaxes a horse, gives him confidence, commands, praise, and occasionally reprimands him. Speak in your normal fashion, not too loud or too soft; you want him to become accustomed to your regular voice. If he's nervous or upset, use your voice to reassure him. Calm a nervous horse by speaking softly, humming, or whispering — give him something soothing and monotonous on which to focus. Don't use a loud voice even if you get upset. Never shout a firm command, such as *Whoa;* just speak more strongly than you would for the soothing *Easy.* Always give the same command in the same tone and the horse won't become confused. Remember, a horse responds more to the tone of your voice than to the words you are saying.

Whips, Spurs, and Other Artificial Aids

Draw reins, spurs, whips, and other devices are called *artificial aids*. Some of these may be beneficial when used correctly, but used incorrectly they can make a problem worse or cause new ones.

The Whip

A whip is effective only as a training aid. If used for punishment, its value is lost because the horse will associate it with pain instead of simply as a cuing device. He can't concentrate on what you're trying to teach because he's worried about the whip.

A whip can be a helpful training tool. It enables you to reach parts of the horse you otherwise could not touch, as when encouraging him to move forward (touching him on the hindquarters) while standing by his shoulder, or when asking him to stay out in a circle while longeing instead of cutting corners or coming to you. Used properly, the horse will think of the whip as an extension of your arm.

FAMILIARIZING A HORSE WITH THE WHIP

Run the whip gently over the horse's back, neck, legs, and belly. If he is afraid of it, simply place it across his neck and let him stand. Once it is resting there and motionless, he can relax. He needs to learn it's nothing to fear — that it's a cue, that its tapping is what he is supposed to move away from. You should be able to place the whip anywhere on his body without frightening him. Approach a

horse matter-of-factly or casually, not aggressively or timidly. If you are worried he might be afraid of the whip and then go about your instruction timidly, the horse may sense your worry and think there *is* something to fear.

Spurs

Use spurs as an aid to better performance. A spur is an extension of a rider's heel; use it to gain an instant response from the horse when an instant response is needed. The purpose of a well-designed spur is to tickle a horse into response, encourage him into obedience, and guide him in a certain direction. The horse moves away from pressure, and light application of the spur gently prods the horse to move. Proper use of spurs produces a quicker response and enables the rider to maneuver his mount with more precision.

The dressage horse, cow horse, or jumper often performs more effectively with the use of spurs than without. The rider merely has to touch the horse lightly or bring his heel toward the horse. The signal is crisp and instant, with less leg and heel motion than if the rider had to squeeze with a leg or press with a heel.

If a horse is sluggish or doesn't turn quickly when asked, the spur gets his attention more emphatically than does a squeeze. The horse understands that the rider means *now*. When precision and instant response are important — to block a dodging cow or to leap a ditch or fence without hesitation — spurs give the rider an edge in communication. A good rider, using spurs judiciously, can make a good horse even better.

Improper use of spurs defeats their purpose; they serve only to irritate the horse or deaden his responses. Spurs are like whips: They are an aid — an extension of your hand (whip) or your leg (spur) — and should never be used for punishment or to cause pain.

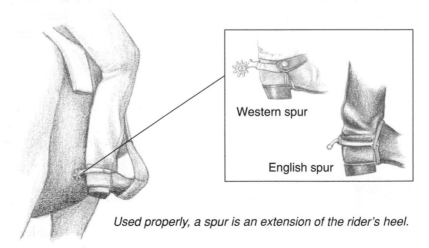

Western spur

English spur

Used properly, a spur is an extension of the rider's heel.

JUDICIOUS USE OF SPURS

Spurs need not be large or sharp. A small blunt spur is less apt to hurt the horse. A novice rider should not use spurs; an inexperienced rider may touch heels to the horse when he doesn't mean to. Nor should you use them for early training on a green horse. The green horse must first learn the basics of response to leg pressure, without being startled by the inadvertent use of a spur. The spur is added later, as a refinement of leg pressure, used in instances when the rider desires a certain response or to reinforce a signal, as when teaching the horse to respond to the lightest leg pressure. (If he doesn't respond to light pressure, firmer pressure is applied. If he doesn't respond to that, he feels the touch of the spur and soon learns to respond to the lighter pressure.)

Other Artificial Devices

A number of devices, among them tie-downs, martingales, and draw reins, are sometimes used to position a horse's head or to keep him from raising his nose. Whenever an artificial aid is used to set the horse's head or for more control, the rider may depend on this rather than trying to develop better balance or control with training. The horse may come to depend on it, leaning into it for part of his balance, and thus cannot truly balance himself. Usually these should be used only as a last resort on a spoiled horse, never in training a green one. A young horse can learn proper head carriage and collection with step-by-step training. The horse that won't collect or carry his head properly usually had poor training.

A *tie-down* or a *martingale* is often used as a shortcut instead of spending time to teach the horse how to balance and flex, or is used on a spoiled horse that already has developed bad habits in trying to avoid the bit. *Draw reins* are used for more control, but if they are used as the only set of reins, they pull down the horse's head to his chest and he is out of your control.

Training Basics

Training is primarily a matter of communication. The better able you are to communicate your wishes to the horse, the easier the training will be.

Develop Mutual Respect

Minimize conflicts by striving to earn respect from the beginning. This is best accomplished if you are consistent in how you handle a horse. He will know what to expect of you and understands that you are the leader in the relationship. By the same token, you must respect him as an individual, take into account his natural reactions, and allow him to be a horse. Don't expect more from him than he can offer.

A horse is most secure when he knows what to expect. He'll trust you more if your consistency is kind but firm. If you're lax, he'll try to bully you; if you're too strict, he'll fear you. Mutual trust is something you develop with a horse as you get to know each other and interact while training.

Treat Each Horse as an Individual

The temperament of your horse should dictate how you handle and train him. Some horses need more firmness to keep them in line; others need more reassurance and encouragement. What works well for one horse might be too harsh or too lax for another.

Progress at His Own Speed

When a horse learns quickly, challenge him with something new. Drilling on things the horse already knows gets boring. A horse needs to be given something new to do or he loses interest and gets into mischief. For a slower learner, provide careful step-by-step lessons and do not move on to anything new until he has mastered each one. Pushing too fast results in confusion on the part of your horse. If he exhibits signs of insecurity, drop back to a lower level and work on things that are more familiar and secure until the horse is ready for the next step.

Not All Methods Work for All Horses

What has worked well for you in training one horse to lead, tie, change leads at the canter, or get into a trailer may not be effective with another. Tailor your training methods to suit each horse.

There are many ways to do things and they all work — sometimes, and with some horses. No method works without fail on all horses. Different horse trainers have different philosophies. Numerous books, articles, videos, and clinics give advice on how to train horses, handle various problems, and teach horses to do specific things.

Some of these methods are similar but others may be quite different. This doesn't mean any one of them is either right or wrong. Some may be better, however, for a certain horse. Occasionally, you encounter a horse that frustrates your best efforts; you have to find some other way to get around a problem. This is why it's good to keep an open mind and continue studying, learning, and searching for the best way to handle your particular horse. Even the best trainers continue to learn; there is always something more to know about handling and training a horse. But don't expect the same method to work equally well everywhere.

LESSON SAFETY

An important rule for evaluating a training method or lesson is to ask if it is safe for both you and the horse. If something were to go wrong, would it put you at risk? Could the horse hurt himself trying to avoid the lesson? If there is *any* risk involved, find a different method. The use of force, which might backfire, or trying to hurry the horse and skip some steps will prove counterproductive and even dangerous in the long run.

Make the Desired Action Easy and the Undesired Action Hard

One of the basics of good training is to make it as easy as possible for a horse to perform a desired behavior. Thinking through your lesson ahead of time and knowing how he reacts to various cues will help you set the stage for making desired actions easy to do and undesired actions more difficult. How you apply and release pressure (on the halter, on the bit, with a leg) will influence how your horse learns. Horses prefer the path of least resistance.

Discourage Unwanted Behavior

To thwart unwanted behavior, anticipate the horse's actions. If you know your horse, you can usually predict how he will react to certain things. There are several ways to head off unwanted behavior; the tactic you choose should depend on the horse, his training level, and the situation.

A horse can't concentrate on two things at once. Sometimes the best way to discourage an action is to distract him by giving him something else to think about or do. If he wants to bolt toward home, make him concentrate on going in little circles instead. He can't run while he's circling. If he's impatient and won't walk slowly while following a herd of cattle, zigzag behind the herd in a fast walk. This gets his mind off prancing. If you can distract his mind and ask him to perform an alternate maneuver, he won't be able to continue the bad behavior.

Don't Intimidate

A horse is stronger than a human. We must influence him to do what we want by using his mind — that is, making him want to do it — rather than resorting to physical force. He must be comfortable with what we ask and able to understand what we want. Because we understand how he reacts to our actions, we work with his mind and his emotions.

Most of us recognize that to train a horse, we must be dominant in the partnership. Often, however, we try to force this dominance by physical means, by improper use of whips or spurs or a chain over the nose. This can lead to more conflict. If a horse becomes afraid or excited, which is likely when we resort to physical persuasion, he will be in no frame of mind to listen to us.

Almost all horse-human conflicts involve physical contact — jerking on a lead rope or bit, for example, or kicking the horse in the ribs

---

ADDRESSING TRAINING CHALLENGES

There's often no single answer for a particular problem. Every solution or method has some drawbacks. To find the best way to handle a certain problem with a particular horse, evaluate the positive and negative aspects of each method and choose the one that seems most appropriate. Always think about a problem before actually addressing it. Try to consider all the repercussions of how you'd deal with it, so you won't inadvertently create more problems. For instance, if you try to teach a horse not to bite by swatting him, you teach him to jerk his head away from your hand every time it approaches his head. This makes future tasks like bridling, treating a head injury, and clipping extremely difficult.

while mounted. In contrast, almost all horse arguments within a herd are resolved simply with body language. As trainers, we must find ways to resolve conflicts and maintain our "boss" status in the social order without physical contact or force, and this means working with the horse's inclinations and emotions.

Be a Teacher

Remember, you are working with the horse's mind as well as his body. Teach him a proper response by having him figure it out. He soon will learn that the more quickly he responds, the sooner he will be rewarded by release of pressure. Some methods that rely on physical manipulation, such as tying a horse's head around to the side to teach him to be "supple" and give to the bit, may work, but it is easier on the horse if you are more subtle. Again, physical methods are best used as a last resort on an already spoiled horse. With a green horse, you'll make better and more lasting progress working step-by-step with his mind.

BE CLEAR

Always try to present a lesson in such a way that the horse understands what is being asked. You can't expect him to respond properly if he doesn't understand your request.

Plan the Lesson

Approach every goal with a plan on how to achieve it. If our aim is to teach the horse to go smoothly into a canter from a walk, for example, or to change leads at your signal while cantering, you can't just ask for this action without some step-by-step preparation.

Be Realistic

Your goal should be reasonable. As you would with a lesson plan for a class at school, have some thought-out steps for how to get there. Start the lesson where your horse is now in his training. The more steps you can give the horse between where he is now and arriving at the goal, the more likely you will achieve it without needing to correct problems along the way. Like the fabled tortoise, a trainer who goes slowly and consistently, without shortcuts, will arrive at the goal sooner — not having to correct problems that occur when rushing the training.

STAY FLEXIBLE

Don't be so rigid in your methods that you can't change them if necessary. You may have to go to plan B or even C or D. The method you decided on may not work with this horse. The time frame you laid out may be unrealistic. Your horse, like you, may have good days and bad days. Feel out his attitude as you prepare for a lesson, and shift gears accordingly. Have an alternative plan for bad days. Perhaps you can review some easy things he already knows and save the next step until later. Even if you don't make much progress on a given day, if you keep his attitude positive and prevent a confrontation, you still will have gained ground. Flexibility is key in any training program.

Decide When to Correct and When to Be Patient

Often a horse does something unwanted because he is afraid or does not understand a cue. Usually his behavior is not the result of disobedience. Understand your horse well enough to know the difference, and don't punish if he's acting badly out of fear or confusion. The last thing you should do is lose your temper with a horse that does not understand what you want. If he's afraid, he needs comfort, reassurance, and patience to get past the fear. If he's confused, you may be rushing him or asking too much. Slow down and simplify what you are asking; until he regains his confidence, go back to something more basic that he understands and can perform.

If your horse understands what is asked but is being lazy and uncooperative, perhaps testing your authority, he needs a firmer cue to reinforce the one he did not obey. Cease the correction as soon as the horse responds properly, and reward him with release of pressure for complying and moving past the bad attitude.

Don't Rush the Horse

Lessons should be progressive. Build step-by-step on earlier lessons. Expand the basics, broadening the foundation of his abilities, both mental and physical. Be in tune with the horse so you can recognize his strengths and limitations. Don't expect him to perform a new maneuver perfectly the first few times you ask, and don't drill him on it excessively.

A mistake many trainers make is getting caught up in a timetable, feeling they must accomplish a certain goal in a certain time. Never have a rigid training schedule, and don't assume that today is the day you will introduce a certain movement and tomorrow you will move on to the next. This puts you and the horse under pressure. Take things one step at a time, moving forward only when you feel he is ready. As always, don't ask him to do anything he doesn't understand. Sometimes a novice trainer punishes a horse for not performing when the problem is that he doesn't know what is expected. The horse generally overreacts to the stimulus or refuses to respond at all.

Most things a horse must learn to do under saddle will require physical conditioning as well as mental learning. He will use muscles differently, develop new muscles, and acquire the ability to perform various maneuvers as he gains the strength and dexterity to do so. If you push him too fast, he'll be physically sore as well as mentally soured; if he hurts, he will resent the lessons.

A horse may develop a bad attitude about training, especially for an athletic or performance career, because he is worked too long and hard; pain will actually teach him avoidance behavior. A training schedule should allow time for him to recover from previous strenuous activity. The greatest progress is made when you take ample time to get there. Shortcuts generally end in failure of one kind or another.

Keep thinking. Plan the next lessons. Evaluate previous ones. If today a horse did not react or respond the way you wanted, look closely at your methods and cues. Consider how you might alter things for tomorrow; be flexible. Most of all, try to understand the horse and why he reacts the way he does. Every horse is different, and each one will be teaching you while you are teaching him. If you are willing to learn as much as the horse does as you go along, you will become a good trainer.

Give the Horse a Break

Now and then, horses benefit from a vacation from daily training drills. The good news is that horses continue learning even during a break. Often after a vacation day, the horse will be sharper in his responses or will perform a new or difficult maneuver better than when you left off. A good training program combines daily sessions with an occasional break. This allows the horse time to soak up the training and come back to his lessons refreshed. Intermittent schooling also helps prevent overdrilling, sourness, and boredom. If a horse becomes bored with his lessons, he will stop trying to learn. Strive to keep training interesting, and give your horse a break when he needs it.

Enlist the Aid of a Role Model

Inexperienced horses learn some things more quickly if a trained horse sets an example. Just as a foal learns to eat grain by mimicking his mother, so does a green horse tackle a new challenge — crossing a stream and stepping over a log, for example — more willingly if he sees another horse doing it. When taking a young horse on a first ride across country, it's easier if an experienced horse accompanies him. By observing his companion, he will learn that mud puddles, barking dogs, cars along the road, and birds that fly up are not so scary.

Horses take cues from other herd members. If one panics, they all prepare to flee. In the wild, horses depend on one another's acute senses for survival. A spooky horse communicates fear to the herd. A calm horse traveling with a youngster, on the other hand, communicates security and steadiness. When encountering unfamiliar or threatening situations, the green horse takes courage from the fact that the other horse is

---

CHECKLIST FOR TRAINING

❑ Keep in mind the basic nature of the horse and how horses communicate and learn.

❑ Take into consideration the temperament and personality of the horse you are working with.

❑ Make sure rewards and corrections fit the situation and the individual, and that they follow immediately after the good and bad actions.

❑ Arrange to have lessons take place in an environment free of distractions.

❑ Use repetition (building step-by-step on earlier experiences) to reinforce proper actions and good habits.

❑ Use logical lessons that last no longer than a horse's attention span; try to end on a positive note while he is doing things right.

❑ Always try to improve your communication with the horse and take advantage of the way horses learn.

not afraid. With this reassurance, the youngster is able to settle down and relax, and again pay attention to his rider.

Take advantage of role models. Load a young horse into a trailer with a buddy for first lessons, or pony (lead) him from a calm horse. Tie the foal next to his mother, or tie a young horse next to a dependable stablemate for his first tying lessons. You can also ease a horse through a trying situation if he trusts and accepts you as he would a dominant herd member. If you project a relaxed attitude, your horse will more readily lose his fear, just as he would take cues from a more experienced and secure herd member.

How to Avoid Conflict and Common Pitfalls

No matter how far you progress when training your horse, there will be times when you and your horse are just not on the same page and experience a conflict. This is partly because we (horse and human) are not machines, but rather individuals with moods and feelings. Sometimes we simply fall out of sync. Also, the horse-human relationship is always in flux. The young horse is growing up, learning more, going through phases and stages. Like parent-child, teacher-pupil, boss-employee relationships, the horse-human relationship is never static. Training is a dynamic process that we can't take for granted. We must constantly work at it.

The horse's mind is like that of a young child. He wants to please you, but he may also want to exert his independence. If he can get away with some little thing, he may try to get away with more. Some conflicts arise because you have not communicated well enough with the horse and he does not understand what is being asked. But other arguments happen because he is trying to see what he can get away with. Perhaps you were too lax in tolerating small infractions and now he is becoming bolder in his disrespect or disobedience. You must be able to pinpoint the cause of a conflict and deal with it appropriately.

Know your horse. If it's a horse you are just getting acquainted with, you must feel your way along. Sometimes you'll have to revisit more basic lessons to regain the horse's respect. (See chapters 3 through 5 for more information.)

Above all, try to understand the horse and why he reacts the way he does; evaluate your own actions. Every horse is different and will teach you while you are teaching him.

Control Your Temper

Anger and frustration have no place in training a horse. At times you'll become impatient and frustrated, but try to avoid or at least minimize these reactions as much as possible; they get in the way of proper communication. If you are angry, you are no longer in tune with the horse and the communication is gone. Horses learn through repetition and in response to positive or negative reinforcement. If you punish in anger, the horse will react in fear or confusion and is less likely to give you the proper response. Anger will usually make your horse afraid to try that same movement again because he will associate it with punishment. And this can undo all previous training.

TAKE A DEEP BREATH

When you lose your temper, you lose control of yourself and no longer think clearly. The rush of adrenaline can make you overreact and put you at risk if you forget the rules of safety and good horsemanship. If you are physically fighting with a horse, you may do things you would never do otherwise, again putting yourself in danger. In addition, the horse will be reacting blindly, from alarm or pain and possibly in self-defense — perhaps trying to bolt, buck, or kick — and this too puts you at serious risk.

When you lose control, you lose the key to staying in sync with and in control of your horse. If you tend to get impatient or lose your temper easily, beware of the early warning signs and head them off.

Begin each training session with a positive and open attitude. If you are already upset about something that messed up your day, don't work with your horse until you have put that behind you.

If you feel anger coming on during a lesson — when the horse reacts improperly or does something that displeases you — try to calm down before it escalates and damages the training session. If he is doing something wrong, stop what you are trying to teach him and go back to something he can handle better, so he can do it right and you can both feel a sense of accomplishment. Remember your priorities and goals. The horse's welfare comes first. You want a well-trained horse and you don't want to ruin your progress by losing your temper. Tell yourself to relax and focus.

If you absolutely cannot halt the impatience and anger, stop the lesson. Let the horse walk freely on a loose rein (or let him stand, if you are doing ground work) while you calm down. After you are more relaxed, think about what triggered the anger. Don't blame the horse; take responsibility for your temper flare-up. You will have to work on training yourself as well as your horse!

Dealing with a Bad Attitude

If you know your horse is having an off day or has a certain phobia, how you handle him and what you ask him to do can make a difference in whether or not you have a confrontation. Sometimes the best way to deal with a problem is to avoid it; in this case, do something that will create a positive effect for that lesson rather than a negative one.

On the other hand, if you get into a situation in which the horse exhibits a bad attitude, how you deal with it makes a difference in whether it is resolved positively or becomes a hindrance to further progress. If the horse is aggressively disobeying, such as breaking into a trot when he's supposed to walk, or backing or rearing instead of standing still, it's important to let him know immediately that his behavior is unacceptable.

Correction should be instant and appropriate: Check with the bit to keep him at the lower gait, exert leg pressure to hold him in position and keep him from backing, use firm leg pressure along with a pull to one side so he cannot rear, for example. When you are consistent in your correction, he will eventually stop trying to make these decisions on his own and instead wait for and respond to your cues.

DISTRACT WHEN NECESSARY

Avoid situations in which you are at odds with a horse and cannot win. If there is a problem and you are not able to resolve a particular conflict, find something else for the horse to do that will make him think you have won and are still dominant. Don't fight; shift gears and ask him to do something else that you know he can or will do, so the result will be his willing obedience rather than his victory. This enables you to keep the psychological upper hand and can help prevent further conflicts or attempts at disobedience.

Use Proper Timing and Patience

Instant response (correction when needed, reward when he does it right) enables the horse to understand what you want (or do not want) him to do — that is, he will associate your response with his action. If you don't correct a bad action immediately, he'll assume he can do it whenever he wants.

If your conflict with a horse is due to his inability to understand your signals, patience rather than correction is needed. Try again to communicate, this time more clearly. There must always be some compromises. If you can't get him to do something using one method, try another. If you get stuck, don't struggle. Do something else for a moment, end the lesson on a positive note, and ponder the problem before you try again. It often helps to sleep on it. By taking some time away, you might come up with a better way to handle the problem or to resolve the communication impasse.

If you can think about it for a while, you should be able to figure out ways to "explain" it differently or create different circumstances that will make the lesson easier for the horse or the problem less challenging. The horse will also have a chance to give things a fresh look. Some lessons soak in better after a horse has a little time to absorb them.

3

Handling and Training the Foal

It's easier to train a horse if he has been handled properly as a foal. It's likely that he trusts and respects people and is accustomed to being restrained and accepts it. The older horse that has not been gentled — that is, he has not yet learned to accept people and trust them — can be difficult to work with because he is so much larger and stronger than you are.

When training a foal, the wrong kind of handling will cause him to be afraid of people or aggressively disrespectful and spoiled, both of which will make later work difficult and dangerous. Always handle and train a foal with conscientious care.

Learning Begins with Imprinting

A newborn horse's brain is more developed than is a human baby's. The foal's critical period for learning and soaking up new information occurs immediately after birth. He must learn who his mother is, get up and find the udder, nurse, and follow his dam *and* be able to run alongside her to flee from danger. The foal does these things instinctively. A foal who isn't able to do these things at birth won't survive in the wild.

Because of this urgency, the equine baby learns much faster than a human baby. Human infants are primed for "instant learning" for the first few years of life, a much longer period than the foal. The foal, in contrast, is a sponge for learning during only the first hours of life and

A young foal is able to run with his mother soon after birth.

becomes like an adult horse in temperament within a few days, suspicious of changes in his environment, wary of new experiences. The most advantageous time to make a lasting good impression on a foal, therefore, is right after birth. The newborn does not innately fear humans, but if the first time he sees a human is after he is a few days old, he may be afraid of this unfamiliar presence.

What Is Imprinting?

To survive, prey animals such as horses, deer, and geese are genetically programmed to absorb a lot of information, to attach and bond with objects seen immediately after birth (usually the mother), and later to flee from anything unfamiliar. This process is called *imprinting*. Imprinting can only be done in the first hours of life, when the brain is programmed for maximum receptivity to experiences. *Conditioning* and *desensitizing*, on the other hand, are training methods that are used later. The only time a horse can be imprinted is immediately after birth.

Just as newly hatched goslings follow the first moving object they see, identifying it as their mother and ensuring a good parent bond (a survival instinct that helps them stay close to the parent bird), a newborn foal reacts to what he sees and experiences immediately after birth. If you can control what the foal sees and experiences, you can have a lasting, positive effect.

Acquaint the foal with humans and the various things that will be done to him by humans so he won't fear them. If he sees only his dam during the first hours of life, he may later need to be taught that humans are not something to be feared. If his first experience with humans is when he is cornered and caught and given injections — and a worried, protective dam gives him the idea that people are something to avoid and flee from — it may take much patient handling and training before the foal begins to trust you. If, however, you are there when he is born and his first impressions include you and the things you acquaint him with, his brain soaks up this experience. He will accept it and be more receptive to handling by humans.

History of Imprint Training

The first Europeans to encounter horses of North African tribes in the 1700s were amazed by the horses' responsiveness to their owners, a result of early handling. These people lived with their horses, carefully tending pregnant mares and paying immediate attention to each newborn foal. Foals grew up in constant contact with humans. They were even allowed in the tents, which created a strong bond between them.

Horsemen in many other cultures, including Europeans, simply ignored the early training of foals. Training manuals, both ancient and modern, are typically geared to the education of two- and three-year-old horses. Only recently have some people come to recognize the value of educating foals.

Imprint training, training during the first hours of life, was formalized by Robert M. Miller, a veterinarian and horse breeder who has been using it since the 1960s. Done properly, handling of a foal at birth makes later training easier, because the foal has an early memory of being handled and accepts it more willingly. Thus, he is less anxious and fearful about things you do with him later. He has already experienced having his feet and ears handled, for example, and is comfortable with it; he remembers the experience as nonthreatening.

Goals of Imprint Training

There are two basic goals for imprint training.

1. Immediately after birth, the foal should have the smell, touch, and appearance of humans imprinted on his brain. The memory of this experience will assure him that future encounters with humans are to be accepted and not feared.

2. The foal should be exposed to experiences, on a limited basis, that he will be expected to tolerate later, such as foot handling, grooming, touching under the tail, and mouth handling, and become submissive to humans.

Imprint training is usually accomplished in several sessions soon after a foal is born: the first while he's still lying down, before he gets up to nurse; the others, after he has nursed. Whatever you expose the foal to in these brief sessions will be accepted later. It's important to do the imprinting properly; mishandling at this time can leave a permanent bad memory about humans.

PLAN AHEAD AND BE VIGILANT

If you plan to imprint the foal, have the mare foal in a pen or stall where you can observe her and then attend the birth. If she foals at pasture some night and you miss the event by an hour or more, you will have lost the opportunity for immediate imprinting.

Session One

Following is a list of what you'll need:

◆ One or two helpers
◆ Very small foal halter
◆ K-Y jelly lubricant or petroleum jelly
◆ Hoof-pick handle or something similar for tapping feet
◆ *Optional:* plastic grocery bag, electric clippers, spray bottle containing warm water, anything else that he will be exposed to that you want to accustom him to

Imprinting works best if the mare is a mellow, trusting individual who is not worried about you handling her baby. It is not recommended if she is a nervous first-time mother and your presence interferes with proper bonding of mother and baby or if she is an aggressive, protective mother who may become dangerous if you work with her foal. Know your mare; use good judgment. If she is upset about your handling the foal, forgo imprinting. Even if she doesn't attack you, she will transmit her anxiety to the foal.

If the mare trusts you, begin as soon as the foal is being born. Kneel behind the mare, making sure the amnion sac is off the foal's head as he emerges. Touch and stroke his head and neck as he lies there with his hind legs still within the mare as she rests before getting up. While she is resting, and as the foal is still gathering strength and organizing himself to try to get to his feet, touch and rub much of his body. If the mare trusts you, this will not disturb her, especially if you are kneeling or squatting low to the ground. If your presence alarms the mare, back off. You don't want her leaping up and breaking the umbilical cord prematurely. After she gets up, disinfect the foal's navel stump and proceed with the imprinting session, as detailed in the accompanying illustrations.

Sometimes circumstances change your priorities. For example, imprinting is the last thing you'd do if you were faced with the more critical task of dealing with a potentially life-threatening problem at birth. But in most instances, with a normal birth and good conditions, you'll have a chance to imprint a foal.

Halter the mare after she gets up and have someone hold her. Allow her to stand by the foal, to lick and nuzzle him, thus creating that important mother-baby bond. She and the foal need to interact as soon as she gets up, to "lock in" each other's identity, but this can be accomplished as you imprint the foal.

Restraining the Foal

The first lesson should begin while the foal is still lying down — right after the umbilical stump has been disinfected. Wait until the mare gets up before you begin, if you are not sure the mare will allow you to handle the foal. (Don't startle the mare into getting up too quickly, as this will break the umbilical cord too soon). The foal may be trying to get up by that time, but it is important that the first session be done while he's still down. This works best if he is born during mild weather, with no urgency about getting him dried, up, and nursing (or into the barn) before he becomes chilled.

Most newborns will be struggling to get up, but you must keep the foal lying on his side for the first session. Kneel behind him and hold him on his side by putting one arm across his upper body, holding on to his topmost front leg and keeping it bent against his body. Cradle his head with your other arm, with his neck bent around so that his muzzle is pointed at his withers. (See illustration on page 78.) Hold him in this position until he stops struggling and relaxes. If at any time during the session he tries to get up, hold him again in this position so he is unable to, and wait until he relaxes again before you continue. The mere fact of not being able to get up creates submissiveness in the foal.

Imprinting Step-by-Step

It can be difficult to hold down a strong foal by yourself when you do the imprinting. Be especially careful not to hurt the foal while trying to restrain him. Three people may be needed — one to hold the mare, one to hold the foal, and one to do the imprinting.

While the foal is gently restrained in the down position, perform each step until he is at ease and accepts what is happening. Speak softly, comforting and reassuring him so he won't panic and struggle. Work on one side of the foal at a time.

Stimulate each body part until the foal submits. Thirty to one hundred repetitions may be necessary. If you aren't sure whether you've done enough, continue a bit longer. You can't overdo it, but if you stop too soon, the foal may not be completely desensitized, which will defeat the lesson.

1. **Face and head:** Gently rub your hands over his face, head, and poll until he relaxes completely. Put on and remove the halter repeatedly until he is oblivious to this procedure.
2. **Ears:** Gently rub the outside of the ear. Insert a finger partway and rub the inside of the ear.
3. **Nostrils:** Gently rub each nostril one at a time. Insert a finger up to the first knuckle and gently wiggle it, until he accepts it.
4. **Mouth:** Gently lift the foal's upper lip, then rub and press on his gums. Put a finger into his mouth and rub the bars on either side.
5. **Neck:** Rub his neck from top to bottom, on all sides, including the poll, and gently bend his muzzle around toward his withers.
6. **Body:** Rub and gently pat all exposed parts of shoulder, ribs, chest, and upper legs. Do *not* rub his flank or the girth area where you would apply leg pressure when he is an adult being ridden. These places must remain sensitive to touch. Also, the foal's dam will nudge or nip him in those areas when she wants him to move.
7. **Legs and feet:** Gently flex and extend the joints of the leg. Next, pat and rub the bottom of the hoof, then tap gently with a hoof pick, until he ignores it.

8. **Belly:** Gently but firmly rub the underbelly and groin, sheath, or udder. Lubricate a gloved finger with K-Y lubricant and insert it gently into the rectum.

9. **Optional:** If you wish, rub the foal all over with a plastic grocery bag; touch him on his ears and body with electric clippers to accustom him to the sound and vibration (but do not clip him); aiming away from him, spray with warm water to get him used to the sound of the spray (but don't get him wet).

10. **Repeat on the other side:** Turn over the foal by grasping his forelegs and gently rolling him onto his back until he is lying on the other side. Take care not to be kicked as you turn him. Repeat all steps on the other side.

If you must stop before finishing all the steps, be sure to complete the step you are working on and go back later — as soon as possible — to finish up. To pick up where you left off, wait until the foal is lying down for a nap after he nurses, kneel behind him again, and complete the various steps. You can interrupt his nap to finish session one, but then let him nurse and nap before beginning the standing sessions.

COLOSTRUM

The foal should nurse within an hour or two of birth to get the antibodies he needs for disease protection. These antibodies come from *colostrum,* or first milk. If you have not completed the first imprint session by the time the foal is an hour old, finish with the body part you are working on, let the foal get up to nurse — and be sure he *does* nurse — then continue with the imprinting after he has napped.

Don't Stop Prematurely

The most common mistake people make when imprinting a foal is stopping too soon — that is, not completing the lesson. Continue each lesson until the foal stops struggling and relaxes completely, seemingly oblivious to any manipulation of feet, ears, mouth, whatever part of his body you are working on. Any new stimulus must be given repeatedly until it no longer alarms or even interests the foal. If you stop too soon, while he is still resisting, he won't be completely desensitized to the action and will remember it as something to be resisted. At this critical

IMPRINT TRAINING

When imprinting the foal, continue each activity until he accepts it and submits to it.

A

B

A. *Kneel behind the foal and restrain him on his side, holding onto his foreleg. With your other arm, gently bend his head toward his withers, until he relaxes and no longer tries to get up. Rub his face, poll, and head until he relaxes and does not resist. Then put a foal halter on and off his head at least ten times, until he is utterly passive about it.* **B.** *Rub his ears, then put a finger partway into each ear and wiggle it around until he completely accepts it.*

C

D

C. *Rub each nostril, one at a time, inserting a finger to the first knuckle and wiggling it until he accepts it.* **D.** *Handle his mouth, rub and press on his gums, and insert a finger into the corner of his mouth and wiggle it.*

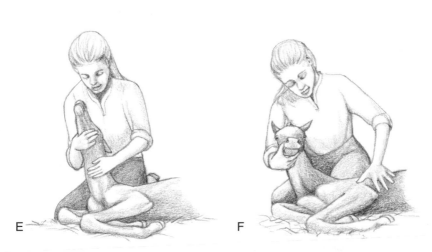

E. Rub all parts of his neck until he no longer resists. **F.** Rub his body — shoulder, rib cage, chest, upper legs — everywhere except the flanks and just behind the girth area.

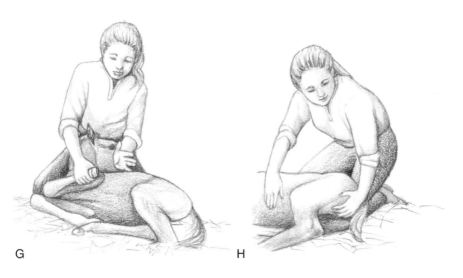

G. Handle and flex all leg joints. Rub and pat the bottom of each hoof, then tap it with a hoof pick until he ignores it. **H.** Rub his belly and groin area, including sheath, or, if a filly, udder. Insert a lubricated, gloved finger into the rectum.

stage, the foal learns what to fear just as easily as what to trust, so take care never to hurt or scare him. If he is scared, he will struggle. If he struggles frantically, halt the lesson and hold him until he stops struggling, taking care not to hurt him. In this way, he learns to submit and learns that he is not and won't be hurt.

Remember, too, that if imprinting is not done correctly, it does more harm than good. If the foal is flighty or independent and you don't continue the desensitization process long enough, you will reinforce his flighty or strong-willed determination to resist humans. For example, if you are working with an ear or flexing a leg and halt before the foal submits, he will have learned to jerk it away whenever you try to handle it later. Always continue a session until the foal is passive and accepting.

Session Two

In session two, you'll undertake the same ten steps as you did before, but this time with the foal standing up. If you managed to complete the first session before the foal got up to nurse, allow him to get up and nurse (remember, make sure that he nurses within 2 hours of birth so he receives the colostrum) and take a nap before you begin. If it took several rounds of work between nursings to complete session one, let the foal nurse and nap a few more times before beginning session two.

In this session, plan to work in intervals of no more than 15 minutes. Let the foal nurse and nap (napping at least 15 minutes) and nurse again between steps. Allow the foal to stand on his own before beginning so you don't wear him out. Because the foal tires easily, it usually takes several rounds of work to complete the standing lessons. Be patient, and take as much time as is necessary for him to accept each step. You may be able to accomplish two or three objectives per standing period; the actual amount of time you'll need will depend on how long it takes the foal to accept each step.

Have someone restrain the foal while you work with him, this time by holding an arm in front of his chest and an arm behind him. He should be held loosely, meeting resistance only if he tries to move forward or back. Once again, work on one side of the foal at a time. Continue working until the foal is utterly relaxed and accepts what you are doing.

In these standing sessions, you can teach control of movement, encouraging the foal to move forward, back, and sideways on command, to produce a responsive, obedient, respectful horse.

The safest way to restrain a foal is with your arms.

If You Miss the Delivery

If the mare and foal are already up before you get a chance to imprint, first make sure the foal nurses so he gets his colostrum. Then halter the mare and have someone hold her facing the foal. Speak to her reassuringly and let her know you are not harming the baby. The mare may be worried; her instinct is to protect the baby and to teach him to follow her example only. If the mare is cooperative, go ahead and work with the foal. If she is alarmed, she will transmit her anxiety to the foal, defeating your purpose. (See Gentling a Timid Foal on page 83 for more suggestions if imprinting isn't possible or if you've missed the chance to do it.)

If the foal is on his feet before you discover him, you can do the imprinting while he is standing, but instead of a single, hour-long session, break it into four 15-minute increments, letting the foal nurse or nap in between. After a nap, before he gets up again, enlist the aid of two helpers: one to hold the mare and one to gently restrain the foal in the down position. Then perform the ground session, reassuring the foal as you work with a soothing tone of voice.

Days Two and Three

If done properly, the lessons a foal learns through imprinting during the first hours of life will remain with him always. He will be more accepting of having all areas of his body handled — for bridling, saddling, trimming, and shoeing; having his temperature taken; receiving injections and medical treatments. The few hours it takes to complete this work immediately after the birth of a foal will definitely pay off later.

After the initial imprinting sessions have been completed, you can conduct additional sessions over the next few days to reinforce the learning. A common mistake is to omit or inadequately do the subsequent lessons. During these sessions, emphasis is on control of movement. Horses establish dominance hierarchy by controlling the movement of their peers. Omitting these subsequent lessons in control of movement will produce a foal that is tolerant of body handling but disrespectful.

PRESERVE THE MARE-FOAL BOND

In all you do with a young foal, consider the mare-foal relationship and how your actions will affect or be affected by that bond. If you work with the foal in such a way that you never alarm the mare, she will not transmit her worry to the foal. Work with the foal in an area close to the mare. Use the mare to help catch him, to be an example and reassuring presence when you first lead or tie him. Tie her when you tie the foal. Have someone lead her as you lead the foal. If he has Mama to follow, he will lead readily. Using the mare as a partner in the training process will make your job much easier.

Handling a Foal during the First Weeks

If you had a chance to imprint the foal at birth, he should be easy to handle later, even if you work with him only occasionally. When he and his dam are at pasture, you may not have a daily routine for handling them, but if the mare is well mannered, it will be easy to catch and work with the foal whenever you have an opportunity. If the mare and foal are in a confined situation or you are feeding them, you'll be able to develop a daily routine for handling the foal.

Gentling a Timid Foal

It's best to handle a foal every day, particularly if you didn't imprint him. If you handle him frequently as he grows up, you can gentle him and do fundamental training, thus making him easier to work with when he is older. The young foal learns quickly. You want to keep his experiences positive and build good habits and responses. As long as you keep lessons short, a young foal is often easier to train than an older first-time pupil. And it is usually easier to gain his confidence when he is young. The first time you catch and handle a foal, especially if you didn't have a chance to handle and imprint him at birth, it's best to use a small pen or a box stall, in which you can quietly corner him next to his dam. You'll be able to catch him without alarming him if you can move up to him without the mare moving away. Then he can't get away from you or get up enough speed to hurt himself crashing into a fence or wall. If the mare is elusive, catch her first and have someone hold her while you gently corner the foal.

Get Acquainted

The first step in gentling a foal is to let him get acquainted with you so you can gain his trust. Work in a pen or stall. Many foals are curious and will come right up to you. Others are timid; it will take more time to win them over. Tie the mare, brush her, clean her feet while you talk to the foal. Let him get used to you and realize you're part of his daily routine. As he grows to accept your presence and is no longer afraid, you can handle him and get him accustomed to the feel of your hands all over his body.

USE YOUR ARMS

The best way to restrain a foal at first is with an arm around his chest and the other around the hindquarters. (See illustration on page 81.) It's better to restrain him with your arms rather than with a rope or halter; you'll be less likely to hurt him if he struggles. Foals can move swiftly, and you don't want him running into a wall or fence. When holding him with your arms, block his movement but don't use pressure as long as he is standing still — just keep him from going forward or back. If he relaxes and stands, release all pressure so your arms are just gently encircling him.

Handle the Foal

Before you try to catch a youngster in a pasture, he should be handled in a small pen or stall — enough that he accepts you without resistance. If he willingly accepts you in the small area, it's usually easy to catch him at pasture, as long as Mama is cooperative. If he is still a bit skittish, however, and discovers he can elude you and keep his freedom, he will try to get away from you in the pasture, so perfect his catching in a small pen first.

If the mare and foal are at pasture and the foal is hard to catch, create a small pen in the pasture, using portable panels. When you let the mare into the small pen for grain once or twice a day, the foal will follow her. Then you can gently corner and catch him.

Handle him from both sides and get him used to being touched and rubbed on the neck, back, abdomen, and legs. Gently handle him all over, including head, ears, and tail. Talk to him softly: Constant soft talk has a soothing effect and will help him relax.

EACH FOAL IS UNIQUE

Foals are all different. Some require more handling to gentle. Some are curious, friendly, and never hard to catch. One may be your buddy, always coming up to you and following you around. Another may want nothing to do with humans and must be gently cornered for several weeks before he realizes that being caught and handled is nothing to worry about. A few foals require months of handling to get them over their fears or to bend their independent nature to include you in their lives (you'll wish you imprinted them). But patient and diligent daily handling pays off; the independent foal will eventually resign himself to your dominance.

Catching a Timid or Elusive Foal

Spend time with a timid foal, touching him, talking to him, getting him accustomed to your presence and actions so he is no longer suspicious. This may take days or weeks, especially if Mama is overprotective and tries to take him to the far corner of the pen or pasture every time you approach. Foals learn attitude and behavior from their dams. If Mama conveys the idea that you are scary and not to be trusted, your job will be difficult unless you convince the foal otherwise.

If the mare herself is elusive, catch her first and tie her in a corner or have someone hold her so you can quietly corner baby next to her — between you and the mare and the stall wall or fence. She won't be able to move off at just the wrong time or station herself between you and the foal. Give her some grain while she's tied so you can work with baby. If he learns to eat grain, you can use this to your advantage, to help him associate you with good things.

SEEK MAMA'S HELP

Halter the mare and have a helper lead her into the corner of the pen or stall and stand at her head. Follow the foal as he moves up next to the mare. If he tries to get away from you by ducking under the mare's neck to go in front of her, your helper gently blocks his way and you can block him from behind. The mare's body and the fence or stall wall form a place to gently corner him.

Restraining a Foal

If a foal needs an injection or enema, one person should hold and restrain the foal while another administers the treatment. He can be held more securely by your arms than with a halter and rope. This not only is safer, but it also makes an impression on his young mind; he realizes he can't get away from you and must submit to and respect restraint.

A large foal may be too strong and lively to be restrained with just your arms. Use one arm around the neck and grasp the base of the tail with your other hand, lifting it firmly but gently

Use the tail hold to help restrain a larger foal.

upward. With tail held straight up, most foals won't try to back up. Raising the tail upward and forward — but not too forcefully — will generally quiet a foal and immobilize him, and this triggers the reflexes that straighten his hind legs. This hold should keep him from kicking and struggling.

Be careful how you hold a foal when restraining him for medical care or in any other situation in which he may struggle. Don't hold him around the rib cage or pick him up by the rib cage if he falls down; a foal's ribs break easily. If you wrestle with him in a stressful situation, don't put pressure on or lift him with your arms around his abdomen, as this could injure him internally. On a small foal, you should be able to immobilize him adequately with an arm around his front and the other above his hocks. If you give him lessons in control of movement soon after birth, he will respect the gentle pressure of your arms.

Beginning Leading Lessons

Before you try a halter, the foal should be accustomed to restraint by your arms. You can teach him to move with you, also. With one arm around his chest and the other around his hindquarters, push him along from behind and guide his direction. This is the easiest way to control him when you take him and Mama somewhere before he is halter-trained; someone can lead the mare while you keep the foal contained and under control, behind or alongside her. In this same manner he can have his first "leading" lessons: Someone leads the mare as you and the foal follow her. Thus, he'll become accustomed to being restrained by you as he follows Mama in a controlled fashion, and it's an easy step from this to being led at halter.

USE A PROPERLY SIZED HALTER

Because a foal grows quickly; it's difficult to keep his halter the proper size. Use one that is adjustable for the smallest foal and that can be let out and made larger until he is a few months old. Then you can use a regular foal halter, adjusted to its smallest size, enlarging it until he outgrows it. A yearling halter can then be made small enough to fit him.

Beginning Halter Lessons

After the foal is gentled and accepts handling, put a halter on him, small enough to fit properly. A halter too loose may pull off his head or snag on something. He could get a foot caught in it if he paws with a front foot or scratches an ear with a hind foot. Never turn a foal loose with a halter on. If he is hard to catch, work on that aspect of training. It's better to put mare and foal into a stall or pen for catching than to leave a halter on a foal.

Haltering the Foal

The first few times you halter the foal, gently corner him between you and the mare or have a helper hold him so he can't back away or resist. Put it on carefully and gently, so as not to startle him. Plan every movement you make to make sure it provides a good experience rather than a frightening one. It takes a lot of work to undo a bad experience!

Putting a halter on is a simple matter if the foal was imprinted and had this done at birth. If not, take care not to alarm him. Don't bump his muzzle or ears. Hold him securely so he can't avoid the halter. Try to put it on in such a way that he won't struggle. Slip the noseband up over his muzzle and then fasten it behind his ears in a fluid, gentle movement

First haltering: Slip the nose-band up over his muzzle, then fasten it behind his ears.

that does not threaten his eyes or ears. After he has worn it a few minutes, take it off and put it on again, getting him used to the procedure so he'll realize it's nothing to be afraid of.

When you work with a foal, move slowly and deliberately; sudden movements may startle him. If he is touchy or jumpy about certain things, slow down and work gradually on that area again — whether it's a touchy leg or his ears when you halter him.

Leading in a Stall or Pen

When you start lessons with a halter, don't try to hold the foal just by the halter and rope; he may throw himself down. Use a rump rope behind him — an extra length of lead rope works for this, or use a separate rope. If he tries to rush backward, you can stop him with that instead of having all the pressure on his head and neck, which he will resist more.

Foalhood is the best time to teach a horse to lead, before he is in the habit of hanging back on the halter rope for instance, or refusing to lead at a trot. Most foals learn to lead easily.

USING A RUMP ROPE

Create a "come-along" loop with a soft cotton rope that is long enough to go around the foal's hindquarters. Hold the ends of the rope in your hand, so you can pull on it to encourage him to move forward, or make a loop that goes around his rump, with the free end long enough for you to hold in your hand or to pass through the halter.

When leading the foal, allow him some freedom of head and neck. If he goes too fast, he runs into pressure of the halter; if he goes too slow, he feels the pressure of the rump rope. If he responds properly, reward him instantly by releasing pressure, thus setting the stage for good response in the future.

The youngster should learn to walk quietly beside you, pacing himself to your speed. At this stage you're as strong as he is; he can't get away from you if he tries to take off. Walk beside his left shoulder, lead rope in your right hand, fairly close to the halter for good control. This way, he cannot charge over the top of you, and you also have control of his head if he decides to go too fast. Many foals lead naturally, walking beside you — until you try to lead them away from Mama or out of her sight.

Until he is further along in his lessons (and older, so he and Mama won't be so worried if they get a short distance apart), stay close to the dam or limit the lessons to walking along with her — one person leads her while you lead the foal alongside or behind her. The foal will go willingly wherever she goes. A daily routine, such as leading mare and foal in from pasture to a small pen for lessons or back and forth from pasture for part-day grazing, is ideal training for the foal. It gives him regular handling along with good leading lessons, and when he is a little older, more independent, and not so insecure, you can start leading him farther away from Mama.

Some bold, curious foals won't protest when you lead them away from Mama, enjoying the adventure, but many refuse to stray from her

Leading the mare and foal.

DON'T PULL THE FOAL'S HEAD

Never try to pull a foal by the halter. He's bound to fight it, thus creating a tug-of-war. Restraint on his head may alarm him, and he may rear up to avoid it, with the risk of falling over backward and injuring himself. Instead, use a rope around his hindquarters to hold him in place if he balks or pulls back. With a rump rope, you control his movements better and the pressure is not all on his head if he fights restraint. If he tries to rear or throw himself down, you can keep him in place. If he falls, control his head with the lead rope by pulling on it so he won't hit his head on the ground. Always work with him in a safe place where the ground is soft.

---

and will balk. The best way to avoid a fight is to use a loop around the foal's hindquarters; this encourages him to move with you. Place it loosely over the rump so you can pull on it if necessary. It always works better to have the encouragement come from behind when a foal balks, rather than trying to pull on his head.

When he feels pressure on his hindquarters, he'll want to move forward to get away from it. He soon learns that when the loop begins to tighten the easiest thing to do is move forward. Keep it slack when he moves out freely and willingly; use pressure only when he hangs back. Be sure the loop is not so slack that it hangs down and bumps him on the hocks as he walks. After he leads well, you will no longer need the rump rope.

Once he gets the idea of walking beside you, take longer walks. Some foals never need the rump rope; others need it for quite a while if they are stubborn or insecure without Mama. You may need a refresher lesson with the loop if a foal decides to challenge your control. Your well-mannered youngster may suddenly think he wants to have his own way and resist during a leading lesson. Go back briefly to using a rump rope until he gets through the rebellious stage and realizes he still must obey you.

Beginning Lessons in Foot Handling

While working with the foal, teach him to pick up his feet. He must learn how to balance himself and not resist. Handle his feet often so he accepts this as part of his daily routine and won't try to fight when you need to clean his feet or trim them. He will be well mannered later for the farrier, too.

FIRST DESENSITIZE THE LEGS

If a foal was not imprinted and is ticklish about having his feet handled, touch his legs a lot before you try to pick up a foot. While someone else holds the foal, run your hands down his legs. If he resists or kicks at a hand, just keep patting the leg until he realizes you are not hurting him. Don't stop patting the leg if he kicks or it will teach him he can avoid having his legs handled. Touch the leg until he is relaxed and will stand quietly for you as you run your hands up and down each leg.

Picking Up a Foot

Until the foal is halter-trained enough to tie, always have someone hold the foal for you while you handle his feet, so he gets in the habit of standing still. A calm and steady influence at his head, or restraint with arms when he is young and not yet haltered, can make matters much simpler.

Before picking up a foot, run your hand down his leg a few times and make sure he is not afraid of having it touched. If he picks up the foot, go ahead and hold it briefly. If he fights, keep holding it, then put it down as soon as he relaxes. If he does jerk it away, pick it right back up again so he won't think he can take it away whenever he wants to. At this age, never punish him for taking a foot away; just keep working at it and be patient. You want to instill trust and relaxation, not fear or resistance.

When you first pick up a foal's foot, he won't know how to balance on three legs. Lean into him a little to help him to shift his weight. A foal that gets off balance may panic and try to take his foot away to put it back on the ground. You can prevent most struggles by helping him shift his weight so he can balance more easily on the other three legs. Help him along by picking up a foot he isn't already standing on with most of his weight.

If at first he is reluctant to give you his foot, tickle the back of his heel a little or press on the back tendon, between cannon bone and

---

BE PERSISTENT

When a youngster throws himself down in a tantrum, let him. But as soon as he's back on his feet, pick up the foot again. Soon he'll realize it is a lot better to let you hold the leg for a moment. As soon as he stops fighting, put it back down. He'll learn to relax when he discovers you're not hurting him, and that when he doesn't fight you, the foot is released. If he stands quietly, the foot is picked up and put down. He will then learn to tolerate having each of his feet picked up briefly and repeatedly.

---

tendon, just above the fetlock joint to encourage him to pick up the foot. He'll soon learn to shift his weight to the other three legs and pick up the foot when you give it a little tickle or soft pinch. If he has trouble keeping his balance, lean into him to steady him and keep hold of the foot until he relaxes. He must learn that you are the one who decides when to put the foot down. Do not reward him by letting him put down the foot when he struggles or resists.

It's easier for the foal to balance with a hind foot up than with a front foot up, as he carries more weight on his front legs. But don't hold a hind foot too high or too far out from his body, or he may become unsteady and try to take it away. Hold it quite low at first, and only for a short time, and the foal will be less apt to resist.

Handling the foot while a helper restrains the foal.

Start the foal on lessons in foot handling early, so he'll be at ease with this aspect of care. All training sessions and grooming should include routine handling of the feet. Most foals need to have their feet trimmed at least once before weaning (and many need two or three trimmings, depending on how their feet grow and wear), so get him used to having his feet picked up.

Some horses develop bad habits about feet. Perhaps the feet weren't handled early enough, and the horse's first experience with foot trimming was traumatic. Perhaps a timid owner tried to handle feet too late, when the horse was a big weanling or yearling, and the horse learned he could rear up and take away a front foot, or jerk away a hind foot. Maybe the horse had an injured leg and now associates foot handling with painful wound treatment. The best way to avoid these scenes is to handle feet from the time the horse is a baby.

THE ADVANTAGES OF STARTING YOUNG

If you handle his feet as a young foal, you can teach him he cannot take a foot away from you; at this stage, you are stronger than he is. He learns that you are in control and that you are not hurting him, and comes to accept foot handling. He will learn to relax and give you his foot instead of fighting, kicking, or trying to throw himself on the ground.

Beginning Lessons in Tying

After the foal knows how to lead and respects the halter and your control over him, it's time to tie him. But before you tie him, simply loop the halter rope around a post or pole. Have the foal stand as though tied while you hold the other end of the rope and give slack, if necessary. He will probably stand there quietly at first, thinking you are holding him. Then you can gradually move a little farther away, still holding the rope; eventually he'll realize that you are not holding him — the fence is. He will come to understand that the fence can restrain him just as you do.

When you tie him, use a strong halter and rope that he can't break, and always tie to something solid. Never tie to a wire or net fence of any kind; he might injure himself if he paws at it and gets a foot caught. Don't have too much slack in the rope; your foal might get a foot over it

By tying the foal next to Mama, you put him at ease. Stand behind him the first few times to help keep him from setting back hard on the rope.

or go over backward in his struggles. Tie him high and short. Stay with him at first so you can get him out of trouble if he has a problem.

And keep these lessons brief. The foal has a short attention span. He will do better with frequent short lessons rather than occasional long ones. A young foal needs to nurse and nap often and should be tied for only a few minutes.

Tie Next to Mama

It helps if the mare is well halter-trained. Tie the foal next to her. She will be a good example, and the foal will be more at ease near Mama. If you tie them both briefly every day, you can gradually lengthen the lessons until mare and foal are standing tied for 20 to 30 minutes. If the mare doesn't tie well, give her some hay or grain to eat in the corner of a pen or stall while the foal is tied next to her.

Precautions for the Young Foal

When tying a young foal the first few times, loop the rope around a pole or post, without tying him hard and fast. Then you can monitor his reactions and give him a little slack if necessary so he cannot hurt himself, then take it up again after he becomes calm. Once he resigns himself to this flexible restraint, you can tie him solidly, but don't leave him unattended.

If he fights the rope and falls down, he'll need help if the rope holds him in such a position that he can't get up. Always tie with a knot that can be undone quickly, no matter how tight it gets (see page 26). With a small foal, you can keep him from setting back or throwing himself down by standing behind him. If he sets back, he bumps into you and you can help keep the pressure off his head and neck. This is safe to do with a young, small foal, but not always safe with an older, large foal.

Tying the Older Foal

If the foal is large and strong by the time you begin tying lessons, make sure he does not injure himself or you during first experiences. Dally the rope around the post (wrap it around the post a few times), tying him solidly *only* after he is accustomed to flexible restraint. Use a body rope or tie to something that will give, like an inner tube, so he won't hurt himself if he pulls back. A foal that pulls back may hurt his neck if he is tied solidly by the head.

Using a Body Rope

In order to avoid injury, and to help teach the foal he cannot pull free, tie with a cotton body rope, so most of the strain comes on his body instead of just his head and neck. Put the rope around his girth and tie it

If using a body rope in addition to a halter rope, allow some slack in the halter rope so the body rope will take the strain if the foal sets back.

under his belly with a nonslip knot so it cannot tighten and put pressure on him when he pulls back. Run the free end between his legs and up through the halter ring under his jaw. Tie him to a post or something else secure that will hold him. If you are using a separate rope in conjunction with the halter rope, tie with both ropes, making sure that most of the strain will be taken by the body rope rather than the halter rope.

Usually the young horse will pull back strongly only a few times. Once he tries it and discovers he cannot pull free, he will respect the restraint. Some headstrong individuals don't give up easily, however, and need more lessons — tie them daily with the body rope until they learn the rope is stronger than they are. Then they will resign themselves to standing patiently when tied.

Using an Inner Tube

Another method that works is to use a deflated inner tube fastened securely to a stout post. Use a new tube without tears, holes, or rot. Tie the foal to the inner tube rather than the post. When he sets back, the tube stretches and has some give, so he is less apt to hurt his neck, but goes back to its original shape when he comes forward. No matter the method you use, stay nearby during the first tying lessons in case your foal gets himself in trouble. Do not leave him unattended until he is well halter-trained to tie.

Avoiding Problems

The rope should be the final master. Never reward the foal with freedom during his struggles or he'll think he can free himself by fighting and pulling back. If you have to untie him or cut the rope to get him out of a bad position, tie him right back up again. After he submits to the fact that he must stand there, then you can untie him. Most foals halter-train quickly and easily, but a few stubborn youngsters take more time and effort. Routinely tying a youngster every day will eventually produce a well-mannered, halter-trained foal.

Grooming Lessons for Handling and Patience

You don't need to do much actual grooming with a young foal, but you should accustom him to grooming tools and having various parts of his body touched and rubbed. This is especially valuable for later training, particularly if you did not have a chance to imprint him at birth. After he is trained to tie, it's easy to groom him while he is tied. If you think he might be ticklish about grooming, which might cause him to set back, tie him with a body rope.

Rub Him All Over

At first, rub just with your hand. When your foal gets used to being touched all over his body, try a soft cloth, then a very soft brush. It will upset him least if you start at his withers and along his back, then gradually work up the neck. Most foals are sensitive about their faces, so go very gently there. Work around his rump, under the abdomen, and down the legs. Each time you groom him, it will get easier; he will be more reassured that you are not going to hurt him and will soon come to enjoy the rubbing and brushing.

Prevent Bad Habits

The foal will readily develop bad habits if you let him — things that seem cute when he's young won't be cute later. Don't let him chew on

TALK TO THE FOAL

Horses like to hear you talk; a constant stream of soothing tones or humming will help keep them relaxed and focused on you. Talking softly and humming can also calm a nervous horse; a loud voice will make him nervous.

Tone of voice is important in training any horse, especially a foal. Your voice should be soft, soothing, and reassuring. *Easy* is a good word for him to learn, used to encourage him to relax. This will come in handy later when you need to calm him.

A strong, severe voice should be used only for reprimands. With many horses, tone of voice is sufficient reprimand; no physical correction is required. Like a child, the horse knows from your voice whether he has pleased or disappointed you. A horse also learns his name by hearing you repeat it often and directing your voice to him. Most horses quickly learn to come when called, particularly if you use the same tone each time.

the rope or nibble at the brush or on you. When he is not tied, don't let him wander around as you work with him. He must learn to stand quietly in one place as you groom him and handle his feet. A tap on the chest and the command *Whoa* whenever he starts to move forward will usually cure him of wandering.

Advanced Leading Lessons

When you are handling the foal every day, you can perfect his responses in all lessons, including leading. As he gets older and more independent, you can lead for longer periods and farther away from Mom.

Leading Solo

After a foal is several weeks old, most mares become less worried about baby and he's less worried about being apart from his dam. You'll be able to leave her tied and take him for longer walks, gradually switching his dependence and obedience from her to you. As long as you are still within her sight, such as leading the foal around in the pasture while she grazes or is tied eating grain, she will generally tolerate these separations. If she worries or if the foal is insecure, take things more slowly — go just a little farther each day as the foal gains confidence. Don't create a stressful situation by suddenly taking him out of sight. A screaming mare and a frightened foal will defeat your purpose.

Leading at the Trot

You can teach a foal to lead at the trot after he leads well at the walk, or this can wait if he's rambunctious. For the calm foal, however, it's the next logical step. Give him the cue to trot by making "clucking" or "kissing" noises and by moving your own feet faster (trotting in place), then begin to move faster. He'll soon catch on and begin to trot as soon as you give this signal. Eventually, you can cue him to move right into a trot from a standstill just by "clucking" or giving the trotting signal with your feet. He also learns that clucking always means "move."

Carry a little switch or crop in your free hand in case he balks. Simply reach back and gently tap his rump as you begin to trot; this will usually make him move faster. Take care not to alarm him, of course, or he may jump forward faster than you want. A switch or training whip is often necessary when teaching older horses to lead at the trot, but foals usually catch on more quickly; many will not need this cue at all.

Leading at the trot.

If your foal hangs back and won't trot, use a switch to urge him forward rather than using a rump rope around his hindquarters. The rope loop works well for teaching him to lead at the walk, but may spook him while trotting, especially if it moves around on his rump or hind legs. You don't want him distracted or annoyed by a rope flapping at his hocks or constantly urging him faster by bumping his hindquarters.

Most foals are quite lively; often the main problem is not getting them to move out at a trot, but rather holding them *down* to a trot after they start off. Never let a foal get up too much speed or he may be able to jerk away from you. He can gallop much faster than you can! Your loss of control can undo a lot of training. With some foals, it's best to save trotting until later, after they accept restraint.

With most foals, however, including trotting in the early education pays off later. If a foal learns to respect restraint at this age, he will be more manageable as he gets older. Teaching him to lead at both walk and trot makes him more maneuverable and responsive. He learns to control his exuberance and move at your command, traveling quietly and responsively at your side, at whatever gait you ask for.

Basic Lesson Planning

The nice thing about starting a horse's training when he's a baby is that you have a chance to minimize bad experiences and teach him all the things you want him to be able to do when he's most receptive to learning. Early lessons can be challenging but are also a wonderful opportunity.

Goal-Oriented Lessons

Always work with a purpose and keep long-range goals in mind. Even if you didn't get as far today with a lesson as you had hoped, you still made progress if your horse had no bad experiences to set him back. A good horseman has both time and patience. Don't be disappointed if you see little progress at certain points. Time is on your side.

Everything you do with the foal is training. You teach him either good habits or bad habits. Curb bad habits when they begin. You want a well-trained, obedient yet affectionate horse, not a spoiled one that does whatever he feels like. If you let him get away with breaking rules when he's young, he'll grow up to be a bad-mannered or headstrong horse. Cultivate his respect and trust, using tact and patience; and always insist on proper behavior.

Keep Lessons Short

Frequent short lessons are always more beneficial than occasional long ones. Never keep working with a foal after he becomes tired or bored; he will begin to resent and resist the lessons. He has a much shorter attention span than does an older horse; don't stretch it too far. Gauge his mood and end the lesson on a positive note while he's still trying, learning, and doing things right. Don't wait until he's tired and fighting or doing things wrong. One of the most important aspects of being a good horse trainer is knowing when to stop for the day.

Avoid Spoiling the Foal

Most foals are inquisitive and sassy, which is part of what makes them so lovable. Unless these tendencies are properly channeled, however, they will become bad habits and bad manners. A youngster may be feeling frisky and you may be tempted to play with him, but don't. The small foal who nips at your clothes or kicks up his heels won't seem so cute when he's larger than you are and takes a bite out of your arm or kicks at you on the way by. Curb these playful tendencies before they get out of hand.

Spoiled Foals Are Hard to Handle

The spoiled foal is much more difficult to train than is a timid foal. The more fearful individual can be controlled and taught that nothing will hurt him as long as he behaves, but a spoiled foal is not afraid of anything or anyone. He thinks no one will hurt him, no matter what he does, because he has gotten away with this behavior before. If someone tries to discipline him, he may fight back or think it's just a game. A spoiled youngster usually grows up to be bad mannered or mean-tempered.

Prevent Bad Actions in the Beginning

Exercise good judgment, and you can usually keep from giving a sassy foal an opportunity to bite or kick. This is important because there are times when you won't be in a position to properly punish or correct him. It is very bad business to let him kick or nip at you and run off; he feels as though he is in control and can do as he pleases.

If he nuzzles you, make sure he gets only to your hand and not to your body. Be one jump ahead of him and tweak his nose or meet his muzzle with your elbow, a flip of your fingers, or the poke of a finger if he starts to get nippy. (For more on maintaining personal space, see page 420.)

Foals are quick; they can get in a fast nip and be away again before you can punish. This is how they play games with one another. You have to anticipate a nip and prevent it. Don't give the foal an opportunity for naughtiness. Try to see or sense it coming and act quickly. You must be able to outsmart the foal.

Some foals go through exasperating stages of being nippy, hard to catch, or in some other way uncooperative. They are testing you to see if you really are boss. If a foal is handled with good judgment, tact, patience, firmness, and kindness, however, he will usually outgrow these little rebellions and become well mannered.

ACT QUICKLY AND WITH RESTRAINT

To be effective, correction must immediately follow the misdeed, or the foal won't know why he's being punished. If you must reprimand him, don't overdo it. Never strike him above the muzzle or around the ears; this will make him head shy.

Dealing with Challenging Personalities

Some foals are easy to work with from the time they are born; others are frustrating and take more care and thought in handling and training. Try to figure out the best way to handle the difficult foal. This way, you avoid mistakes that would hinder your progress in training him.

The Aggressive Foal

A bold foal may consider you a playmate to roughhouse with rather than an instructor to be respected. It can be a challenge to determine how much patience and how much firmness are needed to keep him in line. As always, firmness and consistency are very important. As your foal grows up, he may continue testing you if he feels he is in control or if he is determined to gain control. A foal with this personality is easier to handle and train if he was properly imprinted at birth to accept restraint. If you begin his lessons later, his training will take much diligence until he learns respect.

Precocious Colts

Many male horses become aggressive at an early age. Some will act like stallions, mounting their mothers, playing rough-and-tumble games with other foals and trying to play rough with you, unless you impose rules of behavior as soon as possible. Colts tend to be more nippy and aggressive than fillies, often trying to dominate you and other horses with their teeth and feet. You must thwart this behavior before it gets out of hand. The young colt has to learn he cannot get away with nipping, kicking, rearing, or striking.

Castration during Foalhood

Unless a colt is destined for stud duty, he should be gelded. Young males are easier to handle, work with, and train if they are gelded at an early age. Many horsemen wait until a colt is one or two years old before gelding him, but this isn't necessary and can be dangerous. A young stallion is a nuisance — he must be kept separate from mares and fillies after weaning — and is more apt to get in trouble, perhaps trying to go over or through a fence to get to fillies or mares.

Even if he has a wonderful disposition, the young stallion is more unpredictable than a mare or gelding. You'll avoid many problems if he is gelded as a baby, and surgery is much easier on him at that age than it will be after puberty. Unless you plan to keep a colt for breeding purposes, have him gelded as a foal.

Handling and Training the Colt That Will Be Kept as a Stallion

A colt with a domineering attitude and aggressive disposition should *not* be kept as a stallion. Even the young male with a good disposition will have his "male" moments, and care and consistency should be paramount in handling and training so that he always respects and trusts you. A good rapport and complete understanding of each other will make the training more successful and enjoyable for both the colt and you.

He needs to be handled with kindness rather than force — force from you is generally met by retaliatory force from him — and with firmness and consistency. There must be no doubt in the horse's mind that you are the leader in the partnership. If you can work with him regularly, starting from when he is young, you can check bad habits before they become established. If he tests you continually, however, he needs to be handled by someone else who can better command his respect or he should be gelded.

The Timid Foal

Some youngsters are insecure and afraid. These take much patient care and handling to ease their fears so they learn to relax and trust you. It may take many lessons and gradual stages to achieve the same progress that was gained in a shorter time by the easygoing or bold foal. There are no timetables for certain lessons. It may take days or weeks of gentle encouragement to win over the timid foal, but he may progress swiftly once you've built a solid foundation of trust.

The Headstrong Foal

Perhaps most challenging is the independent foal who is resistant to your efforts. The headstrong youngster may be exasperating in his attempts to avoid being caught or his stubborn determination to fight restraints. Regular handling and diligence are necessary to convince him to accept your leadership.

Some independent individuals need firmer handling and more force-ful methods, such as daily tying with a body rope, use of a rump rope while being led, and harsher punishment for nipping, to bend them the right direction and create willingness rather than resistance. Try to determine the best way to handle each individual. Your goal is to aid progress in training, not hinder it.

Learning Ability of the Foal

Every foal requires an individualized training program. Your foal may be a fast learner, but don't make the mistake of trying to teach him too much at once. Don't confuse him. Make sure he has learned one thing well — leading freely at the walk, without balking, for example — before going on to something new, such as leading at the trot. After you have worked with him several times on a particular cue or movement, he will become more responsive or cooperative as he discovers what you want him to do.

If you confuse or frighten him, or if you let him get away with bad manners, it's easy for his negative reaction to become habit. Be a perfec-tionist in training him, with total flexibility and in tune with his needs. Try to do everything properly and at the right time for that foal. Think first of him and anticipate all possible reactions to whatever you do.

No two foals are alike in personality and learning ability. Some foals need very few lessons on several aspects of training and more drilling on others. For instance, one foal may learn to tie up without ever pulling back hard, yet be sassy, independent, and headstrong while being led. One may have no qualms at all about having his feet handled but has trouble learning to lead at a trot. Another may pull back hard on the tie rope again and again before he resigns himself to being halter-broken, yet be easy to handle for other lessons. What works well in training one horse may not work at all for another. Be flexible, exercise good judgment, and use a great deal of ingenuity. Treat each foal as an individ-ual and tailor your training program to him. Progress at his pace and emphasize the things he needs work on.

In your enjoyment of a foal, don't neglect his basic training. Handle him wisely, often, and well, setting the stage for later training. All of your work with him will be easier if you have laid a strong foundation for learning from the beginning.

4

Handling and Training the Weanling

Weanling age — the period from weaning, at five or six months, to yearling — is a good time for training. The foal is on his own, no longer influenced by Mama. His young mind is curious and he is capable of learning a great deal. Proper training now can make a big difference in his later progress. If he is halter-trained and accepts you as leader, it will make subsequent training easier because he has the basics — good manners, respect, and "learning how to learn" — already mastered.

If he is not properly handled at this age, however, and becomes spoiled or has bad experiences that leave him fearful, later training will be more difficult. Weanling age is an extremely formative period; it's your chance to set patterns for future progress. You want to make sure those patterns are good ones. If he wasn't handled as a foal, you'll begin setting those patterns now. If he was handled as a foal, you'll build on his early lessons.

Weaning a Foal

Weaning is a stressful time for any foal (and for the mare), but this stress can be short-lived and less intense depending on your weaning methods. The trauma of separation is more emotional than physical, so take this into consideration. The younger the foal, the more stressful the weaning, in most cases. A five- to six-month-old foal is usually more independent and better able to handle it, both physically and emotionally, than is a

---

THE OLD WAYS AREN'T NECESSARILY THE BEST WAYS

The traditional way to wean is to separate mare and foal completely. The foal is put in a strong pen or box stall where he can't jump out. Foals run themselves to exhaustion and whinny themselves hoarse in frantic efforts to get out and find their mothers. After a few days, they resign themselves to life without Mama, but this abrupt and traumatic separation is always very hard on them.

---

younger foal. Some must be weaned early, if a mare doesn't give enough milk, for example, or gives too much, which will cause joint problems in the foal because of too quick growth. But there are measures you can take to make the transition go smoothly.

A Low-Stress Weaning Method

As we learned in chapter 2 (see page 44), horses are herd animals, and they are happiest when they are with other horses. Young horses feel very insecure without adults. Even if you wean several foals together in an isolated pen, they will all be frantic. The least stressful way to wean is to put mare and foal in adjacent pens where they can see, hear, and smell each other. They accept this better than a total separation because they still have each other for company. The foal at five months no longer needs milk if he is on good feed. But he is still dependent on the mare emotionally, and if you take her clear away it leaves him panicky.

When a foal and his mother are placed in adjacent pens a few days, he still has the emotional security of her nearness. Foals weaned in pens next to their dams usually spend time near the fence, but they are not very worried. There is little whinnying or pacing. They nuzzle each other at the fence, but the foals cannot nurse, so the mares dry up their milk. After a few days, they can be moved farther apart with little fuss, especially if the foal has other horses for company.

Minimize Stress at Weaning Time

Too much stress hinders the immune system and leads to illness. Traumatic weaning (such as sudden separation of mare and foal) produces stress. Weaning time is not a good time to deworm, vaccinate, or halter-break a foal. Either do these before weaning or wait until a few weeks after weaning, when he has adjusted to his new lifestyle and won't be so upset and stressed. Treatments and training are easier on him if he is accustomed to handling. With more experience, deworming, vaccinating, and trimming feet won't be traumatic for him or for you.

WEANING TIP

Build a fence for weaning in such a way that the foal can't reach through to try to nurse. Make sure it's high enough that neither mare nor foal will try to jump it; 6 feet is a good minimum. V-mesh or diamond-mesh wire works best. The foal cannot put a foot or nose through it but can still see and smell Mama, which eases the transition.

Handling a Foal at Weaning Time

Wait until a few weeks after weaning to begin serious training. But weaning time can be an ideal time to gain his trust if you haven't worked with him much before. Some foals grow up on pasture with their dams, without supplemental feed, and do very well nutritionally (this is the most natural way to raise a foal), but these foals often don't get much early handling unless a person makes time for a daily training program.

At weaning, however, with Mama gone, the foal has to look to you for food and comfort. Even if a foal hasn't had much handling earlier and is still either timid or very independent, you can handle him regularly after weaning and soon win him over.

The Untrained, Hard-to-Catch Weanling

It may be a challenge at first to gentle a foal that is in a pen or pasture with room to run and get away, especially if he was not imprinted at birth. (For more on imprinting, see page 72.) Feed him in the same place daily and stay nearby for a few minutes or until he relaxes while he eats. Each day, gradually move closer to him. As he comes to tolerate your presence and your nearness, he will eventually let you touch him and, finally, halter him. Then you can make the transition to catching him before you feed. Most weanlings will learn to come at your call, stand quietly for haltering, then eat the food you brought.

GET HELP WHEN YOU NEED IT

If a youngster is impossible to catch, you may need several people to help you corner him in his pen against a solid, safe fence or to ease him behind a gate, or to quietly herd him into a barn stall where he can be more easily cornered. Then put a well-fitted halter on him — snug enough that he cannot get a foot through it if he tries to scratch an ear with a hind foot — that you can grab when you get close to him until he gets over being elusive.

Occasionally, however, you encounter a youngster who refuses to let you catch him. You may have to resort to more creative methods to get your hands on him, then leave on a close-fitting halter, in a small pen, for a while so you can catch him. Do this until he learns he cannot play games and run off.

As a last resort, if he won't let you get close enough to his head to get hold of his halter, let him drag an old halter rope for a few days. Keep him in a small, safe pen where there are no protrusions or hazards that could catch or snag a halter or entangle the dragging rope. Letting him drag a halter or rope is always risky; no matter how careful you are, it can still get caught on something. Generally, it is not safe to leave a halter on a horse, but it can be a useful temporary solution for the challenging youngster who has not had proper handling and gentling. Using a halter with a trailing lead rope for a few days, in a safe pen, can help you catch him while you are getting him accustomed to you and daily handling.

Even if he is evasive at first, you can always catch him in the small pen by quietly moving about until you are able to pick up the trailing rope. Never chase the youngster, especially if he is afraid. Walk around in the pen ignoring him, so he won't view you as a threat, and eventually work close enough to get hold of the rope. He soon learns he can't get away from you.

If you always handle him with quiet confidence and understanding, he will have no reason to run. If he is independent rather than afraid, he learns he cannot avoid being caught. If you catch and handle him several times a day, gaining his trust, he realizes there is no reason to try to get away and becomes easier to catch. Soon you can dispense with the trailing lead rope and then the halter.

Train Consistently

Ideally, you have handled the foal and he respects you. But even if you worked with him a lot and he's trained to lead and tie, keep handling him as a weanling to further his experience. The result will be a well-mannered horse when he grows up.

Polish the Leading Lessons

Your foal will need ongoing lessons as a weanling. Often as a horse grows up, he becomes more independent; some aggressive youngsters will challenge your authority. It's a good idea to keep doing some regular lessons so he continues to be well mannered and manageable.

If the weanling did not have leading lessons as a foal, you must start at the beginning, using a rump rope or whip as a training aid (see page 88). At this stage he is larger and stronger than you are and can be a difficult pupil if he decides to pull back or resist. Avoid a contest of wills, especially a pulling contest with the halter rope; use the rump rope to encourage him to move forward when you ask and to keep him from halting or balking.

For the youngster who doesn't know how to lead by walking alongside you (or suddenly decides he doesn't want to), use a whip as an extension of your arm, so you can reach back and touch him on the hindquarters. If he's hanging back on the rope or refusing to walk briskly

---

MAKE THE HORSE "AMBIDEXTROUS"

Handle the young horse equally from both sides. If you do this from the beginning, he won't favor one side. A horse that is always led from the left does not lead readily from the right. There may be times when you'll need to lead him from the right — in an emergency when you must lead two horses, for instance, one with each hand — or must approach and catch him from the right, or mount from the right (in steep terrain you'll mount from the uphill side, whichever side it happens to be). If you teach him to lead equally well from both sides, he will be more versatile and you'll be less apt to have a problem later.

beside you, a tap on the hindquarters will usually encourage him to move forward. You can also teach him to move forward in response to clucking ("kissing" noises). This will come in handy later, when you want him to move forward on the ground or under saddle. Give a clucking cue and immediately follow it with a gentle touch with the whip if he does not respond. He'll soon get the picture and move forward at the sound of clucking alone.

Anything you can do to encourage him from behind instead of pulling at his head makes a more positive and lasting impression, and he will be less apt to resist. With a large youngster, a pulling contest is one you can't win. Judicious use of a rump rope, stock whip, or a long willow stick will get much better results.

DON'T PULL

Some horses become pullers because of the way they are led. If you constantly pull at a horse to try to keep him from going too fast, he responds by lugging harder into the halter. A handler may unintentionally cause this reaction by active pulling or even by just a heavy hand on the lead rope. Always give a led horse some slack so he can accompany you in a relaxed manner. It takes two to have a tug-of-war. Any pressure you exert on the halter to slow or stop him should be intermittent, not continuous. Well-timed short tugs are more effective than a steady, hard pull.

Leading the Overeager Youngster

Some young horses aren't content to walk quietly beside you; instead they try to pull or drag you faster than you want to go. The pressure of a halter noseband is not enough to deter their charge-ahead enthusiasm. In these instances, you may need a lead shank with a short length of chain at one end. This method can be helpful if the young horse didn't have enough handling as a baby and is a large, strong weanling or yearling before you teach him proper leading manners. Remember, you don't ever want the leading lessons to become a pulling contest. Diligent and careful lessons can resolve most "charging" problems, but a few individuals — especially young stallions — may require the use of a chain. Never overuse a chain, however, or you might damage the cartilage in the horse's nose.

The chain is used over the nose and fastened to the halter so it puts pressure on the nose only when you want it to; at all other times it is slack enough that the horse will not feel it. To attach a chain properly, start at the horse's left side. Insert the end of the chain from the outside and put it in through the ring on the side of the halter. Once it is through the side ring, run the chain downward — you don't want it wrapped around the noseband — then drape it over the nose and clip it onto the inside of the ring on the right side of the halter. All the links of the chain in the nose area should be touching the horse.

Put a knot in the lead, just back of where the chain fastens on to the rope. This way, when the horse tries to bolt or go too fast, your hand should be able to hold the line when it meets the knot. If you are using a nylon lead that tends to slip through your hand too easily, tie a knot at the end of the lead with additional knots spaced about a foot apart, along its length. These will keep the lead from slipping through your hands if the horse tries to pull away.

Fold the extra length of line and hold it with your left hand. Your right hand holds the lead where the chain meets it. Walk beside or just ahead of the horse's shoulder, with about a foot of slack in the lead. There is no set rule for body position, as the size (and neck length) of each horse will vary.

If the horse tries to bolt or travel too fast, the lead will be pulled through your hand to the first knot and stop there. This sudden jerk will cause the chain to meet his nose. Like a choke collar on a dog, it is engaged only when he goes too fast. Most horses stop or slow down when the chain puts pressure on the nose; then the chain will automatically loosen, release pressure, and give an instant reward for stopping or slowing. It should not exert pressure unless he starts going too fast or unless you ask him to stop.

To engage the chain at other times to make the young horse

A chain can give better control than just a halter if a horse needs another reminder about restraint.

behave — if he is fractious and trying to rear or drag you along — give one firm downward jerk on the lead, then immediately slack off. Never keep steady pressure on the chain or its effect will be greatly reduced.

Teach Him Whoa and to Stand Quietly

A young horse is often rambunctious, but he must learn to stop and stand quietly. Teach him that *Whoa* means stop and stay stopped. If you combine the voice command with the signal from his halter, telling him *Whoa* each time you put pressure or let him run into pressure with the halter, he will soon make the connection.

Each time you halt, make him stand a moment before you begin any new movement. Let him stand on a slack lead with no pressure. Even if he's fidgety, don't keep pressure on the halter. Standing still should not become a contest of strength. Every time he starts to move, let him bump into halter pressure. You are a "post" to which he is tied — he meets resistance only when he moves and is then rewarded with slack when he stands. Don't let him move around after he halts. If he fidgets, insist that he stand. If he learns from the beginning that *Whoa* means stop and stay stopped, you'll have fewer problems later.

Demand Good Manners When Grooming

A horse also must stand quietly for grooming. If you are consistent in making him stand — and never tolerate infractions such as moving about; rubbing on you, the fence, or the stall wall; nibbling at the rope — he'll have good manners as he grows up. Bad habits devel-

A young horse must learn to stop and stand quietly.

oped at this age become frustrating or even dangerous when he is older if he dances around stepping on your feet or shoves you into a wall. Insist on good manners while he is young, with reprimands when he misbehaves and praise when he is good.

Take Him on Long Walks

As you progress with training, lead the youngster on longer walks. This reinforces his respect for restraint so he is manageable when you take him farther from his stall or pasture. He will be less likely to balk at going away from home or to try to drag you back at a gallop. As you gradually go farther in your walks, he learns about new sights and sounds and becomes more familiar with things he may encounter later, when you start riding him. Lead him down the lane, or on a back road, or out in open country behind the back pasture so he can learn about bushes, rocks, gullies, and wildlife.

The more you do with the horse while he's young, the more tolerant he'll be of new experiences as you continue training. The youngster who has never been out of his home pasture or barnyard can be very insecure when you start riding him. Training is much easier if he has had some gradual acquaintance with new things all along.

Training for Ease of Handling

Much of early training is desensitizing the horse to things that would otherwise be alarming to him, things he must cope with for the care he will need during his life. An untrained horse is difficult to vaccinate and deworm and won't stand still for trimming or shoeing. Procedures we must do for health care and general management are well accepted by most horses only because they have been trained to tolerate them.

Lessons for Routine Health Care

These tasks are more pleasant for him and you (and the farrier or veterinarian) if the youngster is accustomed to routine handling. If he's been gradually prepared, less restraint will be needed and a task like deworming won't turn into a battle of wills and strength.

Mouth Handling for Deworming and Bitting

Mouth handling is an important part of the young horse's routine. He should get accustomed to having his mouth and ears gently touched. Most horses like to be rubbed, and even timid ones learn to accept having their ears and face rubbed if you work up to those areas gradually: Start at the neck; they are not so ticklish there. In the process, also rub the muzzle and mouth area, and stick a finger into the corner of the mouth.

Most horses don't mind a finger in the mouth if you don't startle them with it. Rub the side of the mouth, then put a finger into the corner. As he gets used to this on a daily basis, or whenever you groom him, wiggle the finger around in the interdental space (where there are no teeth). Do this periodically, and the young horse will accept it readily. This will be a tremendous advantage when you put a dose syringe into the corner of his mouth for deworming or other medications. He will also be less suspicious of having a bit put into his mouth later.

Vaccination Ease

Some horses are needle-shy because they were not desensitized as youngsters or because an inept person caused them pain during vaccination. But you can train the young horse to tolerate injections.

A properly administered injection causes no pain (the horse doesn't even know he's had it), but advance preparation ensures that it's accomplished easily, with no alarm for the horse. Usual injection sites are the side of the neck (a few inches above the shoulder blade, midway between the mane and the underside of the neck), buttocks, and pectoral muscles (on either side of the breastbone), and occasionally the large muscle covering either side of the rump.

These locations should be touched, rubbed, and pressed during grooming, so the young horse is not sensitive or ticklish there. A common method for giving injections is first to press the target area firmly before inserting the needle; pressing desensitizes the spot and also ensures that the horse is aware of and accepting the touch.

On areas with mobile skin, such as the neck or breast, sometimes a twist of the skin will help desensitize the spot before insertion of a needle. Do this periodically as you groom the horse; as it is painless, he will tolerate it readily and thus will not be alarmed when a needle must be slipped in.

> ## BE FOREWARNED
>
> It can be a challenge if a horse is a weanling or older before he has routine foot handling or had a bad experience with his first foot trimming. The young horse is larger and stronger than a foal and can be a difficult pupil if he already has his mind made up about not letting you pick up his feet.

Foot Handling for Trimming and Shoeing

Every young horse should have his feet handled as part of daily grooming or training routines, so he will be comfortable with this when they are cleaned or trimmed. It's best to start handling feet when he is a baby, but if this wasn't done routinely (or if you purchased a weanling or yearling that didn't have his feet handled as a baby), pay attention to this part of his training.

If the youngster is well halter-trained, you can do foot handling lessons with him tied. Often it works better, however, when an experienced person holds the young horse and reassures him as you attempt to handle his feet. Whether he should be tied or held for these lessons depends on the individual, his stage of training, and his personality.

Some young horses jump around more when held; they are better off tied. Others will set back when tied and are better off held. If the young horse is a real problem about his feet, first refresh his lessons at halter (to stand patiently when tied and when being held) and have him become more dependable in that phase of training before you try to wrestle with his feet.

Start as you would with a foal, getting him relaxed about having his legs touched and brushed. Pick up his feet as described in chapter 3 (see pages 91–92), encouraging him to shift his weight to his other legs to keep his balance. If one foot seems to give you more trouble than the others, spend extra sessions on it.

If he kicks or tries to jerk away a hind foot, determine whether the horse is reacting from fear, distrust, habit, or annoyance. Knowing your horse and his usual reactions can help you determine the reason for a particular response. If he is truly afraid, you must get him to relax and trust you. Never punish a timid horse for kicking or taking away a foot. Be persistent and keep working with the foot until he realizes that no harm will come to him.

CORRECT WHEN NECESSARY

A horse that is reacting out of stubbornness or trying to dominate you needs more firmness. Correction, such as a swat, may be in order if he jerks away the foot or takes a swing at you.

Whenever you're patient and persistent, you'll make progress. Don't worry if at first you can't hold a foot very long. Hold it briefly, then put it down at your discretion. Don't allow the horse to determine when to put it down. If you can hold it for an instant and put it back down without him attempting to take it away, this is progress. Praise him, then go on to another foot, working around to that one again. As you work with him daily, pick up each foot more often and for a longer period. Every time you do anything with him, pick up his feet as part of the routine.

If he does try to take away a foot, hang on. Don't give in. As soon as he relaxes, put it down. Soon he'll learn that he cannot take his foot away and will realize that you're not hurting him and that you'll put it back down when he's calm.

After the young horse accepts having his feet picked up, don't just pick up a foot for a couple of seconds and put it right back down again. This does not prepare him for having his feet worked on. Hold the foot in shoeing position. Hold a front foot between your legs, hold a hind leg across your thigh. Hold up each foot for a longer time each day. Carry a hoof pick in your back pocket and when he is comfortable about having his feet handled, take a few minutes every time you work with him actually to clean his feet.

Get the horse used to foot handling. Hold the front foot between your legs and use a hoof pick to clean or tap on it. Hold the hind foot fairly low at first, in a position comfortable for him.

Even if there's no dirt in his hoofs, go through the motions of cleaning the feet. When he gets used to having his feet held in a working position, as for cleaning, he won't be so impatient or upset when the time comes for trimming or shoeing.

Tap on the hoof with the hoof pick to get him used to the feel of it: you want him to accept having nails pounded. Add rasping to the routine, a few swipes with the rasp on each foot. The more tools you acquaint him with, the more at ease he'll be with how his feet are handled by the farrier. Most foals need their feet trimmed once or twice as babies and weanlings and several times as a yearling. It's best if you can handle the feet ahead of time so trimming is not traumatic. Set the stage for trimming and shoeing to be good experiences, not something the horse resists.

Hold a foot in a position that does not cause discomfort or anxiety, not too high, not twisted out to the side. The young horse must learn about having his legs held in working position, but take care not to put any strain on leg joints.

TRIMMING TIPS

Gradually extend the time you handle feet, keeping your horse comfortable with it. If feet need trimming, whoever does it should never hold up a leg so long that the youngster becomes impatient. To avoid a wrestling match, trimming should be done swiftly, so the leg can be put back down before the horse starts resisting. This may mean doing the job in stages: working on one foot and then going on to the next, coming back to each one again to finish after the horse has had a chance to stand on it. If you work this way, he will be more comfortable — and cooperative — because his legs get a rest.

Fly Spray and Other Bugaboos

A few lessons getting your horse used to the sound of a spray bottle in practice situations will pay off later. Give him as many lessons as needed to get him over his fears. First use the spray at a distance, then, as he comes to tolerate it, come closer. Because you have prepared him, when it comes time to use fly spray or wound spray, your horse won't go into orbit.

Life Lessons

Horses don't reason like we do, so training is based on progressive, step-by-step lessons — conditioned responses to various situations. They learn to tolerate things that don't hurt only after they discover they are not hurt by certain actions, and learn to respond to our requests by finding they are rewarded by release of pressure when they do what we ask.

A horse can be taught to tolerate activities he instinctively fears if we show him there is no cause for fear. Just as a cavalry horse was taught to tolerate gunfire and action of battle, a horse can with patient practice, learn to tolerate many unnatural situations. What you teach him will depend on your plans for his future. If he will be a ranch horse or rope horse, he must get used to cattle and to ropes touching his body. If he's a show horse, he must tolerate being clipped. Think ahead and, while he is young, plan to give him some "life lessons" for his particular career.

If you will be clipping him, for example, give him lessons before you have to do it for real. First get him used to the sound of clippers. After he realizes the noise is nothing to be afraid of, hold them closer and let him feel the vibration. Once he tolerates that, you can do a little clipping, starting on the areas that are least ticklish and sensitive, then finally clipping him wherever you need to.

Despooking a Future Trail Horse

If you'll be riding the horse in open country, there are many things you can do to help facilitate this transition when he is too young to ride. It's best not to start riding a horse until he is a long two-year-old or preferably a three-year-old. During leading lessons, take him to interesting places so he encounters new things gradually and at his own pace as he becomes less fearful and more confident. Lead him through gates that you open and close, mud puddles (this may take several patient lessons if he is not familiar with walking through water), ditches, and gullies.

Around the barnyard, accustom him to walking calmly past wheelbarrows, tarps, vehicles, and other scary objects. Have someone drive slowly past him as you hold or lead him until he gets used to the noise and movement of cars and trucks. There are many things you can acquaint him with before he must encounter them out on the trail — bridges, dogs, a clothesline in the wind, anything you think he needs to become more at ease with. Many of these can be presented to him in

USING A STABLEIZER

For a young horse that is nervous about clipping, one way to get him used to it at the same time you have to clip him is with the Stableizer (Wheeler Enterprises Inc., Ellendale, Minn.). This tool is a more effective, more humane version of a war bridle, and often works better than a twitch, lip chain, or tranquilizer for calming and immobilizing flighty or unruly horses. It's a loop that goes over the head, putting pressure behind the ears and under the top lip next to the gum. When it is tightened by pulling the cord handle, it stimulates pressure points behind the ears and beneath the lip. This causes the release of painkilling endorphins and blocks adrenaline production, causing the horse to relax and feel good. He doesn't worry about being clipped, having his feet trimmed the first time, or any other new experience that might be frightening — and remembers feeling good about it while it was being done. The Stableizer can make the experience a pleasant one, and after a time or two the horse will tolerate the clipping without it.

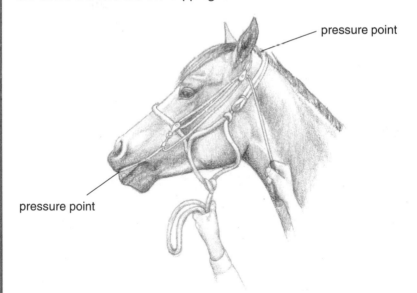

pressure point

pressure point

The Stableizer applies pressure behind the ears and under the top lip next to the gum. When tightened by pulling the cord handle, it stimulates pressure points in these areas, releasing painkilling endorphins. The result is a relaxed horse who feels good.

nonthreatening lessons in his own backyard, where you can give him plenty of time to check them out and approach them at his own speed and comfort level. (See Crossing Water on pages 147–149.)

How to Love Your Young Horse without Spoiling Him

Lots of handling is good for the young horse if it's done properly. The neglected, unhandled youngster is easier to train in later life than the spoiled horse raised as a pet and allowed to develop bad habits.

Practice Consistency

A young horse must learn what to expect from you, to know how to react to what you ask of him. His reactions will be consistent only if you are consistent. If you are not consistent in how you handle him, he will become confused and may stop trying to do the right thing. Make a practice of doing things a certain way, giving signals a certain way, so he will always know what you mean. Without confusion or doubt, he can give the correct response.

If you're consistent in what you expect from him and don't allow him sometimes to get away with poor behavior or bad manners, he will grow up knowing his limits and be more obedient than will a spoiled youngster that had no limits. A young horse that is always handled with consistency is also more comfortable with you and knows where he stands in the relationship. Inconsistency creates insecurity.

Be the Boss, Not a Buddy

A common mistake in handling young horses is to treat them as playmates rather than horses. If a youngster thinks of you as a peer rather than a leader, he'll try rough-and-tumble games as he would with a young herd mate. He needs to learn, early on, that humans are the boss and to be respected. Don't allow him to invade your space, bump you, or bite you, and he must never threaten to kick you. He would not dare do that to a dominant adult horse. If he accepts you as boss from the beginning, life will go much more smoothly for both of you.

Prohibit Game Playing

It is often tempting to play with the young horse, but this usually leads to trouble. He soon becomes larger and stronger than you, and innocent games of foalhood can become dangerous. Insist on correct and well-mannered behavior from the start, and your horse will be much easier to train and won't become inadvertently dangerous in his disrespect.

Dealing with an Aggressive Weanling

Occasionally you encounter an aggressive weanling who wants to play games in order to challenge and dominate you. This is common in young stallions, but even a gelding or a filly may test you too much. Some youngsters become increasingly hard to handle if this behavior is allowed to continue.

The best solution, especially if the youngster is living by himself and you are his only outlet for game playing, is to put him with an older horse that won't tolerate his pranks. An older horse will soon set him straight. This horse will teach the youngster to be more humble and submissive, thus making your training tasks much easier. It's easier to let another horse teach him discipline than for you to have to do it all.

Other horses can teach the youngster more about behaving himself than any human can. A herd situation is the healthiest environment, mentally, for the young horse, as he is able to interact with his elders. Sometimes, however, you have no choice but to keep a horse by himself; in this case, you must be firm and consistent in your handling of him so he learns to respect and accept your role as the dominant herd member in his life.

Be Flexible

Don't assume each horse will react the same way to a certain training method or way of handling. The more foals you raise and train, or the more young horses you acquire, the more you too will learn as you attempt to teach them.

Treat every youngster as an individual, and alter your training methods, if you have to, to fit the needs of each one. Every young horse you train will teach you something new about handling horses.

5

HANDLING AND TRAINING THE YEARLING AND THE TWO-YEAR-OLD

The time to undertake ground work in preparation for riding and driving the young horse is when he is between one and two years old. Up to this point, you have been gentling, handling, leading, tying, grooming, picking up his feet, and getting him accustomed to basic care.

GELD THE MALE

A male horse should be gelded well ahead of formal training, so he'll be fully recovered from surgery before you begin. All colts not intended for breeding should be gelded as yearlings if they were not gelded as foals. (See *Storey's Guide to Raising Horses* for more information.)

Build a Good Foundation for Later Training

If you started training when the horse was a foal or weanling, you can now build on those early lessons. If, however, he spent his early life at pasture with Mama and his adolescence with a group of other youngsters with no handling, now is the time to start his training in earnest and to gain his respect, trust, and obedience.

Crash Course for Unhandled Yearling or Two-Year-Old

Sometimes you don't have a chance to handle a horse as a foal or weanling. Maybe you acquired a yearling or two-year-old with no previous training and must start at the beginning. This can be a challenge, but it's not an impossible situation if you start by gaining the horse's trust and respect and take things step-by-step.

Safety Precautions

Most training books and articles on starting the horse under saddle (or to pull a cart) assume that this is a horse you can catch, lead, and tie up. But if he has not learned these basics, you will have to provide them to a pupil who does not know the proper way to relate to his teacher. If he has not learned the basics of restraint and respect, he may hurt you just trying to get away from you or reacting in alarm to something to which he is not accustomed. In everything you do, consider safety first. Think ahead to what the horse might or could do as you are working with him. (See chapter 1 for more on safety.)

Don't expect him to act like an experienced, well-mannered horse. He's not used to you and doesn't know what to expect from you. He will react with self-preservation instincts (flight or fight). His first reaction to something may be to try to get away from it — perhaps right over you — and if he can't get away, he may try to protect himself. Be prepared, and don't put yourself in harm's way.

Catching and Haltering

If you have lots of time, you can put an untrained horse in a small pen and gradually gain his trust with feed (similar to how you would gain the trust of a weanling; see pages 107–108), at which point you can touch him, halter him, and continue from there. But sometimes you have to get a halter on him sooner than that. With an independent or skittish individual that doesn't want to trade his freedom for food, you must corner him to put on a halter. Then you can leave the halter (or even a trailing rope, if necessary) on for a few days so you can quietly catch him in the small pen and speed up the handling and gentling process (see page 108 for more on this technique).

Generally, the older the horse, the more difficult it is to confine and corner him for the first time. Plan it so neither of you will get hurt; you don't want him crashing over or through a fence or running over

USING PORTABLE PANELS

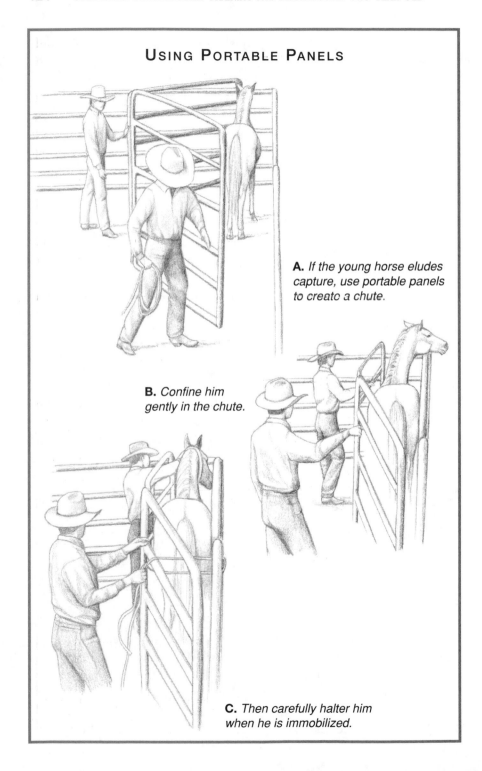

A. *If the young horse eludes capture, use portable panels to create a chute.*

B. *Confine him gently in the chute.*

C. *Then carefully halter him when he is immobilized.*

you. A safe way is to gently herd him into a small pen (with safe, tall fencing) inside of which you have created a catching chute made with portable panels.

If he's in a pasture, make a small catch pen in the corner, using panels, and put his feed and water in it so you can close the gate while he is inside. Use a few portable panels inside the catch pen to make a wing to head him into a smaller alleyway created by the panels, then swing shut the last panel behind him. Then you can slide the panels closer to him, making a "chute" where he can't go forward or back or turn around.

Make sure the panels are securely fastened to the existing fence and that the fence is strong and safe. If it's a wire fence, put panels on the inside of the wire. Work slowly and quietly and don't excite the horse. Once he is secured in the panel chute, carefully lean over the panel near his head and put the halter on him. This method allows you to capture a semiwild horse without trauma, so you can begin to train him. You can then leave him in a small pen where you can easily get hold of the halter without having to corner him.

For a horse that's spooky about being haltered, carry a spare halter when you catch him and slip it on over the one he's already wearing to get him used to having your arm around his neck and slipping the halter up over his nose. If he tries to leave, you'll be able to hold him with the halter and rope he's wearing. If you do this every time you catch him — and repeat several times if he needs to learn not to fear it — he'll soon stand quietly for haltering, and you won't have to leave a halter on him.

SAFETY FIRST

If you leave on the halter for a few days in the small pen, make sure it fits snugly, so he can't rub it off or get a foot caught in it, but not so tight that it cuts into him. There should be no protrusions in the pen that it could catch on. If you leave a trailing rope on the halter, make sure it is long enough that you can pick it up without having to chase the horse; the bigger the pen, the longer the rope should be. Your horse soon learns you can catch him whenever you walk into the pen and that running off or putting his head in a corner won't prevent his being caught, and eventually he'll quit his efforts at evasion. This is safe to do only in a small pen. *Never* leave a trailing rope on a horse at pasture.

Round-Pen Training

Many trainers use round-pen training to build a relationship with a green horse, especially one that hasn't had much handling. Rather than using physical force to dominate the untrained horse, the trainer uses her knowledge of horse psychology and body language to gain control of his mind and to establish a level of communication that allows for quicker acceptance by the horse. Round-pen work can also change the mind of a hard-to-catch horse.

With the horse in a confined area, you can take advantage of how he thinks and reacts to persuade him that a human is not so bad after all. A round pen is perfect for this; it has no corners into which he can retreat. The pen should be 50 to 60 feet across. Any larger and the horse could get too far away from you; any smaller and he may feel trapped. There is also more risk in a small pen of getting kicked. There should be good footing, so if the horse gallops, he won't slip and fall or hurt his feet on rocks. If you don't have a round pen, a square one will do; just put portable panels diagonally across the corners to block them. Don't overdo fast work in a small round pen, especially with a young horse, or it may put too much strain on his legs.

Advance and Retreat

The horse is a flight animal; his first reaction to anything strange is to run. Round-pen training takes advantage of this instinct. The horse can travel in circles until he gets tired and starts thinking he'd rather not run; soon he'll realize the human in the center of the pen is not so scary after all and that it would be much better just to figure out what the human wants him to do. The round pen gives you a position at the center of his activity; you have control over his motion and direction.

Early trainers who started using round-pen methods with wild or untrained horses found that with patience they could change a horse's mind from fear and flight to trust and respect — and willing obedience — within a very short time. They were able to accomplish in days or even hours what it took other trainers weeks to do. A few hours of patient round-pen work transformed a wild horse's attitude much more quickly and completely than traditional methods of "breaking" that involved force; it also provided solid groundwork for a trusting relationship.

People who first tried this type of training had noticed that when Native Americans caught wild horses, they started by chasing and driving them away, often for days — never letting them rest, keeping the pressure on. Then abruptly, the horsemen would wheel around and ride slowly away. As soon as the pressure was gone, the wild horses would turn and follow them back the way they had come. The Native Americans would travel back to a large enclosure, with the horses following them, and lead them into it.

This technique of advance and retreat is an example of the way a horse responds to pressure. Relieve the pressure and he reacts. Relief from pressure — the method of give and take, press and release — is a more efficient way to train a horse than with force. Trainers call this response by the horse *joining up*, or *locking up*, because the horse comes into a willing and obedient relationship with a human.

In this way, training can be accomplished with no violence; the horse responds on his own. The objective is not to conquer him physically with ropes, but to engage his mind and make him a willing follower. This is first accomplished by driving him away (around and around the pen) until he gets tired of it and wants to renegotiate the relationship.

The Goal

One aim of round-pen work is to make the horse want to follow you rather than run away. The ultimate goal is to create willingness: The horse should be willing to have you touch him all over, pick up his feet, put on his saddle and bridle, and start mounted lessons. Once you can engage his mind, you can control his body. A round pen lets you gain this control because in this confined area the horse can't avoid you.

How to Do It

Round-pen work not only instills willingness in the unhandled horse, but it also helps make a better relationship with a horse that already knows you, especially one that is unsure or hard to catch, or one that doesn't accept your leadership. All you need for this training is a whip, rope, or lariat — something to swing to encourage the horse to move.

Bring him into the center of the pen and rub his forehead with your hand. Then turn him loose and move away from him and to the rear, out of kicking range. When you are behind him, toward his hip, or when he takes off, drive him away with a rope or whip, raising it toward his hindquarters. He should take off at a trot or gallop. Some trainers throw a light rope (or a halter, still holding the end of its lead rope) at the hindquarters — something that won't hurt him but instead will startle him into taking off. The horse is retreating, so you advance, quietly keeping pressure on him, swinging your arm, waving the whip or coiled rope at him — whatever it takes to keep him circling the pen, but not so much that it scares him. Make him keep going, but don't get within kicking range. Walk in a smaller circle in the center of the pen, your body opposite his hip.

Maintain an aggressive stance — look as if you will chase him if he slows or stops — so the youngster will trot or gallop around the pen at least five or six times, then make him go around it in the other direction. To have him change, step across the pen to a place in the circle ahead of him; this blocks his forward motion so he'll turn and go the other way. Keep him going in the new direction, moving toward his hip to encourage him if he slows down. Practice changing directions at different locations around the pen until you can work the horse consistently in either direction and have him change direction when and where you request it.

After he has gone around the pen several times both ways, he should be looking for an excuse to slow down. Start asking him if he'd like to stop all this hard work, paying attention to his body language to detect when he might be willing to listen.

Coil your rope or lower the whip and take a more submissive body position. Don't look at your horse directly. This is your invitation for him to stop retreating and come to you. Watch his ears. His outside ear will still be monitoring his surroundings, but soon his inside ear will stop moving, his head will tip toward you, and eventually he will put his head down toward the ground. He is rethinking his flight reaction and wants

---

TURNING TIP

You can dictate whether the horse turns toward the fence or toward the center of the circle. When you move in his direction at a sharp angle, he'll probably turn toward the fence. If you give him more space, he'll probably turn toward you as he changes direction. This puts you in control of his actions.

ROUND-PEN TRAINING

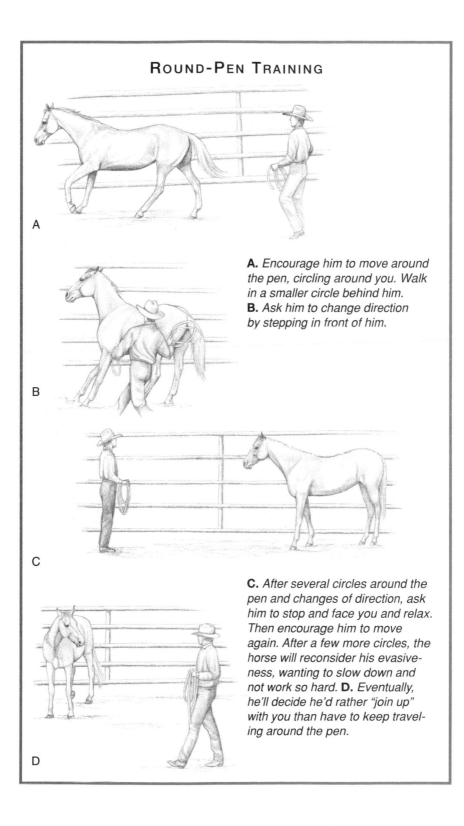

A

A. *Encourage him to move around the pen, circling around you. Walk in a smaller circle behind him.*
B. *Ask him to change direction by stepping in front of him.*

B

C

C. *After several circles around the pen and changes of direction, ask him to stop and face you and relax. Then encourage him to move again. After a few more circles, the horse will reconsider his evasiveness, wanting to slow down and not work so hard.* **D.** *Eventually, he'll decide he'd rather "join up" with you than have to keep traveling around the pen.*

D

to stop. He wants you to take off the pressure. He's ready to listen. Have him make his turns toward the center of the circle, toward you. Back away from him as he turns, to draw him toward you. If he stands and faces you but doesn't come, walk toward him in arcs. Don't come at him directly; he'll view that as confrontational behavior and retreat.

If he lets you walk up to him, fine. If he takes off before you get to him, make him go back to work for a few more laps around the pen, then let him stop again. As you move closer, do so with your body a bit sideways, as if your attention is elsewhere. This is not as threatening as a direct and focused approach, as often taken by a predator. Keep a non-threatening, matter-of-fact, neutral stance — that is, show more of your back to him than your front. He should move toward you (join up) and touch your shoulder with his nose.

When you can approach his head, rub his forehead, then walk away, moving in circles. He should follow, or at least keep his head turned in your direction. If he doesn't follow, your circle will bring you back toward his rear. Put him to work around the pen again, staying out of kicking range. After a few lessons, he'll follow you when you let him stop and invite him to follow. He should voluntarily follow you to the center of the pen and stand relaxed as you rub and touch him all over both sides of his body.

This is the point at which an experienced trainer can take a previously unhandled horse and start picking up feet, progressing to saddling, and so on, because the horse is now relaxed and willing and trusts this person who has given him relief from pressure.

RELEASE TIPS

Release of pressure may be accomplished by backing up, stepping right or left (away from the horse), or just relaxing your shoulders or leaning back. It depends on the horse; see what works to have him feel the release and give a response, so you can tell he is approachable and no longer nervous or resisting.

Fine-tune Your Communication

Done properly and patiently, round-pen training has many uses. First you use body language to control his speed and direction; he responds to you, stopping and facing you when you ask him to. You create a foundation for a higher level of communication, such that he will follow you without a halter as you control his movement around the pen.

With body language and practice, you can direct his turns to the inside or outside to change direction and walk up to him wherever he is without him attempting to walk away. You can teach him to step away from you on cue, so he'll never crowd you when you lead him; step sideways away from you; turn forward again to face you. Soon you can turn him in any direction, on cue, with just your body language. It takes time and patience and the ability to read the horse's body language to know how much pressure — or release — to put on him with your own body language, and when.

If a horse is scared or nervous, go easy and don't make too many continuous circles; he may become immersed in the idea of flight and unable to think. Make changes of direction and changes of speed to keep his attention on you. Let him stop and make his turns toward you whenever he wants to. Later, after he's not scared, let him stop only when you ask, so you stay in control. Some horses get lazy. Keep in tune, adjusting your approach as the horse changes his attitude, so you can keep proper timing and the proper amount of give-and-take, pressure and release.

Leading Lessons for the Yearling and Two-Year-Old

The most basic principle in training a horse is that you control his movements. He must learn to go forward freely when asked to. Teaching a horse to move forward at halter is usually simple; the harder part is teaching him to move forward just at your command, how fast to go, and in which direction. He must learn what *Whoa* means, and that he can't drag you, run by you, or pull back. You usually want a brisk walk, not a prance pulling you along or a dull plod with you pulling the horse.

Unity between horse and rider starts with good leading — the rapport gained from ground handling. The horse that refuses to lead properly will be a frustration, whether you are trying to show him at halter, lead him into a trailer, or trot him for a soundness check at an endurance ride. No matter how good he is under saddle, if he doesn't handle well from the ground, there will be times he'll disappoint you.

BE PATIENTLY PERSISTENT

If a horse backs up in an attempt to get away, go with him, keeping the same light pressure on the rope. If you are working in a small area, he can't pull clear away. Patiently stay with him; he'll stop backing up eventually. Keep the same light pressure on the rope until he finally takes a step toward you. Be patient. The last thing you want is a tug-of-war. Even if it takes several days for first lessons to sink in, it's better to avoid a fight. You want good leading manners for the rest of his life.

Beginning Leading Lessons

When teaching a foal to lead, you have his mother for him to follow. Following is not the same as leading, but it's a place to start. Teaching to follow is also a way to start the older youngster. If you've worked with him enough to be able to catch him, you can start leading lessons in the pen or even in a stall. If he's still a little scared, use a small area so he can't get up much speed or get clear away from you. Progress to leading lessons after he's sufficiently accustomed to you and won't be trying to charge over you to get away.

To teach him to lead, use light pressure — not a hard pull, or he will brace against you. Stand several feet in front and to the side (it's easier to make him take a step to the side), and put just enough pressure on the lead rope to take out the slack and make him a bit

To teach the horse to lead, stand several feet in front and to the side, and put just enough pressure on the lead rope to take out the slack (enough to make him uncomfortable, but not enough to fight it), and eventually he'll take a step toward you.

uncomfortable. He may fuss, but eventually he will take a step toward you. It may be on purpose or by accident, but that doesn't matter. When he makes that step, instantly release the pressure.

Keep asking for another step until he learns he can get release of pressure by stepping toward you when he feels a light pull on the rope. Teach from both sides. Don't ask him to come straight ahead until he comes to a light pull from the side. A few steps is enough for a first lesson; quit before he gets tired of it and balks.

Rump Rope

Use a rump rope on a horse that is exceptionally balky or lazy, just as you do when teaching a foal to lead (see page 88). First give him a chance to learn to give under pressure of just the lead rope. If he refuses, use a rump rope rather than trying to pull harder with the lead rope. He is much stronger than you are; getting into a pulling contest will always end in your defeat.

For a rump rope, make a loop that won't get larger or smaller; you don't want it to tighten when you pull on it. Put the end of it through the halter, so it pulls in the same direction as your lead rope; if the horse runs by you, it won't slip off his rump.

Always ask for a step first with light pressure from the lead rope and follow up with pressure on the rump rope if he fails to respond. Your horse will find that it's more comfortable to move when you ask with the lead rope; he'll know he'll feel the rump rope if he doesn't. As soon as he steps forward, create slack with both ropes. Once he knows what to do, you won't need the rump rope. When he responds to light pressure on the halter and moves forward when you ask, he has learned to follow. Later you can teach him to lead properly, but first he should learn *Whoa*.

Teaching *Whoa*

Learning to stop on command is one of the most important things to teach a horse; it keeps you in control of his actions whether you are leading, longeing, driving, mounting, or riding. You must get him leading, however, before you teach him *Whoa*. Otherwise you may confuse him if you ask him to stop; he needs time to learn one command before you teach another. If you give a light tug on the rope immediately after you say *Whoa*, he'll get the idea and will soon stop without the tug. Let him stand a moment each time you stop him, so he realizes that *Whoa* means to stop and stay stopped.

THE POWER OF THE VOICE

Every horseman uses the voice in training and handling a horse, to soothe, encourage, calm, stimulate, and, sometimes, reprimand. The horse who knows you can be easily encouraged or soothed just by the sound of your voice; likewise, he will be deterred from bad behavior by your tone. If you encourage him to move out freely as he is being led, he will be more likely to do so.

Use of Voice and Whip

To lead a horse properly, walk beside his left shoulder as he moves freely beside you. To accomplish this, you must instill more action and willingness than the horse would normally exhibit, encouraging him to move briskly beside you rather than just following. This can be accomplished with the help of two aids — your voice and a whip.

Many trainers teach a horse to move by clucking with the tongue. The cluck tells him to move, or to move faster. He responds to the cluck, knowing that if he doesn't, he'll get some other form of pressure as follow-up, such as a touch of the whip on his hindquarters, and he quickly learns to move at the sound alone.

The whip, as an extension of your arm, enables you to touch parts of the horse's body you could not reach otherwise. When you are standing beside his shoulder, you can touch his hindquarters and he knows immediately that the touch comes from you, even though you are merely standing beside him. He realizes he is being controlled at both ends at the same time by you.

When teaching a horse to lead, ask him to move forward beside you, clucking to encourage him. Hold the whip quietly in your left hand, the whip trailing behind you so you only have to move it a little bit to touch his hindquarters. He should be able to see your hand. If he doesn't move, tap him lightly on the hindquarters as you ask him to move forward. If he still doesn't move, tap him again. Then, as he understands and takes several steps forward beside you, praise him and let him know by your voice that he has pleased you. If you ask him to move forward by using the whip, keep a little slack in the halter rope, giving him freedom of head — that is, no feeling of restraint — as he walks forward. If he's held too tightly by the halter, he'll be nervous, not understanding the conflicting demands of being urged forward while being held closely.

After he learns to lead properly, you won't need the whip; your horse will understand what you want. It is important to stop using a cue as soon as the horse responds, whether the prompt is a touch of the whip or your leg pressure; use of the reins; pressure on the bit or on the halter. Release of pressure or the halt of a cue is the horse's reward for doing the right thing, and his response will be swifter in the future.

BE GENTLE

If touching his hindquarters with the whip makes a horse jump forward, don't pull on his halter or discourage him; he's just trying to do what you ask. Tap more lightly next time and move forward with him. He'll learn that a touch of the whip will follow your verbal request to step forward, and will begin to move at the sound of your voice or cluck without need of the whip.

Holding the Lead Rope

Your hand on the halter rope should be firm but gentle, like your hands on the reins when you ride. The way you use your hands is one of the factors in determining how your horse will respond and how subtle your cues can be.

Your hand maintains contact with the horse via the lead rope. Your hand should be passive, not active. All too often a trainer pulls at the horse in an attempt to pull him forward or pulls back on the halter if he tries to go too fast. The main purpose of your hand is to regulate the length of the rope. As long as the horse walks quietly beside you, there is no tension on the lead rope because neither of you is pulling. The weight of the rope itself is enough to keep sufficient contact between the horse and your hand. Don't give him too much slack; 8 to 12 inches is usually sufficient to maintain good contact.

If he walks too slowly and drops behind, don't try to pull him forward. Use the whip, pointing it at his hindquarters. A horse

> **TRAINING TIP**
>
> The value of the whip in training depends on how you use it. *Always* use it quietly and sensibly. *Never* use it for punishment. If the horse fears it, its effectiveness as a training tool is lost. If he trusts you and has never been abused by a whip, he will respect it and respond to it as an extension of your arm.

instinctively resists being pulled; the harder he is pulled, the more he pulls back. Avoid the common mistake of trying to pull him forward. Even if he'll lead as you pull, he does it reluctantly, with head and neck braced. He will respond better and more quickly if you just tap him gently on the hindquarters with the whip to remind him to move forward or to do so more swiftly.

Keep your hand passive. Tension on the rope will then be created only if he tries to go faster than you want. If he bolts forward or prances, he pulls on your hand. Resist this pull passively. Don't yield if he pulls at you, but do *not* pull on him. Let him do all the pulling; your hand acts as a fixed point. He'll usually stop pulling — most horses don't continue to pull unless they're pulled at.

The difficult challenge when leading an uncooperative horse is to be unyielding but not to pull at him. It's easy to pull on the horse without realizing it, for to pull when pulled at is just as instinctive for humans as it is for horses. Try to keep your hand steady.

When he realizes he is meeting an unyielding object but is not being pulled at, he will cease his exertions. An exception might be a young or flighty green horse or one that is frightened into shying and bolting. This kind of jerk on the lead rope is hard to meet by a simple fixing of your

Your hand on the lead rope should be firm but gentle. Strive to keep your hand passive. Tension on the halter should occur only when the horse is going too fast.

DEVELOP GOOD HANDS

The same principles apply to leading as to riding. The term *good hands* means a fine, light touch. *Poor hands* means poor horsemanship — sloppy or rough use of a bit, pain or confusion for the horse. Even when teaching him to lead, you need good hands. At this early stage in a horse's training — the ground work before you ride — deep impressions are made in his mind, and experiences he will never forget. Even the most elementary stages of training are extremely important. A horse can become quite spoiled at any stage, just by the wrong kind of handling — and leading is no exception.

hand; you may have to let out a few feet of rope or even travel a few feet alongside him while still concentrating on passive resistance, letting him do all the pulling. Then when he stops, the release of tension on his halter is immediate and automatic; he will realize he is rewarded by release of pressure when he stops. He will come to understand that he is supposed to travel beside you at your speed.

Turn in Both Directions

When leading the horse, turn him right as often as left. It's easier to turn to the left because that's the side you lead from, and many horses are awkward when going to the right simply because their handlers haven't practiced turning them that way. If from the beginning you make him go to the right as much as to the left, he will be easier to lead. You will have much more control; he won't get into the habit of circling around you to the left when he's nervous or walking all over you because he won't go to the right. To turn him to the right — away from you — give short tugs on the halter rope in that direction, turning his head to the right. He will soon get the idea and start turning that way as soon as you extend your arm under his chin.

If a horse constantly veers into you and pushes you to the left, don't let him circle around you to the left; this puts him in control rather than you, and he's pushing into your space. If you've taught him to go to the right as well as to the left, ask him to circle to the right when he starts pushing into you and that will solve the problem. Also handle him from his right side so he won't become one-sided and awkward. Get him accustomed to being led in both directions *and* from both sides.

Teaching a Horse to Back Up at Halter

When a horse has learned to back up, you gain more control over the horse's movements. This is essential if he must learn to back out of a horse trailer. You want him to be able to back up as easily as he leads, without having to jerk on his halter or push on his chest.

Beginning Backing Lessons

To teach him to back up on cue, halt the horse, face him (standing beside his head), reach under his jaw, and use intermittent backward pressure on the halter (never a steady pull). The pressure should be gentle. If he doesn't understand, tap on his chest as you put pressure on the halter. Ask for one step, but repeat this several times during a training session. After a few lessons, your horse will be able to back several steps at your request.

When asking him to back up, use a series of short tugs on the lead rope. Use clucks to encourage him to move, or teach him the verbal command *Back*, if you prefer. Because you always cue in the same tone of

Ask the horse to back up by tapping on his chest and giving short tugs on the halter. Both cues should cease the instant he starts to take a step back.

ANTICIPATE AND REWARD

Release pressure as soon as a horse starts to shift weight or pick up a foot, *not after* he takes the step, or the reward comes too late. Release should come the instant he thinks about backing; he'll get the idea quicker. If he doesn't back away from light pressure, don't add more pressure or he may brace against it and be less able to move his feet. Give a little side-to-side motion with the halter while maintaining light pressure; this won't give him so much to brace against. You want him to back up at a light touch.

---

voice, the horse will associate the command with the proper action, especially when it is reinforced with a gentle tug on the halter.

If necessary, push on his chest or tap gently with the butt of your whip to encourage the first backward step. Slow, rhythmic tapping will make him want to back away from it. As soon as he starts to take a step back, praise him and release all pressure or stop tapping so he knows he did the right thing. Then ask for another step. If you ask for just one step at a time, halting the pressure instantly when he responds, he will quickly get the idea. Soon he'll back with just a tug on the halter and your verbal command, and you won't need to tap on his chest.

Backing as an Aid to Slow Him

If you have taught the horse *Whoa* and *Back,* you have more control if he gets nervous and wants to go too fast. Stop him and make him back up a few steps. Do this quietly and without reprimand, so he learns that if he goes too fast, he'll have to stop and back up. If he wants to charge, dragging you along, stop him and back up repeatedly until he figures out that he shouldn't do that.

Beginning Tying

If the horse wasn't trained to tie as a foal, use a method that won't injure his neck if he pulls back strongly. (See Safety Tip on page 140.) The yearling or two-year-old is large and strong enough to hurt himself if you tie him by the head. Use a body rope or tie him to an inner tube secured to a fence post (see pages 95–96). The rubber tube stretches if the horse pulls back but will still hold him — and minimizes the risk of hurting the horse or of breaking the rope or halter.

Gradual Adjustment to Restraint

In preparation for first tying lessons, to minimize risk of pulling back hard or pulling back while you are still tying the rope, acquaint your horse with restraint by the rope, using a long halter rope looped around a post with you holding on to the end of it. Bring him to the fence post, loop the rope around it, and groom him or work with him without his being tied. He gets used to being next to the post, but if he steps back, he is not solidly restrained and does not panic. He learns the feel of mild restraint but does not become claustrophobic. He is also less apt to pull back strongly because you are there to reassure him.

Once he is at ease with this, gradually move farther away, still holding your end of the rope. The post is still holding him, via your rope; he realizes that he is restrained by the post, not by you. Yet you can still play out rope if he pulls back a little. Usually he'll pull back only a step or two, as the rope is not holding him tightly, but when this results in some pressure, he'll step right back up to the post again.

This gets him used to the idea of standing by the post and not pulling back hard. If your rope is long enough, you can stand quite a bit away from him relaxing or leaning against the fence, perhaps, so he thinks he is tied to the post. You are still there, however, in case he needs reassurance or to be moved back to the post if he pulls back a little. From there, you can progress to tying solidly — either with a body rope or to an inner tube — and he is less

> ### SAFETY TIP
>
> Be careful when tying a horse the first time; you don't want fingers caught in the rope if he pulls back while you're tying the knot. Don't be in the way if he pulls back, then lunges forward or he may smash you into the fence. To be out of harm's way, use a long rope and run it through the inner tube or around the post, then tie it to the next post down the fence, at a safe distance from the horse. If you take a couple of dallies around the post, this will help hold him if he starts to pull back before he's tied.

PATIENCE

Patience while tied makes later training easier. Though he'll need to learn to stand still for grooming, for having his feet handled, and for mounting and dismounting, tying will have taught your horse the patience required for learning these things. It also teaches him to respond and yield to pressure. If he pulls back while tied, the post holds firm. If he moves ahead, he has an instant reward (slack). After he learns to stand tied, he is more responsive when being led; he has learned not to pull away.

---

apt to set back with all his might; he already knows about being restrained by the post.

Don't leave him tied too long the first times; 10 to 20 minutes is enough if he behaves. You just want him to know he has to stand there. If he sets back, leave him tied until he resigns himself to standing quietly for a little while, then reward him by turning him loose. Don't untie him right after he's pulled back or he may think he gets his freedom by pulling back. Gradually tie for longer periods.

To get a horse gradually used to being tied, use a long rope and dally it around a post, so you can give or take up slack as needed. Move farther and farther away in subsequent lessons so he will realize that the post, not you, is holding him.

Perfecting His Tying

Tying is one of the best ways to teach a young horse patience as well as control. By the time he is a yearling, he should be well trained to tie. If he is not, give him beginning lessons as previously described (see pages 139–141).

Make It a Regular Part of His Day

To perfect his tying, so he will always be easy to tie and reliable while tied, do it often. You want your horse to accept tying as part of his day. Even if you don't have time for much training, make it a habit to halter and tie him, and leave him tied in a safe place while you do chores or other things nearby where you can keep an eye on him. This practice will greatly further his training progress.

At first you can tie him in his stall or pen — any familiar place where he will relax. (Just be sure not to tie him where there are any loose horses.) When he is comfortable with that, lead him to a safe spot in the barnyard and tie him there. Leave him tied an hour or so each day, at various places. Don't make the mistake of always tying him in the same place or that will be the only place he will feel at ease while tied. For a while he'll be impatient, maybe even nervous, about being tied in a new place, but soon he'll realize impatient pawing is to no avail. And because nothing is hurting him, he will resign himself to waiting and will learn patience.

If you gradually expand his experience to where you are tying him in a safe spot along the outside of the arena fence or in the barnyard, the horse will get used to the activity going on around him and learn to relax and stand wherever he is tied.

SAFEGUARD THE KNOT

If the horse plays with the rope, nibbles the knot, and unties it — some can undo a knot even with the rope put back through the loop — put the end of the rope where he can't reach it. Dally it around the post and make the knot on the next post, or run the end of the rope down the back side of the fence and tie it again on a lower pole or rail, where the horse can't reach it.

Lessons in Blanketing

Some horses live outdoors and are never blanketed; others are blanketed for a number of reasons: to protect them from cold weather during turnout if they spend most of the time in a stall, to keep them warm if they are clipped for winter, and for protection from sun or mud to keep the hair coat nice for showing. Putting something over his body can be scary for the young horse, so prepare him for it. The blanketing process can be a way to accustom him to the feel of something on his back for later lessons in saddling. If your horse will have to be blanketed, give him some lessons first.

HOLD, DON'T TIE

Don't tie a horse for blanketing lessons; he may panic because he cannot step away from it. Have someone hold him for you at first. If you go about it gradually, however, you can probably give him these lessons while holding him yourself. If at any time he panics, be patient and stay calm; don't force him. Go back to earlier steps he is comfortable with.

Gradual Steps

For his first blanketing, choose a used blanket that already smells like a horse (the familiar smell will be less frightening than would that of a brand-new blanket). You'll want one that fastens at the chest so you can easily put it on and take it off.

Get him used to it gradually. Hang it on the fence or over the stall door so he can smell it while you groom and handle him. Don't let him nibble on it or pull it off the fence or door; the sudden movement may frighten him. Work in a familiar place, so the only thing new and suspicious is the blanket. Let him smell it for several sessions, if necessary, so he gets over any fear before you try to put it on. Once he is at ease with it, you may be able to put it on, in gradual steps, in just one session. It may take several lessons to accomplish, though, so take as many sessions as he needs.

After he is accustomed to the sight and smell of the blanket on the door or fence, hold the halter rope with one hand and slowly bring the folded blanket to his nose. Keep it folded small at first; it will be less scary that way. If he backs away, let him. Wait until he is relaxed again. Once he lets you put it close to his nose, allow him to sniff it until he is satisfied it is harmless.

Keep it well folded so the straps won't flop around, and rub his neck with it, gradually rubbing more of his body as he accepts it. Work over to his shoulder, around his chest, down the front leg. Do this on the other side after he is relaxed about it. Then drape the blanket over his withers and let him feel its weight for a few minutes. If he's comfortable with that, move it around.

Fold the blanket tightly and keep straps from dangling so the horse doesn't perceive it as something big and frightening. Rub his neck with the blanket.

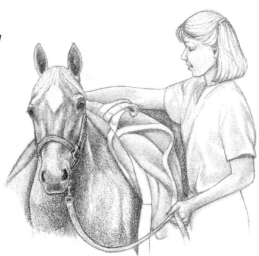

When he's calm about having his neck rubbed with it, drape the blanket over his shoulders and withers to get him used to its weight, then rub it around a little.

After the youngster accepts this, slowly slide it over his back and rump, still standing at his shoulder, out of kicking range, and keep hold of the halter rope to control his movements. If he tries to walk forward or move around, halt him. Work with a slack lead, putting pressure on the halter only when he moves. You don't want tension on the rope or he may feel confined or turn toward you, swinging away his hindquarters.

If he swings away and the blanket slides off, try to catch it — slowly and gracefully, in order not to startle him with your motions — so it won't fall in a heap and frighten him. If it falls off, start over again at his front end. Remember to work from the other side as well. When he seems ready to have the blanket put on, secure all straps so they won't dangle and bump his legs (this first time, you won't be fastening them on him), fold the rear part of the blanket over the front so you can set the front portion over his withers, then carefully unfold it over his hindquarters.

After you put it on this way several times and he exhibits no fear or resistance, fasten the straps, starting with the belly surcingle, making sure he is comfortable about having straps against his body before you actually fasten them. Once he is relaxed about wearing the blanket, lead him around to get him used to the feel of it. The first few times you actually turn him loose with it on, stay nearby and make sure he doesn't get into trouble.

Continue the Leading Lessons

Don't neglect the young horse just because he is not mature enough to start riding or driving. Work on leading lessons. Most horses lead reasonably well without much formal training, just from general handling — knowing they are being taken to water, out to pasture, into a stall for grain, for example. To lead well, however, and to always be under perfect control, a horse must learn a bit more. He must learn not to hang back on the rope, not to lunge ahead, not to gawk and dawdle. He must learn to travel beside you at the speed you want.

Leading at the Trot

After he leads well at a walk, teach your horse to lead at the trot. To trot, start moving faster yourself while giving the verbal command *Trot* or clucking. He will generally start moving faster when you do. It's easy to teach voice commands such as *Walk* and *Trot,* which will come in handy later when teaching a horse to longe, if you always use the same tone of voice for a certain command. *Walk* should be said in a soothing voice to help persuade the youngster to slow down from a faster gait; *Trot* can be said more crisply. Most youngsters quickly learn to speed up when you tell them to trot, especially if you start moving faster yourself as you say it.

Another cue is to start moving your feet faster in place for a second as you give the command. You may have to use the whip to help him realize you want him to trot. If the horse doesn't speed up when you give the command while trotting in place, reach back and touch his hindquarters with the whip.

After he realizes what you want him to do, he should start trotting when you give the command and you'll no longer need the whip. A few individuals are lazy or too independent, however, and don't want to lead at a trot. It may take several lessons, and perhaps more forceful taps of the whip on the hindquarters, to make them realize the easiest thing to do is trot alongside you when you ask.

Advanced Leading

Take the young horse for walks to expose him to new experiences. By this age he should have had enough lessons to be manageable at halter. After you start riding, there may be times when you will have to get off and lead him through or around difficult obstacles, or jump off and hold him the first time he encounters a noisy truck on the road. He must be dependably halter-trained to trust and stay by you in all situations. A few leading lessons out in the big wide world are a good idea if you have not already done this with him at a younger age.

Trail Work from the Ground

Many tasks will be easier for him during early riding lessons if your horse has had some prior experience. If the first time you ask him to step over a log, cross a bridge, or walk through a puddle is when you are on his back, it may be a challenge to convince him he can do it. Horses raised in big pastures have an advantage: They may already know how to walk through water, go through bushes, and get over downed trees without fear. A horse reared in confinement, however, or even in a small pasture, may be afraid of things he hasn't encountered before. Ground work can prevent problems later.

Advanced leading includes walking over poles or posts laid out on the ground. If your young horse balks or tries to avoid stepping over them, cue him with gentle taps with a whip on the hindquarters. Once he figures out he can step over them, gradually raise their height by putting blocks under the ends until he must step higher, as he would if he were going over a log on the trail.

He may be afraid to cross a wooden bridge; simply the noise of his feet on the wood is alarming. Create a "bridge" on the ground, made of wood planks, for him to practice walking across. If this is scary to him, he may resist. Take as many sessions as he needs to learn that walking over it is nothing to fear.

Ask for just one step at a time, using the pressure-release technique. Reward him for even the slightest try by release of pressure and with verbal praise. Don't pull on him. If he steps forward and there is no release of pressure, he will quit trying or become more afraid of the obstacle. Let him realize that moving forward — when you ask him for a step — brings relief from pressure on the lead rope and no more tapping on the hindquarters with the whip and that balking or moving backward brings pressure and tapping. If you are consistent and patient, eventually even a very skittish horse will learn he can step on the "bridge" and walk over it.

Crossing Water

A young horse who grew up in confinement may have had no experience with water. Lead him through puddles or small streams. When he learns it's okay to get his feet wet, he will be easier to ride through water later, and you won't get stranded at a stream that lies between you and your destination. A large rain puddle can be used to advantage when teaching him to cross water, but he may try just to walk around it — he sees no reason to go through it if he can go around. A shallow ditch or stream is often better, as he must cross it to continue on his way. If you have access to a field or pasture with irrigation water running over it, lead the horse around in it. If everywhere he goes is wet, he'll soon realize he can't get away from the water and discover that it isn't hurting him to get his feet wet.

For his first water lessons, be sure you have lots of time and patience. You can lead him to water, but you can't make him cross it. It

CROSSING WATER

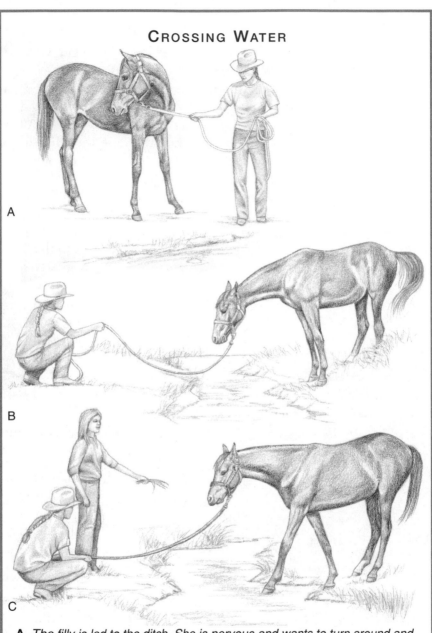

A

B

C

A. *The filly is led to the ditch. She is nervous and wants to turn around and leave.* **B.** *By moving to the opposite side of the bank, the trainer presents the horse with a chance to check out the water. She smells and tastes the water. She is beginning to lose her fear of this obstacle because she isn't forced across it.* **C.** *A helper offers the filly a bite of grass as enticement. Her desire for grass overpowers her fear of the water. There's no pull on the rope, no force to resist, so she decides to come across. Once there, she receives an immediate reward of food.*

has to be his idea. If you resort to force, his reaction is to resist, and he will associate water with punishment. He won't want to come near it at all the next time.

Patience works better than force. If you can lead him to the stream or ditch and hold him beside it for a while, he will eventually relax when he realizes you are not asking him to cross it. After he relaxes enough to nibble grass along the edge, he may become brave enough to sniff and snort at the water. A bold individual may paw at the water and check it out during the first lesson, but a timid one will not want to get that close. Just becoming comfortable enough to nibble the grass beside the stream is enough for one lesson.

In subsequent sessions, you can get your horse to step into the water or across it. One method is to make sure he is thirsty when you take him to the stream. A youngster that has drunk out of a tub all his life needs to learn he can drink from a stream. If he's suspicious and fearful, wait patiently until he relaxes and checks it out. Once he realizes it's water, he will eventually drink. If the young horse is allowed enough time to figure out what water is all about, he'll learn he can get his feet wet with no bad consequences.

Longeing

The word *longeing* (or *lungeing*) simply means to exercise a horse on a long line. Longeing can be a good training tool, but too much work on the longe can be harmful physically to a young horse. When you work him at fast gaits on flat ground, he must lean inward to keep his balance while moving in a circle. This puts a twisting strain on the legs. Footing with some give to it is better than hard ground.

Longeing is probably more valuable as a training tool than as exercise. As a training aid used in brief intervals, it helps teach a horse to balance himself and gain better coordination, as well as become more responsive to the trainer. It can help him become more supple by encouraging him to bend on the turns, striding shorter with the inside legs and longer with the outside legs. And it's beneficial whether you're starting a young horse or reschooling an older one to regain respect and control. Just don't start longeing too young. If he is too young to ride, his bones are still immature and he is too young to longe. It's best to wait until he's two years old.

Equipment

To longe a horse, you need the proper equipment and must know how to use it. It's best to use a longeing cavesson, especially when starting a green horse, rather than a halter. Be aware that a halter that slips can put pressure in the wrong place or damage the horse's eye. Most halters do not provide the amount of control necessary to effectively train a green horse. A well-fitted halter will work adequately, however.

A longeing cavesson is designed to stay in place and has a heavy, padded noseband that gives control if you need it. The line is snapped to the center ring at the front of the noseband, just the opposite of snapping to the ring on a halter (behind the jaw). When pressure is applied to the line on a cavesson, the nose piece presses against the soft cartilage on the horse's nose. You have more control to make him decrease speed or to turn when he is trying to go too fast, for example, or to run out of the circle and away from you.

You need a longe line — flat nylon webbing 20 to 30 feet long — with a swivel snap that can turn as the horse circles, thus preventing kinks and twists in the line. Webbing is less apt than rope to slip through your hands if the horse tries to buck or run off. Some trainers use a heavy, stiff, small-diameter rope like a lariat because if used properly, it gives better control and feel of the horse, but it works best if it has a swivel attached. Most trainers prefer nylon webbing. The webbing won't burn your hands as readily and is easier to handle. It's wise to wear gloves, especially with a rambunctious horse that may pull.

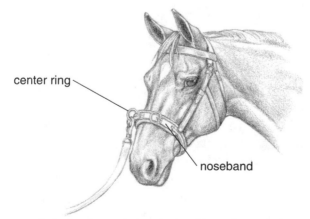

center ring

noseband

A cavesson is better than a halter for teaching a horse to longe because the longe line can be attached to the front.

PRACTICE *BEFORE* YOU WORK WITH YOUR HORSE

Handling a longe line safely and efficiently requires practice; you want to keep fine-tuned control of your horse without getting into a tangle. One way to get a feel for how to handle a longe line before you attach it to the horse is to snap it to a fence and practice with it until you can feed it out (backing away) and take it up (moving forward) without having to look at it, always maintaining a light tension on the line. You must learn to handle the line equally well with both hands.

Using a Whip

Your whip should be about 5 feet long with a 6-foot lash. This makes a good popping noise if you snap it, which encourages the horse to go forward. It is rarely necessary actually to touch him with the whip. He should never fear the whip; you should never lash him with it. Before you try the longeing lesson, a few preliminary sessions might be needed to get the horse used to the whip being pointed at him or waved if he is skittish about it.

If you used a whip properly when teaching him to lead at a trot, the horse should not be afraid of it. If he hasn't seen a whip before, acquaint him with it gradually. Show it to him, let him smell it, move it around, move it slowly and gently over his back. Usually one short session is enough for your horse to accept it (see page 57).

Your voice will eventually become your most effective cue as your horse learns what you want, because you can't reach him with the whip when he's making large circles. The popping of a whip, however, can reinforce voice cues, encouraging him to move forward or change from walk to trot when you command it. The whip is an aid, in the beginning, to help your horse understand what you want.

When teaching him to longe, hold the whip a little behind the horse, about 2 feet off the ground, and not pointed directly at him. To encourage him to move forward, bring the whip up and down behind him, touching his hindquarters if necessary. As soon as he understands, you'll need to raise it only slightly to obtain the desired response. Practice until you feel comfortable using the whip in either hand. It may be awkward to handle the whip with your nondominant hand at first and doing so will take more practice.

Your body position, reinforced with the action of the whip, is what encourages the horse to move around you in a circle. If you "drive" him from the rear, he will move forward or go faster.

Use a Round Pen

Use a small enclosure at first. A 60-foot round pen is perfect for beginning sessions at a walk, although not for extensive work at faster gaits, because the circle is too small and will put too much strain on the horse's legs. If necessary, you can make a temporary barrier with poles set on sawhorses or barrels. A small round ring gives the horse the idea of traveling in a circle without being able to make a wider one or pull away; you can keep him under control. A square pen is harder to use; there he can get into a corner and become confused. With a square pen, block off the corners with panels or poles.

PROTECT HIS LEGS

When working in circles with the young green horse, such as when longeing, it's a good idea to use splint boots on all four legs. These are like shin guards and have padding along the inside of the leg. Most fasten with hook-and-loop closures and can be put on and taken off quickly and easily. If the horse strikes his lower leg with the opposite hoof, as some young horses do when first learning to trot in a circle, the boots help protect the cannon bone.

Once the horse learns to longe well, you can do it anywhere, even out in a big open field. But the green horse may try to pull away from you or buck and play in a large area, or make his circles irregularly shaped. Until he learns, stick to an enclosed area — preferably a ring or pen about 60 feet in diameter.

Beginning Lessons

The first step is teaching your horse to circle around you. Do the first lesson with a halter and lead rope until he gets the idea. If you are using a cavesson, lead him with it before you begin, to familiarize him with it. When stopping, put a little pressure on the rope so he can get the feel of the cavesson pressing against his nose.

Hold the rope or longe line in your left hand when he is traveling to the left and in your right when he goes to the right. The whip is held in the opposite hand. Back away from the horse a few steps and stand slightly to his rear, even with his flank, but out of kicking range. From this position you can encourage him to move in a circle. Most horses want to turn to face you until they understand what you want.

To start him forward, raise the line, give the command *Walk*, or cluck, and tap gently on his hindquarters with the whip. You may have to walk in a small circle beside him in the beginning, driving him in a larger circle around you until he understands. The key to longeing is your stance and position in relation to the horse. From where you stand, you are holding him in position between the whip and the line; you thus have control over his speed and movements. If you get too far forward, he will slow, stop, or turn to face you. If you get too far back, he may be able to pull or rush forward and, therefore, be out of control.

As he begins to move in a circle, gradually play out more line to let him make larger circles. The first lessons will be small circles until he learns the pattern and how to stay controlled. The goal at this point is to keep him moving at a calm walk. After a few minutes of work to the left (generally the easiest way to start him), stop him with *Whoa* and a tug on the line, then start him in a circle to the right. For the first lessons, walking and stopping are enough. Longe him equally in both directions.

In the beginning you may have to stop him, walk to his head, turn and reposition him, then start in the opposite direction. Later you can teach him to reverse direction with voice commands. You may find that he is more reluctant to circle to the right unless you have taught him to lead from both sides. You may have to spend more time perfecting his circle to the right.

GET HELP WHEN YOU NEED IT

If you're inexperienced at longeing or have a stubborn pupil who doesn't get the idea of circling, it helps to have an assistant for the first lesson. One person can work the longe line while the other travels alongside or even behind the horse, if necessary, with the whip, encouraging him to move forward in a circle. After the horse gets the idea, you'll no longer need a helper.

Larger Circles

Don't ask the horse to move any faster than a walk or very slow trot while making small circles. A fast trot or canter can injure his legs (and will make you dizzy). Don't work too long during the first few lessons — 10 to 15 minutes is plenty. A little progress each day is enough. The green horse must strengthen his muscles and joints.

After he travels nicely in small circles, gradually increase the length of the line so he can make larger circles. Keep him out at the edge of the circle by pointing the whip at his hindquarters. Some horses try to cut in on the circles, and you have to be more forceful, directing more aggressively with the whip. Others pull strongly on the line, almost leaning on it — but this is better than having slack in the line and risking a tangle.

Teaching Control of His Actions

Sometimes a lazy horse will stop when out on a large circle, knowing he's too far for you to reach him with the whip. If this happens, reexamine your body position; it is this more than anything else that cues the horse. Always be slightly behind and to the side, almost in a position to "drive" him. In this way, when you encourage with the whip, it is his natural impulse to go forward. If you are even with his head or slightly in front of it, his inclination is to stop or turn toward you. To stop him, then, you will step forward as you say *Whoa* and give a tug on the line. Once he has that figured out, you should be able to stop him from any gait while working on the longe.

KEEP LONGEING LESSONS SHORT

The young horse can be trotted as soon as he learns to walk in a circle and stop on command. But he should not be galloped until he learns better balance and control. To avoid leg injuries, never trot him on a longe for too long a time. Ten to 15 minutes of longeing — walking and trotting — is enough for any lesson. It's better to have a short lesson every day than a long one once or twice a week. A horse is more likely to become bored and misbehave when you overdo a lesson.

Teaching Voice Commands

From the very first lesson, the horse should be taught to respond to your commands. In teaching voice commands, always use a firm, clear voice and be consistent in the way you say each one (*Whoa, Walk, Trot, Reverse*). A horse understands your tone of voice rather than the actual words; you'll confuse him if you give the same command in different ways or different commands in exactly the same tone of voice. He will soon know what you want, however, if you are consistent in the way you give commands and in your body position as you encourage the proper response. Your body language is very important in communicating with the horse and reinforcing your voice commands. Horses are extremely observant; once they understand what something means, they will obey the slightest cue.

Trot

To encourage a lazy youngster to move faster than a walk, you may have to step slightly behind his center for a moment, urging him forward with the whip as you give the command to trot. Insist that he trot and not merely speed up the walk. Don't let him gallop, play, or buck. If he gallops, calm him back to a trot with a soothing voice (*Easy* is a good command for slowing) and apply a little more pressure on the longe line. If he begins to buck, kick, or come into the center of the ring, emphatically drive him forward, even if he begins to gallop. Never allow him to kick at you or cut into the ring; he must learn to stay out on the edge of the circle.

Walk

To slow the horse, give the command to walk. Step slightly forward, giving a few tugs on the line if he is reluctant to slow to a walk. If he tries to stop rather than walk, take a quick step back behind his center and encourage him to keep moving. Work at proper angles with the horse's body to encourage proper response.

Whoa

Make sure your horse learns what *Whoa* means. (It helps if he already knows the command from leading lessons.) If he is reluctant to stop, you may have to direct him toward the ring fence to force him to stop. But most halter-broke horses will readily respond to your *Whoa* and a pull on the nose. Once he is stopped, let him stand a few moments before allowing him to move again. He needs to learn that *Whoa* means to stop and stay stopped. He should not move forward until you tell him, and should not be allowed to take several steps before responding to *Whoa*. He needs to know it means *stop now*. If he is reluctant to stop, go back to reinforcing *Whoa* during leading lessons. If he tries to move forward when stopped, give another pull on his nose and repeat the command.

Teach your horse to stop and stand when you tell him Whoa.

Come Here

Teach the horse to come to you by giving the *Whoa* command, then, after he has stood for a moment, say *Come here* and pull on the line a few times, encouraging him to come. He will get the idea. At all other times he should stay on the outside of the circle.

Reverse

As he progresses, you can teach your horse to reverse directions without having to approach him and turn him around. When he stops on command, he will generally turn his head toward you. Take advantage of this in teaching him to change direction. If he is traveling clockwise, bring him to a stop and hold him there, facing your right hand. Then step to the right, at an angle to his shoulder, while giving him the command *Change*. Change hands with your line and whip at the same time as he starts the new direction.

After he begins to understand, he will watch your hands and how you move. When he sees you change hands with the line, putting your whip in the other hand, he'll know to change direction. At first you may have to step almost completely around his shoulder and move a little behind him to indicate what you want him to do.

KEEP HIM GUESSING

Don't always stop in the same place or ask a horse to change direction at the same spot. Don't give the same commands at the same place in the circle. If you do, he may start thinking he's supposed to start trotting at that spot every time he goes past it or to *Whoa* at the same spot. Avoid this tendency from the beginning to save a lot of trouble later. He must learn he has to obey every command properly, wherever and whenever you give it.

Don't Overdo Fast Work

A good walk and trot are generally all your horse needs to learn on the longe. Most youngsters get very excited at the canter and try to speed up, buck, or pull you out of the center of the circle. Speed also puts more stress on the legs and increases the risk of slips and falls for the horse.

As an aid to prepare for saddle work and in small sessions, longeing can teach patience and control. Your horse will learn to work quietly at a walk and trot, with strong, stable gaits, bending to the circle and obeying voice commands. It will be easier to start him under saddle, as he will have better balance and more strength, and will be more obedient, too. (See page 213 for more on longeing.)

Ponying the Young Horse

Ponying means leading one horse from another. The *pony horse* is the one you are riding; the *ponied horse* is the one being led. Ponying can be a good way to exercise a horse that can't be ridden or to help train the green horse in preparation for riding.

If you have a reliable saddle horse, ponying across country is a more natural exercise for the green horse than is longeing. The latter can be hard on legs because the constant circling creates torque on joints when done at speeds faster than a slow jog. Longeing can also become boring to a horse. Ponying teaches the young horse about the big wide world outside his pen or stall and how to travel on trails, across gullies, and past or through all kinds of natural obstacles.

The young horse learns to accept being led anywhere, and new situations the two of you encounter are not so strange and frightening when you start riding him. A calm pony horse also serves as a good example for him. He gets an introduction to what will be expected of him. You can talk to him, calming or encouraging him, getting him used to voice commands. He becomes accustomed to your being "up there" on the horse next to him — similar to being "up there" on his back.

Ponying is a good way to get the young horse more physically fit before you ride him, especially if he lives in a stall or pen where natural exercise isn't possible. Ponying over the hills with a saddle on will help

---

PEACE OF MIND

Learning to be led from another horse is good for any horse. If you and your horse ever part company out on the trail — on an endurance ride or in the hunt field, for example — it's nice if he's cooperative about being led back to you by another horseman.

get him "legged up," or conditioned, before he has the extra weight and stress of a rider. It can put miles on an exuberant horse to settle him down before you get on him the first time.

Some energetic youngsters try to buck, gallop, or play on a longe line, which increases the risk of injury to young legs if you use longeing to take the edge off high spirits. Instead, pony the fractious horse over hills or pastures in a more controlled manner, giving him as many miles as needed to settle him down before you come back to the training pen to get on him.

USE CAUTION

Ponying is most practical when you have adequate open space. In an area with few trails or pastures, of course, it may not be an option. If there are obstacles or frightening situations around every corner, ponying would be counterproductive for teaching the young horse, and even dangerous. Never pony an inexperienced horse along a busy road, for example, or in places where you don't have much room to maneuver both horses if a problem arises.

Use a Dependable Pony Horse

When teaching a youngster to pony, use a quiet, well-trained mount that responds to your cues. You don't want your mount spooking or shying at the youngster's antics; you'll find it hard to hold on to the young horse, especially if the two start pulling in opposite directions. You don't want your horse balking at obstacles and becoming a bad influence on the green horse, or bucking you off if the lead rope gets under his tail. A few lessons first from the ground — moving the rope across his rump and over his body — are advisable for the horse you will be riding if you've never led another horse from him before.

Your mount must have good manners. The young horse may harass or jostle him, but no matter how disruptive the led horse might be, do not allow your mount to bite or kick him. Your horse must neck-rein and be very responsive to leg cues and remain under your control at all times.

You must be able to maneuver him perfectly as you lead the young horse alongside. Your horse may have to sidepass, back up, or even spin around quickly to enable you to keep control of the led horse. He must be able to walk through, over, and around obstacles at your command. On a narrow trail, he can't be allowed to hog it. You may need him to move out into the rocks or bushes sometimes to give the green horse enough room on the trail.

Proper Equipment

Make sure your saddle, especially the girth or cinch, is in good repair and fits the horse you ride so it won't be pulled sideways if the lead horse spooks and pulls you strongly to the side. Have a strong halter and a long, thick, soft rope for leading the green horse. A large-diameter soft rope is less apt to cause a friction burn on your hand than is a small-diameter hard-twist or nylon rope if the youngster tries to pull away. A leather lead strap will work if it is long enough; it won't burn your hand as readily as a rope can.

Wear thin, pliable gloves to protect your hands, but first make sure you can handle the lead rope without losing it. Tie a knot in the end of the lead rope to give you more leverage for hanging on if you do come to the end of it. The rope or strap should be at least 10 feet long to give you extra leeway; you may lose hold of a short rope if the ponied horse spooks or pulls back.

Holding the Rope

When handling the rope, loop the extra length through your hand. Don't hold it in a coil; if the horse pulls back, a coil can encircle a hand or wrist and drag you off. Hold the extra rope in loops. That way, if the youngster pulls back and you must let go of a loop or two, you still have some loops left.

HE SHOULD KNOW THE BASICS

Make sure the young horse knows the basics of leading before you try to lead him from another horse. He should be very well halter-trained. Then you won't have to drag him with the other horse and he won't try to drag you. He will respond to restraint from the halter and know *Whoa*. If he is reluctant to lead from another horse at first, enlist a helper to follow him with a whip to encourage him.

You can drop a loop or two to give him slack all at once if need be, yet still keep hold of the rope. Sometimes you must play him like a fish, giving him some line, taking in some, letting him have slack if he suddenly spooks or needs more rope to go around an obstacle in his path. When that moment is past, you can then quickly take up the slack again and have him right where you want him, with his head beside your leg.

First Ponying Lessons

The first few times you lead the youngster, do it in a small pen or paddock where he can't get clear away if he does pull free. Do all your first lessons at the walk until you know he is going to cooperate and be manageable. After he leads well at the walk, try some short trotting sessions.

For first lessons, have someone hand you his lead rope after you are mounted and ready. If you don't have a helper, position the green horse to the right of your own horse, parallel with him and with his head about even with your own horse's shoulder. Put the lead rope across your horse's withers where you can hold it as you mount from the left side. If need be, mount your horse from the right. Once mounted, adjust the rope in your hand as you want it and ready yourself to start off.

When you first move off, the led horse may not understand that he must come, too. Start him off at an angle; pull him toward you a little to get him moving. If you have help, the ground person can encourage the youngster from behind until he gets the idea. If the led horse already knows your voice command "Come here" or is accustomed to moving at your clucking signal, this will also help.

Keep the led horse's head even with your shoulder, about level with your knee, to give you good control. Never let him get much ahead of you or he may try to pull ahead more or kick at your horse. If he tries to go too fast, give a few sharp pulls on his halter. Don't let him lag farther back, either: Your horse may try to kick him or he may get the rope under your horse's tail.

Out in the Open

Once he figures things out and behaves, take the green horse out of the pen and into wide open areas and gradually add trotting to his lessons. Do some hill climbing if the terrain warrants it. Tackle new challenges as you feel your horse is ready. If he's spooky about going through brush or other obstacles, don't leave the main trail until he gains more confidence. The last thing you want is a bad experience where he might pull away from you. Progress to new challenges only when you are in full control and the green horse accepts your guidance.

Some horses are more balky than others and require more lessons in ponying before you can take them out of the small pen. A few are so independent that they need some sessions with a stout pony horse and a strong saddle horn so they can't pull away. Some are no fun to pony and not safe. If they are not controllable in the small pen, don't take them out. You are better off skipping this aspect of training until they are older and wiser. They will be easier to pony after a few years of being ridden. Most young horses, however, adapt quickly to being led from another horse. They are sociable and therefore willing to go where the pony horse goes, making this aspect of training a useful transition from ground work to ridden lessons.

Keep the head of the ponied horse about even with the position of your shoulder and about level with your knee.

6

MORE GROUND WORK BEFORE YOU RIDE

There are many things you can do with the young horse in preparation for riding or driving, and the more you do before you actually get on him, the better. This early training will make later lessons go more smoothly when he becomes mature enough physically for mounted lessons or for pulling a cart.

Most horses should not be ridden until they are three years old. You can do some early lessons — mounting, for example, and teaching a horse to move out, stop, and turn — at age two, but it's usually better to wait until he is three years old before you do much riding and four years old before you do a lot of trotting and galloping. A horse is not mature enough physically to handle hard work — many hours and miles each day — until he is five years old. Some equine sports such as racing, cutting, and reining start horses young, but this may be hard on them physically; many do not stay sound. Let the horse become more fully mature before asking him to do speed work.

Remember, training is a step-by-step process, building on what you have already accomplished. If a horse's experience with a new situation is a good one, he will be relaxed about it in the future. If you prepare for each step as you go along, anticipating possible reactions and responses, you can prevent a lot of problems.

Every horse is different. Methods may have to vary with each green horse you train, as you feel your way along and find the best way to gain each horse's trust, confidence, and respect. Some young horses accept new things without fuss or resistance; others are suspicious or fearful, and you must take them more slowly. Others are headstrong; some are lazy. These horses may resist what you are doing just to see if they can get away with it. They're not really afraid; they're just testing you. With these individuals, you may need to be firmer.

First Bridlings

You don't think twice about bridling an old dependable horse. But the first time you bridle the young horse can be a traumatic experience unless you've prepared him for it. Bridling is such a basic part of everything you'll be doing with him in the future that you want to make sure his first experience is a good one. The horse should already be accustomed to having his head, mouth, and ears handled. He should be used to wearing a halter and standing quietly beside you. With these lessons behind you, and because the horse trusts and respects you, it should be fairly easy to introduce the bit and headstall.

Mouth Handling

From the time he is a foal, your horse should be used to an occasional finger in the corner of his mouth rubbing his gum. If not, start doing this before the first bridling lessons as part of his daily grooming routine. Then he will not be alarmed the first time you try to put a bit in his mouth.

Teaching a Horse to Relax and Lower His Head

Teaching a horse to lower his head is not only handy for haltering, bridling, and giving oral medication and dewormers, but it also breaks the pattern of flight or fight if your horse is upset and establishes you as dominant in the relationship. It teaches him to relax. The excited or nervous horse raises his head. If the horse is taught to lower his head on cue, it can also get him under control in an emergency situation. There are a number of cues to teach the horse to lower his head. It's best to be consistent, however. Always use the same one with the horse so he understands what you are asking. Choose the cue that works best for you and the horse you are training.

PRESSURE AND RELEASE

A. *Standing next to the horse's left shoulder, put gentle pressure on the halter. Don't jerk on the rope.*

B. *Keep steady downward pressure and release it as soon as he lowers his head, even slightly.*

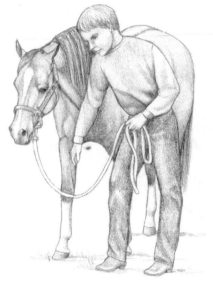

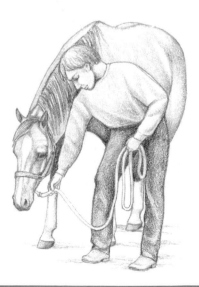

C. *Then ask him to lower his head a little more, rewarding any downward movement with release. Soon he will understand your cue and lower his head as much as you ask.*

Short Tugs

One way to teach him to lower his head is to give rhythmic and regular short tugs on his halter. The tugs should be gentle, more of an annoyance than a pull. He will at first respond by raising his head; this is his natural reaction to get away from annoyance. If he raises his head or tries to back up, let him; just continue with the gentle tugs. It may take several minutes, but as soon as the horse begins to lower his head — even just a little — stop tugging and give him praise. Then continue tugging until he lowers it again.

Soon he will get the connection: When he lowers his head, there's no more annoyance. Within a few lessons he will lower his head on cue, with just one gentle tug on the halter. He can also learn to lower it as much as you request. The key is patience, especially in the beginning, until he figures it out.

Pressure on the Poll

Another cue is to put one hand behind the horse's poll at the top of his neck and the other on the bridge of his nose, applying slight downward pressure with both. When he gives to this pressure, immediately release it and pat him, then do it again. If he doesn't want to lower his head, use your hand on his face to move it from side to side a little, continuing the mild pressure at the top of his neck. Always release the pressure as his reward when he responds.

You can cue the horse to lower his head by putting gentle pressure on the bridge of his nose and at the top of his neck.

If his head comes back up, repeat your cue. If he tries to back up or move away, just move with him and continue the pressure, letting him know he's not trapped but that the pressure stays there until he lowers his head. Never force him to leave his head down. Eventually he will lower his head — as low as you wish — on cue, for the pressure is released when he does.

Ear Handling

Some horses are very sensitive about ears; it takes extra time and handling to get them over their fear of having their ears touched. It's always good to start handling all parts of the horse when he is a baby (see chapter 3 for more on imprint training). If this aspect of handling was skipped or inadequate, now it's time to catch up on it before you bridle him.

If his ears are sensitive, start gently handling them every time you groom or work with your horse, until he realizes you are not going to hurt him. It may take several sessions before you can handle his ears with ease. Work to them slowly and gradually. Rub his neck and work closer and closer to the ears until he no longer is afraid of having them touched. You can tell when he starts to change his mind. He will be relaxed instead of tense and evasive. His head will come down and he may sigh. Once you can touch his ears, gently rub all parts of them until he accepts this readily.

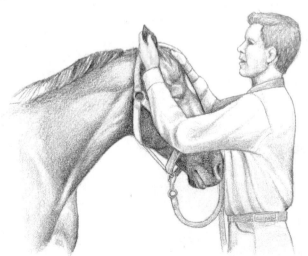

Accustom your horse to having his ears handled.

First Bridling Lessons

Make sure the headstall is of the proper size before you try to put it on the horse. Hold it up next to his head and judge the distance between the poll (behind the ears) and his mouth. Use a headstall that's easily adjustable when on the horse, so if you misjudge the size, you'll be able to take it up or let it out as necessary.

To ensure that the horse will not resist the bit, put molasses or honey on it before you put it in his mouth. Detach the bit from the headstall and put it in his mouth alone a few times, if you want, letting him taste the molasses and get used to grabbing hold of the bit and mouthing it — and also spitting it out.

With a horse that is well halter-trained, ties well, and has no qualms about having your fingers in his mouth or his ears handled, you can leave him tied in his usual place (to the fence or with cross ties in the barn stall or aisle) for his first bridling. Restrain him in a familiar spot until he gets used to the procedure, just in case it takes a few false starts to get the bit into his mouth. You don't want him backing up and then realizing he can avoid the bit by moving away from you.

If he is not well trained to tie, do *not* tie him for bridling. Leave on the halter, and put the headstall on over the halter. Even if you will eventually ride the horse without a halter under the bridle, always leave it on during bridling lessons and early training so you can control the horse *without* having to pull on the bit in his mouth. If you are unsure about how the horse is going to react, enlist someone to hold his halter rope, loosely, on the other side of the horse, the first time you bridle him.

Remove the reins from the bridle for first lessons and use just the bit and headstall. Hold the crownpiece in your right hand (with that arm up around the horse's head to hold his head in place) and the bit in your left hand. Some people bridle a horse with the right hand up over his forehead, but if the inexperienced horse feels too confined with your arm up over his head, use your arm on the other side of the head to keep it steady between your two hands; he won't be able to move around to avoid the bit.

> **SAFETY TIP**
>
> Use a snaffle bit with a thick, comfortable mouthpiece; it will be less apt to pinch or cut into the youngster's gum. A narrow, thin bit puts too much pressure in one small area. (See page 200 for more on bit types.)

Again, when you have prepared your horse, first bridlings should be easy. The bit rests on the fingers of your left hand, beneath his mouth. Bring the crownpiece up to his ears and place the bit into his lips. If there is molasses on the bit, he will probably open his mouth for it. (See page 28 for the basic technique.)

If he doesn't open his mouth, put your thumb or a finger into the side of his mouth, pressing on the bars where there are no teeth. When he opens his mouth, slip in the bit and raise the crownpiece over his ears, being careful not to fold or irritate them. Slip the headstall over the ears one at a time (starting with the far one), then straighten the brow band and forelock before fastening the throatlatch. If you are using a Western one-ear headstall, place the ear gently in its slot, then slide the crownpiece over the other ear. Don't cram in the ears or fold them; ears are very sensitive.

BRIDLING TIPS

If the horse refuses to open his mouth, put more pressure on his gum with your finger or tickle his tongue and he will usually open up. The main things to avoid in first bridlings are bumping the teeth with the bit (a finger against the gum is more effective encouragement) and mishandling the ears; these mistakes can make the horse difficult to bridle the next time.

Unbridling

Taking off the bridle properly is just as important as putting it on properly. If the bit bumps his teeth on the way out, your horse will raise his head to try to get away from the pain. He may then raise his head in anticipation of pain every time you start to take off the bridle, making it impossible to get it off without clanking his teeth.

Thus, it is crucial to make first lessons smooth and relaxed. You want him to lower his head so you can slip off the headstall over his ears and let the bit drop out gently. This is where it pays to have him lower his head on cue. One way to get him to relax is to rub his head or forehead. Only after his head is lowered should you remove the headstall, waiting for him to release the bit. Let him spit it out after his head is lowered. Don't ever try to pull it out; it will hang up on his

---

UNBRIDLING TIPS

If the horse's head is too high when you begin to remove the bridle, stop — the bit will catch on his lower incisors. First encourage him to lower his head. When you take the headstall forward over his ears, make sure the bit stays well up in his mouth (the headstall supports the bit) until he starts to spit it out. Then it can be eased past his teeth without clanking and causing him to toss up his head. Hold him gently by the halter as you remove the bridle, encouraging him to keep his head low, and hold on to the headstall until he releases the bit.

---

teeth. He must be given time to let go of it. After a few careful sessions, he'll learn to lower his head each time you start to take off the bridle.

First Saddlings

Saddle the young horse several times before you actually ride him. This way, he becomes comfortable with the feel of the saddle. If he is already used to being touched all over, the first saddling should be easy. Do his first saddling with him tied in a familiar place, such as where you usually groom him.

It's important that he be thoroughly halter-trained before you progress to saddling lessons. He needs to learn he is to stand still for saddling and not move around. Unless he's not accustomed to being tied, he should be tied rather than held. He must be content to stand tied for these lessons; you don't want him flying back or trying to avoid the saddle pad. This is why you build on each lesson, taking him step-by-step, so he has to concentrate on only one new thing at a time.

Be Confident

Work with the horse quietly and confidently. Take care not to upset him, but act as though saddling is a routine thing. If you're quiet and relaxed, he's more likely to stay relaxed. If you are tense and nervous about the lesson, he will be too: He will feel your mood and become more insecure.

Your horse looks to you for cues on how to behave; act with confidence so he will see you as someone he can depend on. You are the leader in this team effort. If you're unsure or nervous, your horse will think there is something to fear and be suspicious of the lesson.

If you're nervous and your horse is still not convinced that you're actually the leader, he'll become pushy and uncooperative, trying to take advantage of your lack of confidence. In all lessons, you must establish a confident attitude so he knows you are the one in control. This does not mean you have to be forceful; it just means you know you are in control, so he will know it, too.

Get Him Used to Blanket or Pad

Use an old pad that has a familiar horsey smell. While you are grooming, show the horse the saddle pad and let him smell it. Put it on his back slowly and carefully, without a lot of extra movement. Some young horses accept this readily; a nervous or spooky youngster, however, may be fearful. For a skittish horse, it helps to fold the pad or blanket smaller and to ease it onto the back, being careful not to startle him. Put it on and take it off a few times, so he becomes accustomed to it and realizes it is nothing to fear.

Get your horse completely used to the pad so you can eventually put it on in normal fashion — that is, without having to ease it carefully onto his back. After he is relaxed about it, drag it off his back, put it on and take it off from the other side, flip it on with more motion, and raise it up as you put it on (the height you would lift the saddle to get it on his back), getting him familiar with it. When he is no longer worried about it, he is ready for the saddle.

With some horses you can go ahead and saddle them after putting the saddle pad on and taking it off. Others take a little longer to feel comfortable with this new thing; and you'll need to spend a few sessions working with just the saddle pad before you try the saddle.

Use a Lightweight Saddle

For his first saddling, it's best to use a light saddle, such as an English saddle or a child's lightweight Western saddle, rather than a big Western saddle. A light one won't be so apt to alarm him and is easier to retrieve if he tries to jump out from under it before you get the girth fastened. Set the saddle on his back the first time without using the cinch and just hold it there. If he protests or moves around too much, lift it up off his back before he gets overly worried. Once he relaxes and stands still, put it on again. Put it on and take it off several times.

Don't startle your horse when you set the saddle on his back. If it's an English saddle, have the stirrup irons run up so they won't flop around. If it's a Western saddle, hook the off stirrup bow over the saddle horn and lay the cinch across the saddle so it can't flop down on the off side as you put on the saddle. If you never give him cause to startle, he'll be more likely to stay calm, relaxed, and trusting about the saddling process.

Make Sure the Saddle Fits

Select a saddle that fits the horse. An uncomfortable saddle will make him resist lessons with it. If the saddle tree and bars are too wide or too narrow for his back, it will cause pain. A wide-backed horse with low withers needs a wide, flat tree; a narrow horse with narrower withers needs a narrower tree. He can't concentrate on lessons if he's uncomfortable; he'll be thinking about pain rather than what he's supposed to be learning.

Ease the saddle gently on his back, and be ready to lift off the saddle if the horse gets nervous. Don't cinch the saddle this first time unless he is very relaxed about it.

Tightening the Cinch

How far and how fast you progress with the first saddling lessons will depend on the individual horse. On a calm and mellow individual, you may accomplish it all in the first lesson. On a nervous horse, work in stages over several sessions. The first time, you may not want to use the cinch at all. It may be better to wait until the second or third lesson, after he is more relaxed about having the saddle placed on his back.

It's usually when the cinch is being tightened for the first time that a young horse becomes scared or resentful. It helps if you've prepared him by touching him a lot with your hands. During grooming sessions, you will have discovered whether your horse feels comfortable with your touching him. For the sensitive, ticklish horse, extra brushing under his belly and rubbing the girth area and behind his elbows will be necessary until he feels at ease with something touching him around his girth.

CINCHING TIPS

Tighten the cinch slowly and carefully, trying not to pinch or startle the horse. Do not pull it very tight the first time — just enough to keep the saddle from slipping off if he jumps or moves around. If he is nervous, don't leave it on very long. Take as many sessions as necessary until you can eventually tighten it enough to hold the saddle securely in place.

Leading Him with the Saddle

When you have saddled the horse a few times, the next step is getting him accustomed to moving while carrying it. If it fits well and he is comfortable with it, he should never try to buck it off. A skittish horse, however, may buck if it alarms him, especially if you try to lead him with it on before he is used to the idea. Whether you start leading him the second day after saddling or the second week will depend on the horse's personality and his reaction to the saddle.

Before you lead him with it on, make sure it is cinched up properly — not too tight, but definitely tight enough that it will not move or shift if he jumps around. You want to make these early lessons a pleasant experience. If you don't have the cinch tight enough and the saddle

Longe your horse with a saddle so he gets used to the weight and sound of it and the feel of the stirrups flapping.

turns or slips under his belly if he bucks, it will scare him into bucking harder. This frightening experience could make him afraid of a saddle forever. Remember, the young horse does not have the experience and tolerance of an older, trained horse. Prepare for each step carefully to avoid creating a bad experience.

Most young horses will walk around with the saddle on, but others are frightened. Be alert for sudden movements and prepared to keep the horse under control if he does explode. Tie the stirrups to the cinch rings before you lead him so they cannot flop and scare him into bucking harder. If he's skittish, you'll be glad you spent time working on his leading lessons and perfecting his response to *Whoa*.

When you lead him for the first time with the saddle on, do it in a small pen or a box stall so he can't get up any speed or get away from you if he explodes. If he jumps or bucks, halt him and reassure him. Then, after he is fully relaxed, again ask him to take a few steps. Don't lead him in the open until he is comfortable moving around while saddled in a small enclosure.

If the saddle seems to bother him, make sure that it fits correctly. If it doesn't, change saddles. If it does, give him more lessons in moving around while saddled until he learns to relax and accept it. With the saddle on him, you can lead, longe, or pony him. Some trainers turn the horse loose in a round pen so he can move about freely, while others prefer never to give him such freedom to discover that he can buck while saddled. Whether or not he bucks in the round pen, he'll still learn that he can't get rid of the saddle and that it's not hurting him. Give him plenty of time to become accustomed to carrying it; let him get all the skittishness out of his system.

ANTICIPATE YOUR HORSE'S RESPONSE

Some horses learn to trust you very quickly and never try to buck. Others test you every step of the way. Know your horse and be able to predict what he might do. You must outsmart a horse that constantly tests you and prevent bad behavior before it happens. If you feel your way along with each horse, rather than trying to use the same method for each one, you can usually find the best way to handle a particular animal.

Wait until He Is Mature before You Ride

Once he is comfortable with a saddle, you can start mounted lessons whenever your horse is ready, after lessons in bitting and responding to the bridle; see chapter 7. If he is too young to ride, continue his training with more ground work. The bones and joints of a two-year-old are not mature enough to handle athletic work with the weight of a rider. Even extensive ground work such as longeing faster than a walk can be hard on the joints of a young horse. There are a lot of things you can teach him from the ground while you wait for him to mature, however. You can get on a two-year-old a few times and start the basic work of moving, stopping, and turning, but wait until he is three years old before you get out and go.

Ground Lessons in Maneuverability and Control

You can teach the horse basic actions and control before you ever get on him. If he already knows how to stop, turn, and respond to bit and leg pressure, it won't be so confusing for him when he has to try to understand your signals from his back. Leg pressure can be simulated by pressing on his side in the area your leg would be when mounted, as when teaching him to move his hindquarters over (see page 178).

The green horse gains confidence when he understands how to respond to a cue. If he knows he will be rewarded by release of pressure, he will try to make the proper response. If he does not know how to respond, however, your signals will confuse, irritate, and startle him. He may react with a very improper response such as bucking or pulling on the bit. Anything he can learn ahead of first rides will help ensure that they go well.

Pressure and Response

A basic concept of training a horse is that almost everything we ask him to do is in response to pressure. He learns to do the proper thing willingly because his reward for responding is instant release of pressure. When riding, you press with your leg and he moves forward or to the side, depending on the cue. You touch the reins lightly and he feels bit pressure, so he slows, or stops, or turns. These are conditioned responses; he learns how to respond properly because the pressure goes away when he does something correctly. His response becomes automatic, a reflex response to the stimulus, created by training. The essence of training is to use less and less pressure as the horse learns to respond, until he will perform the correct response at just the lightest touch.

If you don't release pressure, he sees no reason to move when you press with a leg or to halt when you use the bit. He responds to the release of pressure — the reward he gets for doing something correctly — rather than to the pressure itself. Remember, horses don't reason; they learn by means of step-by-step training. We ask a horse to do something and when he responds we reward him by halting the signal. We may ask again immediately for another step, but we don't keep pressing or pushing; we repeat the process, asking with lighter pressure until he responds at the slightest cue.

Lessons in Stopping and Turning in Response to Reins

After the horse is accustomed to saddling and bridling, you can lead him a few times with bridle and saddle on (with the halter under the bridle and a lead rope for use if needed). For most horses it's a simple matter to make the transition from signals via halter to signals via the bridle reins as you are leading. When you want to halt, put pressure on the bit with the reins as well as the halter, getting him used to the feel of the bit. Pull and release, pull and release, with intermittent gentle pulls and never a steady pull. Release all pressure as soon as he halts. He will soon make the connection.

When you want him to turn, put a little pressure on the proper rein and pull his head around to the desired direction. If he leads equally well from both sides, you can do this easily as you lead him. If he is more comfortable with you on his left, give him the signal to turn to the right by shifting his head in that direction using your hand under his neck and directing him away from you as you make the turn to the right. The pull-and-release action on the bit works better than a steady pull as you encourage him to turn.

If at first he is confused, use the halter and lead to reinforce what you are asking. Soon he will understand about following his head in whichever direction you pull on the bit, knowing he gets release of pressure from the rein when he makes the correct response. In just a few lessons he will learn to respond to bit pressure for stopping and turning, and these signals will not be confusing to him when you get on his back for the first rides.

Teaching the Turn on the Forehand from the Ground

The turn on the forehand teaches the horse to move away from pressure and helps him become more maneuverable. Stand at his left shoulder, facing him, with the halter rope in your left hand about 10 inches from the halter. Using your right hand doubled into a fist, put pressure with your knuckles on his side, at about the same place where the calf of your leg would be if you were mounted. His response should be to move away from the pressure, swinging his hindquarters away from you. Use the halter to keep his front end from moving too much. You want his left front foot to act as a pivot, with only some up-and-down motion, as he moves his hindquarters away.

Try to get the horse to move only his hindquarters, not his whole body. It is often necessary to pull his head toward you a little in order to accomplish this. At the first "give" of his hindquarters, cease pressing and offer him praise. Repeat the procedure until he does several side steps with his hind legs.

Do the same on the opposite side, so he learns to move away from pressure on either side. After a few lessons, he will move his hindquarters either way whenever firm pressure is applied with your fist and the

TURN ON THE FOREHAND

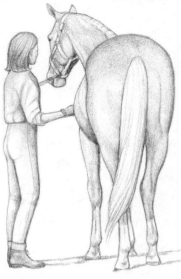

A. *Stand by his shoulder and ask your horse to move his hind legs one step away from you while keeping his front legs in place.*

B. *Ask for one step at a time, pulling his head slightly toward you (but not enough to move his front feet), and press with your knuckles behind the girth area, where your leg would be if you were riding.*

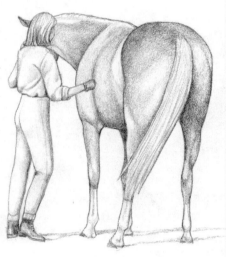

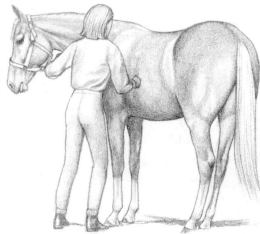

C. *Ask for another step, one at a time, until the horse has turned 180 degrees with his hind legs walking around his stationary front legs.*

halter rope is held lightly to keep his front legs stationary. Use lighter pressure as he learns what you want. If you release pressure as soon as he starts to move his hindquarters, he will respond readily. These lessons will help prepare him for leg pressure when you start riding him and also make him more maneuverable for ground handling and for better control of his actions.

TURNING TIP

If the horse is confused and wants to move forward instead of to the side, position him facing a fence. Now he'll be able to see that he can't move forward and will be more apt to move his hindquarters to the side. Stay far enough out from the fence that you can make a 180-degree turn, however, so that he will be facing in the opposite direction when you finish the maneuver.

Teaching the Turn on the Hindquarters from the Ground

To rotate on the hindquarters — that is, moving his front legs but not his hind legs — it helps to attach two lead ropes to his halter like reins, putting them over his neck. Face the horse, standing at his left shoulder. Take the right rope in your right hand and the left one in your left. Pull them both to the side and a little back to make him move only his front legs to the left. He may refuse at first, not understanding what you want. Be patient. When he makes his first step toward you with one front foot, stop all cues and praise him. Repeat the process, asking for another step. After he makes a couple of steps, cue him from the other side.

Another way to teach the horse to move his front quarters is to push them around, having him turn away from you instead of toward you. Stand next to his shoulder, with slack in the lead rope but pressing gently on his cheek with the hand holding the rope, and put pressure on his shoulder with the other hand. Release all pressure if he moves away. Repeat, walking the horse's front end away from you, a step at a time, pivoting on his hindquarters. If he attempts to step forward instead of to the side, tug slightly on the halter rope, thus shifting his weight back onto his hindquarters. Now ask him to move his front. If he has trouble with this, have him stand facing a fence in order to block the forward movement.

TURN ON THE HINDQUARTERS

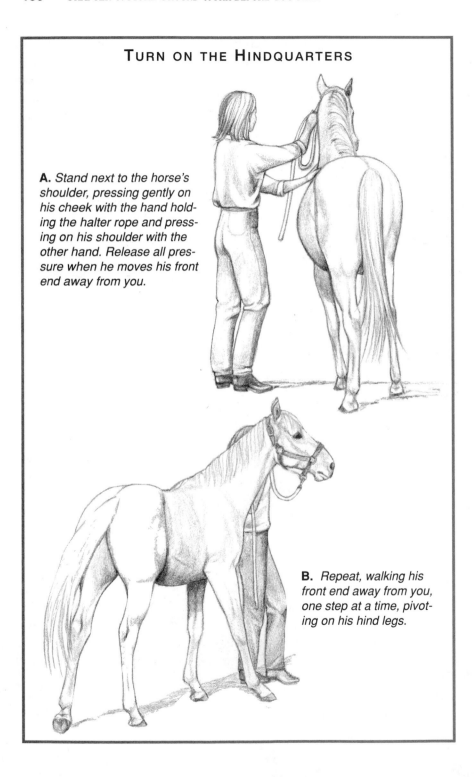

A. *Stand next to the horse's shoulder, pressing gently on his cheek with the hand holding the halter rope and pressing on his shoulder with the other hand. Release all pressure when he moves his front end away from you.*

B. *Repeat, walking his front end away from you, one step at a time, pivoting on his hind legs.*

WATCH HIS LEGS

It is all right if your horse moves his feet up and down on the opposite end when rotating on the hind legs or the front as long as the stationary end does not move to the side. When rotating the front legs around the hindquarters, a backward or forward step or two in the hind legs is acceptable as long as there is no side movement. The legs that are crossing over in making the side steps should *always* cross in front of the opposite leg. This keeps a feeling of better body control and balance, which will be important later when you do this maneuver while mounted. If the horse does not cross his legs correctly, he will be more apt to trip himself. Reevaluate your cues if he does it incorrectly, and try again.

Teaching the Back-Up from the Ground

If you have not already taught your young horse to back up at halter, now is the time to do so. To teach the horse to back up, see page 138. Learning to back up on cue when handled on the ground will be useful to you and the horse for your entire relationship; there will be numerous occasions when you'll need to reposition him for something or keep him from crowding you. If he is maneuverable on the ground — backward and sideways as well as forward — there will be fewer bumps and bruises and stepped-on toes.

The back-up under saddle is something you should wait to teach your horse until after he has the other basics — stopping, turning, moving forward willingly under saddle — mastered (see chapter 16). All these lessons will be beneficial later when you start to ride your horse and will improve his ground manners. The ground work is the foundation for having a well-trained horse.

7

BITS AND THEIR USES

For hundreds of years, riders have communicated with and controlled horses through mouth contact with a bit. The snaffle, the first type of bit, was used on chariot horses. Egyptians were using jointed snaffles by 1400 B.C. These had straight cheek pieces, sometimes with sharp spikes pointing toward the horse.

There are basically only three types of bits: the snaffle; the curb; and the Pelham, which is a combination of snaffle and curb. The type of bit you use is not nearly as important as the horsemanship that accompanies it. When it comes to developing a good mouth on a horse — a mouth that is pliable and responsive to the bit, never pulling or fighting it or trying to avoid it — there is no substitute for light hands, good seat and balance, and the sensitivity to know how to interact with the horse.

It has been said that any rider who is experienced enough and capable enough to use a severe bit doesn't need one, and no one else *should* use one. The factor that determines whether a bit is severe or mild is ultimately the hands that use it. The snaffle is sometimes categorized as a gentle bit, but when used with rough hands it can be cruel. Some snaffles are designed to be as mild and comfortable as possible, for starting a young horse with a tender mouth; others are harsher, for use on an older, resistant horse. The latter are too severe for a young horse.

Snaffles

There are two kinds of snaffle, the bar snaffle and the jointed snaffle. The *straight bar* exerts pressure straight back, directly on the corners of

PRESSURE POINTS

All bits work on one or more of seven pressure points. These are the corners of the mouth, the bars (the toothless area of the gum between the front and back teeth), the tongue, the roof of the mouth, the chin groove, the poll, and the nose. The bars, tongue, and roof of the mouth are quite sensitive because they have many nerve endings. The chin groove, nose, and poll are also sensitive, being thin-skinned flesh over cartilage or bone. The corners of the mouth, composed of thick muscle tissue, are somewhat less sensitive.

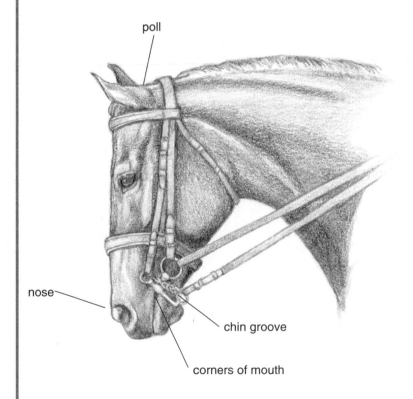

The seven points of pressure for the bit are the corners of the mouth, bars, tongue, roof of mouth, chin groove (curb chain), poll, and nose.

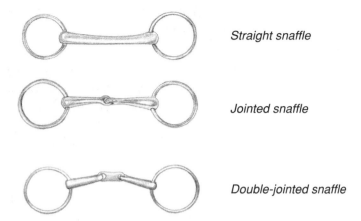

Straight snaffle

Jointed snaffle

Double-jointed snaffle

the mouth, when the reins are pulled. The *jointed snaffle* bends in the middle and comes farther back when pulled; it is usually more severe. A curved, bar snaffle, called a *half-moon,* is a very mild bit, as the horse's tongue has more room to move around.

The snaffle has a ring at each side for attaching to the cheek pieces of the headstall. The same ring is often used for attaching reins, though some bits have a separate ring for each purpose. The snaffle has no leverage action; it controls the horse by pressing on the bars and tongue and by direct movements to the side.

Straight Snaffles

A straight snaffle rests on the tongue rather than on the bars. It moves upward and back when rein pressure is applied and has no squeezing action. These bits have either a straight or a curved mouthpiece and are considered very mild. Although they lack the scissors effect on bars and the corners of the mouth that a single-jointed snaffle has, they still can be harsh. If the mouthpiece is thin and cuts into the bars when pressure is applied, or if it is made of twisted wire, the mouthpiece will have a sharp edge that can be harsh if pulled hard. There are several varieties of straight snaffles, with rubber, metal, or twisted-wire mouthpieces.

Jointed Snaffles

Many snaffles have a single-jointed mouthpiece; in these, the bit bends in the middle. This puts more pressure on the corners of the

---

MILD BITS

A snaffle mouthpiece that tapers — that is, it's wider at the sides and thinner at the middle — is a mild bit: It has more surface area over the bars and won't cut into them, and it gives the tongue more room and comfort. If the horse has a thick tongue, a double-jointed mouthpiece (one with a flat plate in the center) may be most comfortable. The flat plate rests against the tongue, creating a bridge of thin smoothness over the tongue, instead of one joint protruding far forward over the fat tongue and poking the roof of the mouth. A rubber mouthpiece makes a mild bit, not only from the softness of the rubber, but also due to the extra thickness.

---

mouth and on the bars and may also poke the roof of the mouth with the joint of the mouthpiece if the bit does not hang properly when the reins are pulled. A jointed snaffle has a lot of squeezing action.

When merely carried in the mouth, the single-jointed mouthpiece is quite straight, but as tension on the reins increases, the mouthpiece bends in the middle, pushing the joint toward the front of the mouth and squeezing the corners. If it does not fit properly or is of a poor design, it may pinch the tongue or pinch the corners of the mouth between the mouthpiece and the ring. Rubber bit guards are sometimes used to prevent this. *Egg-butt* and D-*ring* snaffles, which have large rings, help reduce the risk of pinching the corners of the mouth and are also less likely than O-rings to be pulled clear through the mouth.

Full-Cheek Snaffle

The full-cheek snaffle incorporates a cheek piece on each side in conjunction with the ring. The action is basically the same as with any other snaffle, except it gives more lateral control of the horse. Even a very slight tug on the rein will encourage the horse to bring around his head and turn, because the cheek piece on the opposite side is putting pressure on the jaw. This makes a headstrong horse easier to control.

Another advantage of the full-cheek snaffle, which is helpful for the green horse, is that the top of the cheek piece can be held parallel to the bridle cheek pieces with leather keepers in order to hold the bit in proper position in the horse's mouth. It keeps the joint of the snaffle from dropping down in the center, as it tends to do unless the bridle is quite snug

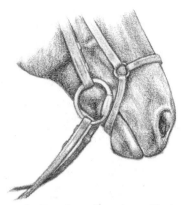 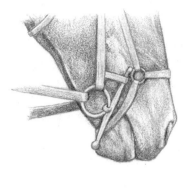

An egg-butt, or barrelhead, snaffle is held in place with an adjustable drop noseband.

A full-cheek snaffle is held in place with leather keepers attached to the cheeks of the bridle, along with a drop noseband.

and the bit is held quite high in the mouth. The full-cheek snaffle with keeper loops can thus be adjusted a little lower in the mouth for the comfort of the horse, without the horse putting his tongue over the bit.

Because the full-cheek design makes it possible to hold the bit at a specific angle in the mouth, special mouthpieces are sometimes used, such as the double-jointed Dr. Bristol. The central link is a flat tab. When using a full-cheek snaffle, this mouthpiece is positioned to lie comfortably flat on the tongue when the horse is carrying his head in the proper position, but it tends to turn at a more severe angle if he sticks his nose out too far.

Western Snaffles

The Western snaffle, or *colt-training snaffle*, is actually a very short-shanked jointed curb or a training curb with a jointed mouthpiece and rings for snaffle reins. Even though it is called a snaffle, this is a misnomer. These bits have shanks and curb straps and are therefore curb bits rather than snaffles: They work on the leverage principle rather than with direct reining. The exception is the training bit with snaffle rings; it has the action of a snaffle when only the snaffle rings are used. When reins are attached to the short shanks, it becomes a *curb*.

A jointed shank snaffle should always have a bar across the bottom of the shank. This helps stabilize the bit and keeps the jointed mouthpiece from folding up too much at the center and possibly poking the roof of the horse's mouth.

SNAFFLE ACTION

A snaffle puts direct pressure on the bars and corners of the mouth and on the tongue; bit severity depends on the horse's head carriage and on the position of the rider's hands. A high head carriage, coupled with hands held low, puts the action of the bit mostly on the bars of the mouth. A low head and high hands shifts pressure to the corners of the mouth. The snaffle does not encourage flexion at the poll or relaxation of the lower jaw, as a curb bit does; mainly it encourages high head position due to its action on the bars and corners of the mouth. The snaffle requires the use of two hands on the reins and direct, or side-pull, reining.

When your horse is in a snaffle and you have a rein in each hand, use a give-and-take pull on one rein to turn his head in the direction you want him to go. Create slack with the other rein at the same time, so his head can turn. The "passive" hand must give as much rein as the "active" hand takes.

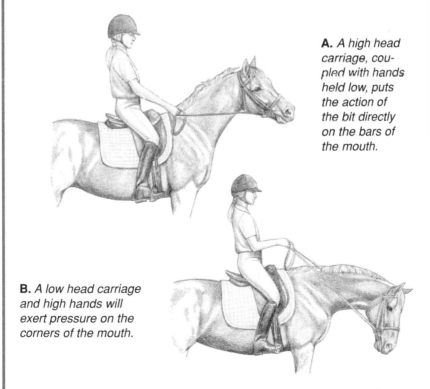

A. *A high head carriage, coupled with hands held low, puts the action of the bit directly on the bars of the mouth.*

B. *A low head carriage and high hands will exert pressure on the corners of the mouth.*

Gag Bit

This bit usually has a snaffle mouthpiece, but it has a hole through the top and bottom of each bit ring for a rounded piece of leather to pass through. The leather piece attaches to the bridle cheek piece at one end and to the rein at the other, and must be able to slide freely through the holes in the metal bit rings. When the reins are pulled, these leather pieces move the bit higher in the horse's mouth and also put pressure on the horse's poll via the bridle crownpiece. Because of the added pressure at the mouth and poll, this bit can be quite severe if used harshly.

Curb Bits

The curb has either a jointed or an unjointed mouthpiece. The mouthpiece may be straight or have a *port* in the center to allow more room for the tongue. The port is a rise in the mouthpiece, in various heights, shapes, and designs, that can affect the function of the bit. When the bit is carried in the mouth (no rein pressure), the mouthpiece rests on the tongue and, if the port is high enough, on the bars. A port gives the tongue more room and helps keep the horse from getting his tongue over the bit. With a low port, the tongue receives pressure first, before the sides of the mouthpiece make contact with the bars. A high port is considered more severe; the bit puts pressure on the bars more quickly when the bit rotates, and the port also may touch the sensitive roof of the mouth.

The main feature that defines a curb is its shanks. Unlike a snaffle, a curb produces leverage action. A pull on the reins tips the bit in the horse's mouth, which affects him in several ways. It puts pressure on the

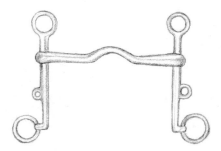

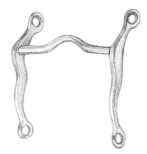

An English curb has straight shanks.   *A Western curb has curved shanks.*

BIT SEVERITY VARIES

Some curb bits are more severe than others. A curb with short shanks and a wide, low port is the most gentle because it has less leverage, allows the horse a lot of tongue room, and does not touch the roof of his mouth. The thickness of the mouthpiece is also a factor; a thick mouthpiece is milder than a thin one, which if used harshly tends to cut into the tongue and bars.

The proportionate length of the cheek pieces also contributes to the severity of a curb bit. The length of shank below the mouthpiece and the length of the upper cheek piece, above the mouthpiece, determine the amount of leverage on the jaw. A long shank puts more leverage on the jaw and on top of the headstall than a short shank, with just a slight pull on the reins. The bottom of the shank is pulled up and back, which tips its upper part downward and forward, pulling on the curb chain or strap and transferring downward force on the poll.

---

bars of his mouth or on the bars and tongue as the bit tips, depending on how it is made. As the bit is tipped, it brings the curb strap or chain tight against the horse's chin. It also puts some pressure on the top of his head with the headstall. The port, if high, may push against the roof of his mouth.

There are many styles of curbs, but they all work on the same principle. There are English curbs and Western curbs. The English curb usually has a straight shank; the Western curbs (and some walking-horse bits) have shanks that curve backward. The curve prevents *lipping*, the horse grabbing the shank with his mouth. English curbs use a lip strap.

Short-Shanked Curbs

The length of the shank is partly what determines the severity or mildness of a curb. A short-shanked bit does not provide as much leverage as a bit with longer shanks; action of the short shanks puts less pressure on the horse's chin via the curb strap or chain and on the top of the head via the headstall. For instance, the grazing curb is a mild bit with a low port and short shanks, short enough that the horse can graze with this bit in his mouth; the shanks do not bump the ground as he eats.

LOOSE SHANKS

A bit with "loose-jawed" shanks that swivel on the mouthpiece is a milder bit than one with solid shanks. The loose shanks give the horse a little more advance notice for rein cues and more chance to balance the bit in his mouth without discomfort. The horse is often more responsive to a bit that has joints between the mouthpiece and the shanks; the rider can communicate with just a jiggle of the bit. A hinged shank also gives more directional function in that the horse can respond to a pull that is slightly to the side, whereas a solid shank is useful only for a pull straight back.

Many bits have a cheek length (total shank length) equal to the width of the mouthpiece. Thus, if the mouthpiece is 5 inches, the cheek (shank) will be 5 inches long. A common short-shanked bit is the Tom Thumb (used as just a curb or sometimes as a Pelham), which has a cheek length of only 3½ to 4½ inches with a 5-inch mouthpiece.

Long-Shanked Curbs

There are a variety of long-shanked curbs, both English and Western. Longer shanks make the bit more severe. The Western "cutting-horse bit" is relatively mild, however, as it has a low port and the shanks curve back dramatically, producing less leverage than does a straight-shanked bit. A Western "grazing bit" with long shanks is mild because the shanks are curved back so far that the horse can get his nose to the ground without bumping the bit.

Curb Chain or Strap

Hooked to the top of the curb bit's cheek piece — either to the loop for the headstall or with its own loop, slot, or hook — the strap or chain hangs behind the bit and connects with the back of the jaw, in the chin groove, only when the bit is tipped by a pull on the reins. The looser the curb chain, the less severe it is; the reins have to be pulled quite hard and the bit tipped quite a distance before the chain puts pressure on the chin groove. In contrast, a tight curb strap or chain will press into the chin more readily, even with a lighter touch of the reins.

Most riders adjust the strap or chain at moderate looseness. If the bit is resting in the horse's mouth, you should be able to fit two fingers between the jaw and the chain. If you pull back the shank of the bit, as a pull on the reins would do, the chain or strap should make contact with the horse's jaw by the time the bit shank is at a 45-degree angle.

For a sensitive horse that doesn't take much touch on the reins to stop, a loose chain or strap is best. When starting the young horse in a curb after early schooling in a snaffle, keep the

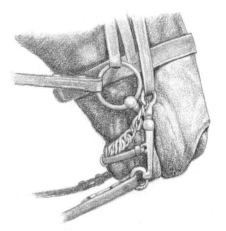

The curb chain is adjusted for moderate looseness and will not contact the chin groove until the curb reins are pulled and the curb bit is tipped about 45 degrees.

curb strap or chain fairly loose until he gets used to it. For a spoiled horse that doesn't respond well to a light touch and needs a firmer hand, you'll want it tighter: This will give him more signal to stop when you put pressure on the bit.

Spade Bit

The spade bit should *never* be used by a novice. It was traditionally used by master reinsmen on well-trained horses that were first started in a hackamore, then put into the bridle to become "finished" horses after extensive schooling. If improperly used, a spade bit can be severe.

The spade bit acts on the roof of the mouth as well as on the bars and tongue. It gets its name from the shovel-like projection rising from the port, the purpose of which is for more total area of contact in the mouth. It lies flat against the tongue when the bit is properly adjusted. The spade usually has copper components, which stimulate more salivation. The high port can injure the roof of the mouth if the bit is used harshly, but a good, experienced horseman uses a light touch on the reins, and the horse is trained to a high degree of responsiveness.

The spade bit is quite heavy, which encourages the horse to carry his head properly, in a flexed position. The weight of the bit hangs from the bridle; it does not rest on the bars of the mouth.

Half-Breed Bit

The *half-breed bit* combines features of a spade bit with those of a standard curb bit. It has a high port that usually contains a *cricket,* or *roller,* and is used on horses that are well along in their training.

CALMING CRICKETS

Crickets are rollers attached to the mouthpiece of the bit, usually a curb bit with port. The horse can roll these with his tongue, making a clicking noise. Nervous horses seem to relax more if they have something like this in their mouths to play with. They also encourage more saliva production, which makes the bit more comfortable.

How to Use a Curb

If you're riding Western with a curb bit or with curb reins on a Pelham, you'll have both reins in one hand to neck-rein. To turn the horse, press the left rein against his neck when you want him to turn right and the right rein against his neck to turn left. Basically, you are

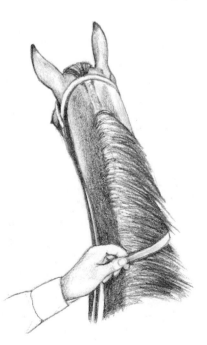

When neck-reining, gently press the indirect rein against the horse's neck. To turn to the left, touch the right rein to his neck; to turn to the right, touch the left rein to his neck.

pushing his neck in the direction you want him to go, using the rein fairly high on the neck, if he is still learning, so he can feel it. On a well-trained horse, a slight touch of the rein a few inches ahead of his withers is adequate.

The curb bit, used by a good rider who keeps a constant light touch on the bit via the reins, encourages a horse to lower his head and flex at the poll, with his face and nose almost vertical rather than stuck out in front. The horse is usually put in a curb after he's well started in a snaffle or after a lot of work in a hackamore (see page 203). Most horsemen don't feel a horse is "finished" in training until he is schooled in the curb, as responsiveness and collection can be refined more fully. Proper use of a curb bit promotes a more relaxed jaw and better head carriage for the collection and balance needed for precision work, whether executing advanced dressage movements or cutting a cow.

The Pelham

A *Pelham,* or combination bit, has attributes of both the snaffle and the curb bit. It has a curb mouthpiece and shanks, with rings at the mouthpiece for snaffle reins and rings at the bottom of the shanks for curb reins. Use of snaffle reins gives the effect of a mild bar snaffle, and the horse's head can be influenced by a direct rein. (There are also a few Pelhams with jointed mouthpieces.) Use of the curb reins gives the leverage action of the curb.

There are many shank lengths available in a Pelham, from short Tom Thumbs and Kimberwickes to some Pelhams with 9-inch shanks. Mouthpieces come in a variety of shapes and materials, including steel,

Short-shanked Pelham bit with straight mouthpiece

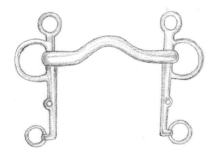

Long-shanked Pelham bit with port

metal alloys, rubber, and copper. There are jointed Pelhams, but these can poke forward in the center and hit the top of the mouth when the reins are pulled. The double-jointed mouthpieces on some Pelhams and Kimberwicke bits lie more smoothly on the tongue and don't form a sharp point in the middle when bent. Mouthpieces can be made of soft or hard rubber. The latter is more durable. The bulky rubber makes a comfortable, mild mouthpiece.

Combination of Effects

The Pelham is a handy bit for all kinds of riding. You can ride a horse English-style or Western in a Pelham, using either snaffle reins or curb reins. It is an excellent schooling bit when using all four reins, as the rider can engage different pressure points for different purposes to help raise the horse's head or collect him more freely.

Using the Pelham as a Training Bit

After a horse is working well in a snaffle, a transition from snaffle to curb is easily accomplished with a Pelham. Use just the snaffle reins until the horse gets used to the bit, then add the curb reins. Use all four reins for a while, then gradually increase using the curb reins and work less with the snaffle reins, until the horse is accustomed to the feel and signals of the curb bit.

This works well for putting the horse into a curb and for further work in teaching him collection and lightness. The snaffle reins are used to help raise his head and the action of the curb helps tuck his nose and set his head, thus enabling the horse to balance himself better, putting less weight on his front legs.

Some trainers use a *converter strap* on the Pelham when making the transition from snaffle to curb. This is a small strap that attaches to both the snaffle and the curb rings on the Pelham. One set of reins is then attached to the converter strap. The reins put some pressure on the curb, but the horse can also be direct-reined (that is, you pull his head to the side with one rein, as with a snaffle) if necessary. After the horse is ridden with the converter strap awhile, remove it and ride him with just the curb reins of the Pelham. The disadvantage of a converter strap is that you don't get the precise effects of either a snaffle or a curb; most trainers prefer the four reins.

> ## MOUTH-WATERING BITS
>
> The metal used can make a difference in whether a horse likes a certain bit. Most horses don't like aluminum or the taste of chrome-plated bits; some don't like a rubber mouthpiece and will try to spit out the bit. Stainless steel, copper, and iron are usually well tolerated. Many horsemen prefer copper or an iron-copper combination because this tends to stimulate salivation — hence, more comfort for the horse. A wet mouth, in which the bit moves around more easily, is preferred.

The Double Bridle

The full double bridle uses two bits — a curb, such as the Weymouth, and a thin bridoon-sized snaffle — each with its own set of reins, which may be used independently to achieve various responses from the horse. The double bridle is mainly used in higher levels of dressage after years of training for the horse and rider, with gaited horses in the show ring, or on polo horses.

The double bridle used for dressage has thicker mouthpieces and relatively short shanks, whereas the gaited show ring bridle has longer shanks and a thinner bridoon. A horse with a shallow mouth may not be able to accept the bulk of a dressage double bridle. The snaffle hangs from a separate strap of leather threaded through the brow band under the crownpiece of the headstall. The curb is attached to cheek pieces of the bridle and hangs below the snaffle.

Ensure Proper Fit

Whatever bit you choose, it must fit the horse. A bit too wide slides back and forth and is annoying and less precise; a bit too narrow can pinch the corners of the mouth. If it doesn't fit, the bit causes discomfort — or acute pain — and adverse reactions to your cues. Your horse may open his mouth, try to spit out the bit, or toss his head. He may try to avoid bit pressure, throw his head in the air, carry his head too low, or get behind the bit, meaning he'll tuck his chin and refuse to engage the bit, leaving you with no control and no communication through the bit.

Pain from an ill-fitting bit may cause your horse to travel with his neck stiff or with tense muscles, losing agility and fluidity as he tries to avoid the discomfort. Some horses pull against the bit or root the nose forward in an attempt to get away from pain. If your horse reacts adversely to a certain bit, check it for proper fit.

How to Tell whether the Bit Fits

Years ago there was an instrument horsemen used to measure a horse's mouth. The exact width of the mouth was measured and a bit was made to fit that horse. Today bits come in specific sizes and many are bought with no thought as to size or width of the mouth they are to fit. Often a bit is selected for other reasons, such as cost, looks, and favorite style or because a friend has one that works well on his horse.

If you are ordering or selecting a bit for a young horse or a new horse, measure the width of his mouth by pulling a piece of string through it where the bit would go. Mark or tie a knot in the string at each side, at the corners of his mouth. Select a bit with a mouthpiece about ⅛ inch longer than your string. The bit must be slightly wider than the mouth, so the hinges of the snaffle rings or shanks of the curb just clear the mouth corners. A bit that is much wider than this, whether snaffle, curb, or Pelham, may shift around or rest unevenly in the mouth.

Size and Balance

Width and fit are just as important in the curb and Pelham as they are in the snaffle, and in addition, the curb or Pelham must be well balanced. For the horse's comfort — and for best communication through his mouth — the bit must hang properly so that it puts no extra pressure on any one spot. An unbalanced bit does not hang properly, lower ends of the shanks tend to come forward when the horse mouths the bit, and points of contact in the mouth are altered, thus interfering with signals through the bit.

THINK BEFORE YOU BUY

When selecting a bit, remember that bits and mouth size vary and that different bits have different purposes. Always consider the horse's training and the rider's ability when choosing a bit.

A balanced bit always returns to the proper position when it is resting in the mouth with no rein pressure. A curb bit must be heavier on the bottom half or it can't do this. You can't get proper action from a bit that isn't balanced.

Many horses like a heavy curb or Pelham bit, and a bit made of steel or iron usually works better than one constructed of aluminum. Not only do they dislike the taste of an aluminum bit, but also it is so light that it floats around in the mouth instead of resting comfortably on the bars and staying in proper position. A horse likes the feel of a heavier bit, and if it hangs properly balanced, it encourages him to travel with his head in a balanced position.

The heavier the bit and reins, the lighter your touch can be on the reins. Your horse can feel the slightest motion of your fingers even if there's slack in the reins, because the weight of the bit and reins maintain a light feel on his mouth at all times. A heavy bit gives a more instant signal and also a quicker release of pressure, thus allowing feather-light communication through the reins. In contrast, a light bit

BIT BALANCE

To check the balance of a curb or Pelham, hold it with one finger, resting the center of the mouthpiece on your finger. If it is properly balanced, it will rest on your finger in the position it should rest in the horse's mouth, with shanks hanging down and slightly to the rear.

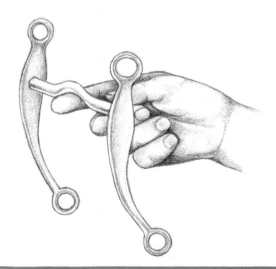

and light reins cannot convey that touch without stronger fingering of the reins. In order to keep proper feel of the mouth with light reins and bit, you must have a constantly taut rein.

Avoiding Bit-Caused Pain

Pain from a bit generally results from harsh use by the rider or improper fit. Rough use of a bit may permanently damage nerve endings in the bars of the mouth. The horse's tongue is sensitive, but it can move around freely and get away from some of the pressure caused by a bit that fits improperly or is used too harshly. The bars, chin groove, and mouth corners cannot escape the discomfort, however, and the horse may toss his head, pull against the bit, or show some other sign of displeasure.

If a horse reacts improperly to a certain bit, the rider's first thought may be to try a different bit. This is the proper thing to do if the bit fits poorly, because it must fit him to be comfortable. All too often, however, a rider goes to a more severe bit rather than trying to find one that fits better, and this aggravates the horse's problem, especially if it was caused by the rider's rough use of the bit in the first place.

PROPER ADJUSTMENT OF THE HEADSTALL

Proper bit fit also depends on the tightness or looseness of the headstall. If the cheek pieces are too tight, the bit may press up against the horse's cheek teeth, especially if he has wolf teeth, and will also rub the corners of his mouth. If cheek pieces are too loose, the bit hangs too low in the mouth and may clank against the incisors or canine teeth, causing annoyance or pain. Also, the horse may put his tongue over the bit.

Teeth Problems That Interfere with Bitting

If a horse has bit problems, check his teeth. Many horses have *wolf teeth,* which are residual teeth located next to the molars. These may cause discomfort when a snaffle hits them. A jointed snaffle is pulled back into the corners of the mouth when the reins are pulled and may bump the wolf teeth or pinch the horse's flesh between the bit and his teeth.

Canine teeth (all male horses have them, and some females have small ones) are located farther forward, a short way behind the last incisor. These generally don't cause trouble unless the headstall is too loose and the bit clanks against them, though sometimes they can make it awkward to put on and take off the bridle. Canine teeth are usually not as much problem as wolf teeth. The latter are often removed before a young horse goes into training so they won't create trouble with a bit. A bit rarely causes pain from contact with molars (as when it is adjusted too high in the mouth or a snaffle is being pulled against the corners of the mouth and hitting the molars) unless it hits a sensitive wolf tooth.

The usual cause of pain from the scissors effect of a snaffle being pulled into the corners of the mouth is due to pinching the flesh of the cheek between the bit and a tooth. When a horse resists the bit, check his mouth before you decide to change to something more severe. If the horse's cheeks have been injured because the bit has been crushing them against the molars, the horse may need dental work to adapt the molars for use of a bit. The flesh of the cheek is a cushion between bit and tooth, and if the tooth interferes with the bit, the cheeks will become sore and perhaps infected. The molars the bit is working against can make even the mildest snaffle an instrument of torture.

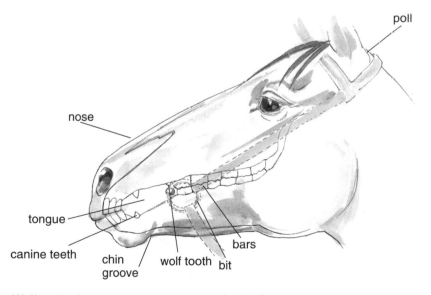

Wolf teeth often interfere with the use of a bit. Canine teeth can be bumped when unbridling and may also be a hazard for your fingers when you put a finger into the interdental space. Be careful.

Choosing a Training Bit for a Young Horse

Most trainers start with a snaffle (straight bar or jointed) and some use a Western snaffle (see page 186). Whatever bit you use should be comfortable for the youngster and fit him properly. The important thing is to make his first experiences pleasant, not painful. If a bit irritates or causes pain, he will focus more on the problem in his mouth than on what you are trying to teach him.

The snaffle is ideal for starting a young horse because of its action. The purpose of the snaffle is to raise the head and enable the rider to use direct reining — that is, pulling the horse's head around in the direction you want him to go. You cannot properly direct-rein a horse with the curb. Until the horse is further along and has learned responses to leg pressure and neck reining, he cannot respond well to the curb. He needs a bit in which "turn signals" can be given with a direct pull to the side.

The snaffle is designed for lateral action — pulling one rein at a time to turn the head to the side. Pulling forcefully with both reins at once for control, when trying to stop a horse, for example, or to make him go slower, leads to resistance. It creates a painful nutcracker action in the mouth, especially with a single-jointed snaffle, and he'll try to brace against it. Your horse soon becomes unresponsive. To control him, use a seesaw action (one rein at a time) or just one rein to pull his head around. Use of two reins in a snaffle is for communication — gentle give-and-take when teaching him to accept the bit, stretch out and walk, stay at the proper speed in a certain gait — never for control.

The green horse must learn to raise his head and balance himself while carrying a rider. Every untrained horse travels uncollected, meaning heavy on his front end. Until he learns to travel lighter in front and to transfer more weight off his forehand, keep him in the snaffle, as its action tends to raise his head.

Importance of a Thick Mouthpiece

An effective snaffle has a fairly thick mouthpiece that won't cut or pinch the corners of the mouth. Most horses do best with a thick mouthpiece, even in a curb, because it rests on the sensitive gums between the incisors and molars. The bars where the bit rests are sharp-edged bones with a thin covering of delicate tissue and nerve endings. The thinner the bit, the more it can press or even cut into that tissue and the more pain it can cause. The ideal thickness for a mouthpiece, however, will depend on the build of a horse's mouth, the size of his tongue, and the type of bit used.

Choose a Mild Bit

The young horse does not know how to respond to a bit. There will be times when you have to pull a little harder than you would on a trained horse, when the youngster shies, for example, or when his attention is suddenly distracted by things an experienced horse is used to. Until a youngster learns what is expected of him, he may be unpredictable and occasionally need a little more control.

You don't want to injure his mouth with a severe bit. You can maintain good control with a mild bit if you know how to use it, such as pulling his head around with one rein in the snaffle if he tries to bolt or buck. Most hard-mouthed horses — those that are unresponsive to the bit and pull against you —are made that way by riders with severe bits who did not know how to use them and abused the mouth to the point where the horse has little feeling left.

No matter what bit you want to use on the horse later, start him in a mild snaffle or a colt-training snaffle using only the snaffle rings. You want him to learn to move out boldly. With a mild bit and thick mouthpiece, he'll be more likely to take hold of the bit and be comfortable with it, without fear of it pinching or hurting him. The purpose of a bit is communication, but you can't establish this if he associates the bit with pain or if he is confused by it. There are several good training snaffles, including the egg-butt snaffle, in which the ends of the mouthpiece join to the rings in a smooth meshing that cannot pinch the corners of the horse's mouth.

An egg-butt snaffle is a mild bit.

Adjusting the Fit

In a snaffle, adjustment for the cheek pieces of the headstall is about right when there's just one small wrinkle at the corners of the mouth. If it is looser than that, the bit can shift around or even bang the teeth. The horse may put his tongue over it. When first starting him in a snaffle, have the headstall a little tighter — two wrinkles at the mouth corners — to make sure he can't get his tongue over the bit while trying to spit it out. If he gets his tongue over the bit, you have less control and little communication; this position changes the point of contact in the mouth.

After he gets used to carrying the bit and no longer attempts to move it around with his tongue and spit it out, readjust the headstall to give him a little more room. Then he can safely move the bit with his tongue and mouth and put it where he feels most comfortable.

COMFORT IS KEY

Some horses can still get their tongues over a bit when it is adjusted high — with the headstall quite tight — and then have a hard time getting the tongue back to its proper place; it becomes stuck over the bit. When the bit is set at a comfortable level, the horse may play with it, but he soon learns how to hold it in a comfortable position with his tongue under it.

The Importance of Good Hands

Your horse must learn to accept bit pressure without fear or resentment. You want him to relax, flex, and be pliable to your hands, giving to the bit rather than fighting it. If he develops a good mouth, he won't have a rigid neck or head, nor will he protest the cues he gets through the bit. A horse with a relaxed jaw accepts your signals; he's not afraid he'll have to endure pain. Proper bitting and a good mouth are necessary before he can advance in his training and become collected and responsive to all the rider's signals.

Horsemanship that accompanies a bit is more important than the bit itself. Too many riders seem to think bits are for making horses do certain things. The novice horseman usually puts too much emphasis on the bit. A good trainer and rider works more with the legs, balance, proper seat, and good hands. The severity of any bit is relative to the skills, or lack thereof, of the rider.

A horse will usually resort to devious behavior like head tossing, tongue rolling, and mouth opening as a defense against rough hands and poor horsemanship. Once he develops these habits, it is hard to get rid of them. It may take much time and patience to retrain the horse whose resistance to a bit has become automatic. It's better to exercise care and patience when training the horse the first time so only good habits become established.

The Hackamore

Some horsemen use a hackamore (not to be confused with a mechanical hackamore; see the box on page 205) when starting a young horse. It provides a way to teach him to respond to the rider's hands without risk of dulling a sensitive mouth. The hackamore is a simple headstall with a noseband, called the *bosal*, made of braided rawhide. Some hackamores have a *fiador*, a rope piece that serves as throatlatch and goes down to tie to the heel knot of the noseband. It provides loops for reins.

The hackamore method was used in California and Mexico by early Spanish-influenced riders, who started all young horses in the *bosal* (a rawhide noseband) and later finished the training in a spade bit (see page 191).

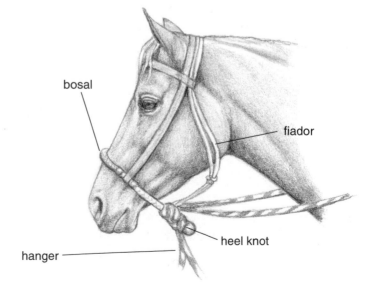

bosal

fiador

Hackamore with fiador

hanger

heel knot

The most important part of the hackamore is the rawhide bosal. Similar in shape to a small unstrung tennis racket, the "handle" is a short rawhide knob (*heel knot*) that rests under the jaw. It balances the hackamore, and keeps the reins from sliding off the bosal. A good hackamore has a rawhide core, not a cable core. The rawhide shapes itself to the horse's nose and jaw and stays properly shaped; it won't put pressure in the wrong place.

The reins (*mecate*) consist of a one-piece, 18- to 20-foot hair rope. One end is left free for use as a lead rope and is attached to the saddle when the horse is ridden. The bosal rides high on the nose where the bones end in cartilage. The heel knot, and the rope wrapped around it to form the reins, adds the necessary weight to make the noseband fall away from the horse's sensitive jaw whenever rein pressure is released. Normal position of the noseband is tilted, with the front part on the nose well up and the heel knot down.

WHY *HACKAMORE*?

The word *hackamore* comes from the Arab word *hackma,* which is the name for the headgear used on camels in some countries. The Moors invaded Spain more than 1,000 years ago, bringing their horsemanship and tack with them. In Spanish, the word *hackma* became *jáquima* (the Spanish *j* is pronounced as an *h*). When the Spanish term came to North America, it became *hackamore.*

Hackamore Training

When you pull a rein on the hackamore, the heel knot comes up under the jaw and makes contact, and the noseband comes down to put pressure on the nose. The bosal rests on the cartilage just below the nasal bone, but can be periodically adjusted higher or lower to keep from making one spot sore. It's also a good idea to make adjustments in order to keep the contact area sensitive to pressure so it won't become tough and callused.

If the bosal is too high — that is, resting on bone — the rider has less control. The lower part of the nose is more sensitive. If the bosal is hung too low, thus putting pressure on delicate tissue low on the nose, a hard or prolonged pull can damage the nose. The bosal on a starting

BITLESS BRIDLES

Mechanical hackamore. Not really a hackamore, this bitless bridle uses leverage points on the nose, jaw, and top of the head. When pulled by the reins, shanks put pressure on the noseband, bring the curb chain against the jaw, and tip the cheek pieces forward to put pressure on the poll. Some riders use this for spoiled horses that don't respond to bit pressure or for horses with injured mouths. The principle of the mechanical hackamore is similar to that of most bits: Longer shanks provide more leverage action on the chin groove and poll; thinner nosebands are more severe than broad, thick ones. Using a mechanical hackamore improperly can permanently damage the cartilage of the nose.

Mechanical hackamore with fleece-lined noseband

Side-pull. This very simple bitless bridle puts pressure on the nose when both reins are pulled and turns the horse by direct pull to the side with one rein. There are several varieties, and some have a curb strap that tightens when the reins are pulled.

Side-pull bridle with curb strap. This side-pull puts pressure on the nose, poll, and chin groove when the reins are pulled.

hackamore should be padded on the sides so it won't skin the jaw on a hard pull. The trainer must usually pull hard at first to take away the horse's head when *doubling* him — that is, pulling the horse around to where he is facing the opposite direction.

The trainer teaches the horse to give to a pull and release. A horse, being stronger than a human, can't be controlled with a halter, leather noseband, or bosal with sheer pull force, so the trainer uses a technique that gives control over the horse's mind, and hence his body. After learning to respond to the hackamore, the trainer can switch to a bit and won't have to pull hard on the reins because the horse already knows how to respond.

Handling the Reins

Always handle hackamore reins with your hands held low. The green horse should be doubled from both sides during ground work, before the rider gets on, so the youngster knows how to give to a pull on one rein. Then, when he is ridden, he learns to follow his nose around when one rein is pulled.

Doubling is the basis of hackamore training. If the horse tries to buck, bolt, rear, or balk, the trainer doubles him. This thwarts his improper action, teaches him always to obey the rein, and to position his feet for agile turns.

Reins are rarely used at the same time. If a young horse starts to bolt, the worst thing to do is pull on the reins — let him run a few seconds on a loose rein. Then take a good, short hold on one rein and pull low and hard, turning him clear around in a series of quick pulls. The pulls should be straight back and low, with give-and-take action — pull and release. After a horse has had his head suddenly pulled around a few times and has been made to double, he soon learns to stop when asked to.

Doubling the horse gives the rider control. After a horse learns to work in a hackamore, he becomes very responsive. One quick pull and he stops. One pull to the side and he turns. In the beginning, however, it will take several give-and-take pulls to bring him around from a gallop.

If the trainer uses reins correctly, the horse becomes lighter on his feet and learns to handle himself for a stop or turn. After the horse works calmly and doubles nicely, he can do circles and figure eights at a canter. He learns better balance. He starts to collect and to change leads at the canter.

DOUBLING

Doubling must always be done from a loose rein. A tight rein enables the horse to disobey by exerting his own strength to resist, pulling against the rider's pull. It is instinctive for the horse to pull against pressure. If the rider uses constant pressure, the horse will lean into the snug noseband and find he has more strength than the rider, and the rider won't be able to pull him around. If the rein is always loose, however, the horse never knows when he is going to be pulled around and has no opportunity to brace his neck muscles against the pull. If the trainer handles the reins correctly, the young horse will never learn his own strength.

Control of the horse with a hackamore depends on being able to "double" him: pulling his head around quickly with one rein and turning him around so that he is facing the opposite direction.

Some trainers keep a horse in a hackamore a year before switching to a bit. It usually takes at least 10 months to prepare a horse for the bridle. If he is not put into a bit when he is ready, however, he eventually may become insensitive to the hackamore and will regress in his training. The hackamore is a great way to start a horse, but it is not the final finishing tool.

8

BITTING AND DRIVING

The information in this chapter can be used to train any horse and will be helpful before beginning a horse's mounted training. It will make it easier for the young horse to respond properly to the bit, making his first mounted lessons go smoothly and safely.

There is no best way to train a horse. Some trainers use a bitting harness to get a horse accustomed to the feel of bit pressure. Some drive the young horse in long lines, walking behind him with long reins attached to the bit, before riding. Every trainer has his own favorite methods, and every horse is unique. It's good to know a number of different methods that work, because what works well for one horse may not work for another. Feel your way with each individual, keeping in mind his personality and his feelings.

TREAT EACH HORSE AS AN INDIVIDUAL

Horses become ready for new challenges at different rates and react to new experiences in different ways. A rigid training schedule or method can eventually run into problems. The youngster you are working with may not be ready for the next step you planned or may suddenly seem ready for something you had not yet planned on trying. You must be able to feel his mood and readiness, to sense when he is or is not ready for what you want to do. You will gain insight to his level of readiness as you work with the horse and get to know him. Be intuitive, and become "partners" with the horse you are training.

Bitting Ground Work for the Riding and Driving Horse

Some horses do best with lots of ground work — leading, bitting, saddling, driving, longeing — taking everything one step at a time. Others are like precocious students in school who can skip a grade and do well; with these horses, you can skip some of the ground work and still be ready for riding. A good trainer can reach the final goal by any number of routes, always using whatever methods are best for each youngster.

Even if you don't plan to drive a horse in long lines, get him used to bit pressure before you start riding him. You can lead him with the bridle on (snaffle bit), using one hand on the reins behind his chin and one hand on the lead rope, signaling him to stop and to turn with the halter and the bit.

When turning, give a little pull on the appropriate rein (pulling him toward you to turn left, pulling him away from you to turn right). Put a little straight-back pressure on the bit as you tell him *Whoa* (he should understand what this means from earlier lessons). Some horses quickly learn to respond to the bit, but others will do better when they have some lessons with a bitting harness.

Using a Bitting Harness

Bitting lessons teach the horse to give to a bit instead of resisting it, playing with it, or pulling on it all the time. Always use a mild bit such as a snaffle. A bitting harness is a good way to acquaint the horse with the actions of the bit; it won't hurt his mouth, yet it teaches him to give to the bit.

Creating a Bitting Harness with a Surcingle

A bitting harness consists of a headstall and snaffle bit, surcingle or bellyband, an overcheck to keep the horse from putting down his head too far, and sidechecks made from an elastic material such as strips of rubber or old inner tubes. The overcheck lines are fastened to a hook at the top of the surcingle, at the withers.

The stretchy sidechecks allow the horse to pull on the bit without hurting his mouth; he doesn't run into a "solid" bit that bumps his mouth when he moves his head. Sidechecks should not be too loose or too

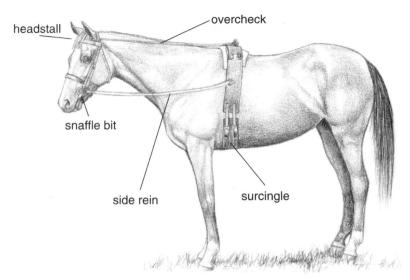

Bitting harness with surcingle. The horse is pulling a little on the bit, stretching the side reins.

tight. There should be no slack when his head is in proper position — that is, above level (wither height is level), nose slightly out, and relaxed. Elastic reins allow him to stretch them while pulling on the bit, yet bring his head right back into place. Soon he learns to relax his jaw and neck, giving to the bit rather than rooting his nose forward.

After a few lessons, shorten the side reins somewhat to encourage the horse to bring down his nose a little more and keep his face more vertical. At first, however, you want the bitting harness to be as comfortable as possible so he will accept the bit and not fight against it.

Creating a Bitting Harness Using a Saddle

If you don't have a bitting harness or surcingle, you can improvise using a saddle. A Western saddle works well because you can attach the overcheck lines to the saddle horn. Attach the side checks to the cinch rings or around the latigo or D-rings if the cinch rings are low.

You can make an overcheck from light cotton rope or a curtain cord. The cords attach to the bit on each side at the top of the bit ring, running to the top of the headstall through the loop where a throatlatch would go. It then goes back to the saddle horn. The overcheck puts no pressure on the bit when the horse's head is in normal position but keeps him from putting down his head too far.

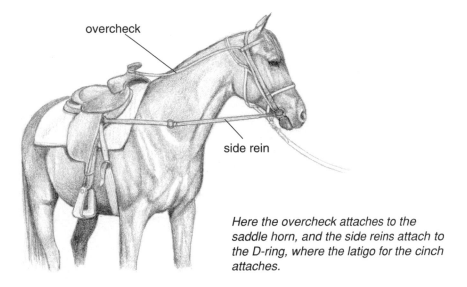

overcheck

side rein

Here the overcheck attaches to the saddle horn, and the side reins attach to the D-ring, where the latigo for the cinch attaches.

How the Bitting Harness Works

The headstall and bit should fit properly, the bit in a comfortable position but sufficiently snug so the horse can't get his tongue over it. If he has a halter under the headstall, for leading or longeing, it should be a close-fitting halter that doesn't interfere with the fit of the headstall.

The overcheck and side checks keep the horse's head in place, putting pressure on the bit when he tries to root his nose forward or put down his head. There is no pressure on his mouth when he carries his head in the proper position, but there is instant pressure when he roots or lugs or throws his head. The pressure is immediately released when he brings his head back where it should be.

The horse's own actions teach him to respond to the bit. This is the first step in teaching him to flex at the poll and give to the bit. He learns to relax his jaw instead of bracing against the bit and discovers that the easiest position to carry his head is in the proper one, with head up — a little above level — and his face nearly vertical, with a slight arch in his neck.

Lessons in the Bitting Harness

It usually takes several sessions with the bitting harness for the horse to learn to give to the bit and relax while wearing it. Wearing the bitting harness for 15 to 30 minutes at a time once a day is enough. Let him wear it in a corral or round pen for a while so he can walk around and try it out fully.

LEADING WITH A BITTING HARNESS

He strains against the bit and harness and stretches the side rein.

By placing his head in the proper position, all pressure is released. The elastic side rein has pulled his head back into place.

When he's used to the bitting harness, usually after several sessions wearing it in a pen, lead or longe the horse with it on. Snap a lead rope to the halter and lead him around the pen or take him for walks out in the open. This lets him discover that he can travel with the bitting harness on, getting used to the bit as he goes. Refresh his memory by revisiting earlier leading lessons — traveling freely at walk and trot at your command, stopping, and turning. He can begin to make the transition from being controlled merely by the halter and your voice to doing all these things while wearing a bit.

Longeing in the Bitting Harness. If the young horse has already been taught to longe, it's a simple step to longe him in the bitting harness. Attach the longe line to the halter or cavesson as usual. Then he can be walked, trotted, and cantered on the longe line. This prepares him for moving at various gaits with a bit in his mouth and lets him feel some pressure on the reins if he gets his head too far out of position or pulls too much on the bit.

If your horse has not been taught to longe, do not try to teach him this at the same time you are using the bitting harness. Work on one thing at a time so he can focus. If you want to longe him in a bitting harness, teach him to longe first (see pages 149–158).

Voice Commands

If you plan to drive your horse in long lines in preparation for riding or to pull a cart, he must develop instant response to voice commands. Give a certain command the same way and in the same tone so he knows exactly what you mean. Commands should be crisp and distinct. Your horse should learn instant obedience; you must stay in control. If he does not obey — he starts to move before you tell him to, for example, he refuses to halt exactly when you ask, or he misses a gait transition — get after him and insist that he does. If you wait a few seconds before correcting him, he will not respect you and will be sloppy in his responses in the future. (See pages 52–53 for more on correcting.)

Driving in Long Lines

The next step after the bitting harness is to drive him with long lines, teaching him to move out, stop, and turn. It is not absolutely necessary to drive a young horse before you ride him, but the more he knows before you get on him, the less confused he will be and the

---

ACQUAINT HIM WITH THE LINES

Before driving a young horse for the first time, get him used to the feel of the lines. Take a lead rope and loop it over his body, touching his hindquarters with it and brushing it against him. He will soon realize it's nothing to fear. Then, when the driving lines rub against his sides and hindquarters or accidentally bump his tail and hocks, he won't be jumpy.

---

quicker he will learn what you want him to do. It's usually easier for a horse to learn to drive if he has already been taught to longe.

Do not try to drive your horse out in the open at first; if he spooks or tries to run off, you will have a serious problem and defeat the purpose of the lesson. A large corral or round pen works well for first lessons. A round pen is ideal, as it has no corners in which the horse can become "stuck."

Snap *driving lines* (long reins used to drive a horse hitched to a cart) or, if you don't have driving lines, two longe lines, onto the snaffle bit and run them through the guides in the surcingle, or through the stirrup bows if using a saddle. This will keep the lines at his sides and in the proper place, with less danger of them drooping to the ground. You want the lines low, so they won't come up over his back, but not too low. If your stirrups are low, attach rings on your saddle pommel to run the lines through. You don't want your horse to get a foot over a line or become entangled. Keep the lines fairly short.

When using a saddle, tie the stirrups so they won't flop and spook the horse and won't be pulled too far back along his sides. Tie them down to the cinch rings or together under his belly. Tie a rope or twine to the inside of one stirrup, run it under his belly and through the little ring on the bottom of the cinch, if it has one, then tie it to the inside of the opposite stirrup. The rope or twine should be tight enough to keep the stirrups from flopping but not so tight that it irritates the horse's belly.

At first you may need someone to help you get started. Your assistant should lead the horse as you drive him. If the horse is not used to having someone following him, he may try to turn and face you instead of moving out. If someone leads him for the first part of the first lesson, he will quickly learn that he must move forward instead of trying to turn around. Once he understands what you want, he will move out freely without the helper. Walk about a horse length behind the horse, with a line in each hand and the whip in your right hand.

Start your horse with a voice command or touch of the whip and allow him to move in any direction he wants for the first few minutes so

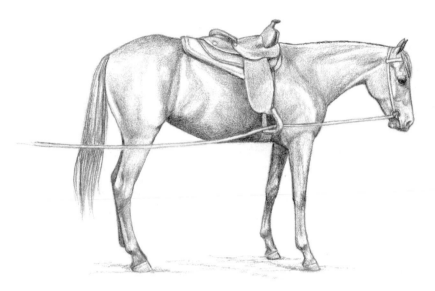

Driving in long lines, with lines through the stirrup bows to keep them at the proper height.

If you don't have a surcingle for driving and your stirrup bows are too low to run the lines through, attach rings to the saddle pommel to run the lines through so you won't be pulling the horse's head down too low.

he can get accustomed to you following him. Some trainers use a gentle slap of the lines against his sides, but it's best not to do this; it puts a jerk on the mouth, which to the horse should mean *Whoa*. Remember, you never want to confuse him. Instead of slapping the lines, touch him with the whip on his side, where your leg would be if you were riding. If your whip is not long enough to do that, tap him lightly on the rump.

To stop him, tell him *Whoa* and use a slight seesaw pull on the lines (squeeze alternately with each hand, applying a little pressure on one rein at a time). Release all pressure the instant he responds. If he has learned *Whoa* from leading lessons, this will make it easier.

Whoa should always mean stop and stay stopped until otherwise directed. Give the lines slack when the horse stops and let him stand awhile each time you stop him, so he remembers this. The signal to stop means to stop completely — whether the horse is being led, longed, or driven. Walk him forward again and repeat the lesson until he learns to stop at your verbal command with very little pressure on the lines.

> ### SAFETY TIP
> Wear gloves when driving to protect your hands if your horse becomes frightened and tries to pull away.

When you want him to start moving, take up the slack in the lines and speak to him. Give him the command to walk or to move out, or cluck, whatever he is accustomed to. If he doesn't move, touch him lightly with the whip. Always keep your command words simple and consistent. One and two-syllable commands like *Walk*, *Get up*, and *Let's go* work best. If you plan to hitch your horse to a cart later, you may want to teach him the commands for turning left (*Ha*) and right (*Gee*).

END ON A POSITIVE NOTE

When driving in long lines, work only at the walk, so the horse will stay calm. You don't want him to become upset and get out of control. He thinks about things more clearly if he stays in a calm, quiet walk. Keep first lessons brief — 20 to 30 minutes is plenty — or even shorter if he has a briefer attention span. If you work with him too long and he becomes bored and frustrated, you will lose ground in the training. Always end the lesson on a positive note, while your horse is still trying to do things correctly.

Driving in Circles

Sometimes the easiest way to start ground-driving a horse that knows how to longe is to make a transition from longeing to driving as he goes in circles in the round pen. Get him used to longeing with a bridle and saddle on (a Western saddle works well for this step), with the stirrups tied beneath his belly so they won't flop. Then substitute driving lines for the longe line.

Making the Transition

Longe your horse first, using one line snapped to his halter, then halt him and snap the line onto the bit (to the near side if you are on the horse's left). Snap the other line to the bit ring on the far side (the off side, in this instance) and run it through that stirrup bow and up over his back, behind the cantle. The line will be held in place by the saddle cantle.

Now you can encourage him to move around you in a circle as if he were longeing, but you will walk a short way behind him. As you follow him, a little to the inside, as he is accustomed to you being at the inside of the circle, keep just enough pressure on the inside line to keep him in a circle.

When your horse has relaxed and is walking in a circle around you, with you following a little way behind him, flick the outside line gently so he knows it's there. When he accepts that, gently slide it off the saddle and onto his hindquarters. Keep sufficient tension on the line to keep it from drooping down to his hocks or lower.

If your horse is nervous about the line touching his hindquarters and speeds up, just go with him and talk to him, giving him the command to walk and controlling his speed with the inside line, just as if you were longeing. Reduce the size of the circle until he calms down to a walk again. Stop him and reposition the line behind the saddle cantle and start over.

After he travels calmly with the outside line in place around his hindquarters, halt the horse and run the inside line through the stirrup bow also. Now you are actually driving him rather than longeing. Start him again, and walk parallel with him but a little behind him, gradually letting out more line so that eventually you are walking directly behind him, dropping farther and farther behind, with even pressure on both lines.

When driving in circles, let the right line slip down over the rump. The left line is still snapped to the halter, so if the horse must be pulled on to stop, it won't hurt his mouth.

Turning

To turn, pull gently on one line and give a little with the other; this way, the horse can turn his head to follow the direction of the turn. If he gets confused and stops instead of turning, encourage him to keep moving with voice commands or a slight touch with the longeing whip, if necessary. Use the bit gently; rough handling will produce a hard mouth or cause the horse to bend too much.

Drive him in a circle in both directions. Get him used to starting with you standing to his right and to his left and behind him. Until he learns to turn in response to a gentle pull on one rein or the other — in the direction you want him to go — use the fence as an aid. When he gets to the fence and realizes that he can't continue to go straight ahead, he'll turn more easily.

At first, just let him go the way he decides to turn, encouraging the turn with a slight pull on that rein with equal give on the other. He'll soon get the idea that a pull on the rein means a turn, because the rein pulls his head around. As he learns to respond to the rein pressure, you can turn him any way you want.

After your horse drives well in a corral, is under control, and understands your signals, you can drive him at a trot as well as at a walk and out in the open as well as in a corral. This isn't necessary, however, unless you plan to take him further with driving training — to pull a cart, for example. The purpose of teaching a horse to drive, as a preliminary lesson for riding, is to get him used to giving to the bit and learning to stop and turn with rein pressure, and all this can be done at the walk.

When he gets to this stage in his training, the horse is ready to be ridden if he is physically mature enough. He just has to make the transition to rein signals when you are on his back. He knows a pull on a rein means stop and that a pull to the side means turn in that direction. He knows the basics of moving out, turning, and stopping on command. If you carefully laid the groundwork for his first riding lesson, that next transition should be accomplished fairly easily.

Overcoming Resistance to the Bit

Many horses will play with the bit and try to spit it out the first few times they are exposed to it, but if it fits well and is comfortable, most horses will soon get used to it. Once they start doing things that keep their interest, such as longeing, driving, and being led or ponied wearing a bridle, they stop fussing altogether. Some horses, however, have a hard time getting used to the bit and continue trying to spit it out or put their tongue over it. Even if the bit fits well, they are not content to leave it alone.

Check the Mouth

When a horse resists the bit, the first step is to have your veterinarian check his mouth to make sure there is no physical problem causing discomfort. An old tongue injury, sensitive wolf teeth, an abnormally shaped jaw — there are a variety of reasons a bit could bother your horse, and the veterinarian can detect what's the matter and give you suggestions. If the mouth and teeth are normal, however, try the following approaches to help the horse accept the bit.

Make His Mouth Happy

It's easier to deal with resistance in a young horse who is just learning about bits than it is in an older one for whom resistance is already a habit, but some of the "remedies" that follow can be helpful in either case.

- Use a bar snaffle with a very thick, comfortable mouthpiece (see page 200). It's not as easy for the horse to get his tongue over a straight mouthpiece.
- Put honey, jam, or molasses on the bit each time you bridle him.
- Rub the bars of his mouth with your fingers before you put on the bridle.
- Adjust the headstall so the bit is very snug (that is, so there is no room for your horse to put his tongue over it), and let him wear just the headstall and bit in his stall or in a round pen for an hour or two each day until he accepts the bit.
- Use a headstall with a brow band and a throatlatch so he can't rub it off, and make sure there's nothing in the stall or pen he could catch the bridle on.

Another way to make your horse more at ease with a bit is to use a *bitting snaffle*, or a mouthing bit. This is a straight bar snaffle with little "keys" that hang from the middle of it. These small metal pieces lie on the horse's tongue and he plays with them rather than trying to put his tongue over the bit. He will chew on the bit, which relaxes his jaw and produces more saliva in the mouth, thus making the bit more comfortable. Many of the old-time trainers used this type of bit to get a young horse ready for driving training. Most horses soon learn to accept it.

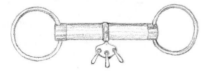

Example of a bitting snaffle. The horse uses his tongue to play with the "keys."

Training the Driving Horse and Pony

You may want to continue the horse's training in driving if he is to work in harness and pull a cart or buggy. You can even give driving and riding lessons at the same time: Teach driving in the morning and under-saddle lessons in the afternoon, for example. You can teach the young horse to pull a cart now and start riding him next year, when he is older and stronger. Or you may decide to teach an older horse to drive. This can be fun and rewarding, especially if you want to take along passengers or enter your horse in driving competitions.

If you don't know anything about driving, take time to learn about the sport before you buy equipment. Find out from experts what you will need and what your choices are. Don't buy equipment that does not fit your horse or is not in good condition. Remember, as always, safety must be your first consideration.

Can My Horse or Pony Pull a Cart?

You can teach almost any horse or pony to drive and pull a cart, but some are better candidates than others for this type of training. Evaluate your horse's mental and emotional characteristics to determine whether he is a good prospect. The nervous or flighty individual poses a bigger challenge and makes many situations more risky than does a calm, easygoing horse.

> **SAFETY TIP**
>
> Every decision you make about obtaining a harness and cart should be governed by safety. A driving horse depends on his driver for reassurance as well as control; if something happens — the harness breaks or the cart overturns — and you come out of the cart, the horse will probably bolt, with the vehicle banging along behind him. You don't ever want this to happen. Make safety your highest priority.

Temperament Is Important

When riding, you have a lot of control over a nervous horse's actions: You can use your hands, legs, seat, and balance as aids, as well as your voice. If he jumps sideways, your leg pressure can control the jump and get him back into position. When driving a horse, however, you have only your hands with the reins, your voice, and the whip, which can be a poor substitute for leg pressure in some situations. You can't just turn the horse around or back him to return him to the trail if he shies. The spooky horse that shies at every strange thing is a poor choice for a driving animal. When choosing a driving prospect, evaluate his mental and emotional attributes as well as his physical abilities.

Age

A horse can be taught to drive before he can be ridden. At two years old, a horse is too young for strenuous work under saddle, but he's capable of pulling a light vehicle without much stress on his back or legs. It is often easier to teach the young horse to drive than the already-ridden horse. The horse that is accustomed to a rider must make adjustments. He has to get used to restricted movement between the shafts and reduced vision wearing a closed bridle and blinkers. He must learn to cross his legs when turning, keeping his body straight, and to obey voice cues as well as rein cues.

Size

Also consider the horse's size and what you expect him to pull. A small pony can handle two medium-sized adults in a light cart on smooth roads with no steep grades. If you want to take your cart across country or to carry more passengers or to use a larger vehicle, you need a larger animal.

Harness

To drive your horse, you'll need a harness, which includes a driving bridle and bit. There are several types of snaffles appropriate for starting the driving horse, among them a rubber snaffle, a loose-ring snaffle, and a half-cheek snaffle. (A snaffle with cheeks prevents the bit from being pulled sideways through the mouth.) Choose a bit with a thick, mild mouthpiece.

It's often better to buy a harness custom-made to fit your horse rather than try to make do with a borrowed one. In addition to fitting properly, a new harness is less likely than a used one to fail in an emergency. Most driving accidents are caused by a poorly fitting or broken harness; it's not always wise to save money by buying used tack.

Selecting a Harness for Training

There are two types of harness — those with a simple breastplate and those using a collar. For pleasure driving — that is, pulling a light vehicle carrying only one or two people — a simple breastplate is fine, and easier to put on and take off the horse. Collars, which are worn by large draft horses, are more suitable for pulling heavy loads because they distribute pressure more evenly over a greater area. A collar can be difficult to fit correctly if you don't have experience doing so.

You can purchase a good leather pleasure-driving harness through mail-order suppliers (you must know what size you need, of course). Have an experienced person help you measure your horse and also advise you on harness quality. The cost ranges from $400 to $1,000. Avoid cheap mail-order harnesses; quality construction and proper fit are crucial for your horse's safety and comfort. By the time you replace pieces that don't fit, your "bargain" will turn out to be quite expensive. A better option, perhaps, is to buy through a harness shop. An important advantage is that if you are inexperienced, someone at the shop can measure your horse (or tell you how to do it) and suggest the best type of harness for your purposes.

Two harness colors are available — black (dyed) and russet (natural). Russet is generally more expensive because it requires a better grade of leather with no blemishes. Imperfections can be covered up if the harness is dyed black. Whatever the harness color, however, reins are always russet, so you won't get leather dye on your clothing.

Nylon harnesses generally cost less than those made from leather (prices range from $200 to $500) and require less maintenance.

> **SAFETY TIP**
>
> Don't buy a cheap used harness at an auction. If it's been hanging in a barn for years or stuffed in a box, the leather is probably dried, cracked, and weak. When leather is bent, it should not show any tiny cracks; if it does, its strength is compromised and it could break under sudden strain.

The material is strong and durable, but doesn't have as much give as leather and usually becomes worn-looking after a few years. Nylon harnesses are often less adjustable, too. Nylon is not always the best choice for a beginning driver or a young horse in training because it will not break as readily as leather if the harness gets caught on something.

Getting the Horse Used to the Harness

Before you hitch up to anything, get the horse accustomed to the harness so he'll be comfortable with it before you add the novelty of a vehicle. There are many parts to the harness; the best way to prevent problems is to get him used to it a portion at a time. Don't try to put on all of it the first time, especially if the horse is green. Step-by-step acclimation will build his confidence.

Putting on the Harness. Have a helper hold him as you acquaint the horse with the harness; this method is better than tying him in case he becomes nervous. Let the horse see and sniff each part of the harness before you actually put it on him.

The first piece of harness you introduce to the horse is the surcingle, or *belly band*, if he has not already experienced it as part of a bitting harness. Then add the *crupper*, which goes back from the surcingle and loops around the base of the tail to keep the harness from sliding forward. Most

Harness with breastplate

Harness with collar

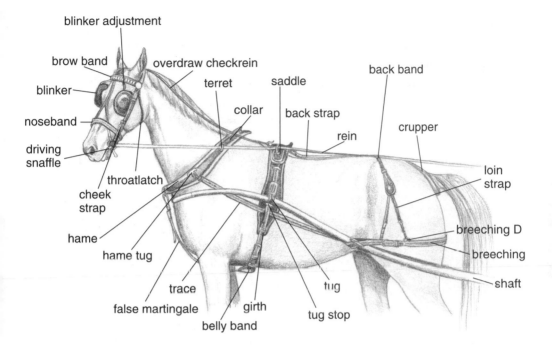

blinker adjustment

brow band   overdraw checkrein   back band

blinker   terret   saddle

noseband   collar   back strap

driving   rein   crupper
snaffle

                                                                          loin
        throatlatch   strap

    cheek
    strap

                                                                      breeching D

hame   breeching

    hame tug

        trace   tug   shaft

    false martingale   girth   tug stop

            belly band

Parts of a driving bridle (closed bridle) and harness

horses don't mind the surcingle, but many are wary of the crupper; it takes several sessions for a horse to get used to this. If your horse is accustomed to being in a stall, add the crupper to the surcingle in his stall and let him wear it in there for an hour at a time, for as many days as it takes him to become comfortable with it. Remove any obstacles the harness might catch on.

If your horse has not been in a stall, turn him loose in a round pen for an hour at a time, so he'll get to know how the crupper feels when he moves around.

Enlist the aid of an experienced person to hold him for you as you put on the crupper, in case the horse is nervous or resists. Be prepared if he explodes when you turn him loose to wear it. When turning him loose in a stall, be close to the door and ready to exit; don't turn your back on the horse. After a few daily sessions, he will realize the crupper is not going to hurt him, but at first he may clamp or wring his tail, kick, run, or buck. Don't put any more of the harness on him until he accepts the crupper.

COMFORT TIPS

The crupper dock, which goes under the tail, should be made of very soft, well-conditioned leather, pliable rather than stiff and hard, and the loop should be large enough to be comfortable under his tail. To prevent chafing and discomfort, it's a good idea to sew sheepskin onto the surface that will rest next to his tail.

Bridle with Blinkers

Next, take a few sessions to introduce the horse to a driving bridle and bit. A horse that has already been bridled may need only one lesson to get used to the new bridle with blinkers. If this is his first bridling experience, however, see chapter 6 (page 168) for helpful lessons. After he is used to the driving bridle, put it on, along with the partial harness (surcingle and crupper), and let him wear it in the stall or lead him with it.

Some horses are nervous when introduced to the driving bridle with blinkers. Take as many sessions as needed to get him used to it. If he has been taught to pony (see page 161), you can lead him from another horse while he's getting used to blinkers, surcingle, and crupper; he can rely on the security of the other horse while getting used to the feel of everything and adjusting to restricted vision. (*Note:* Blinkers are not essential for driving, but they help focus the horse's attention on what he is doing; he is unable to look back at the cart following him, for example. Most horses are less spooky about the cart if they wear blinkers, but a few do better without them. If your horse can't get used to them, don't use them.)

While training, leave the halter on under the bridle, so a helper can lead the horse. The halter should be close fitting. All parts of the bridle should fit well and be taut so there is never any danger of having a shaft slip under the halter or bridle.

When the horse is comfortable with the driving bridle, add the rest of the harness. Practice putting it on and taking it off while someone holds the horse. You won't be turning him loose with the full harness on, but he must get used to the feel of it. The breeching drapes over his hindquarters, and he will definitely need some time to become accustomed to that.

Longeing in Harness

After the horse is at ease with the bridle and partial harness (surcingle, back pad [top portion of the surcingle], and crupper), longe him with them on. (He must first be well trained to longe with voice commands; see page 155). Run the connecting end of the longe line through the inside bit ring, up over the poll, and snap it to the outside bit ring, so when there is a pull on the line it will pull upward on the outside of the bit and not just pull on the inside bit ring. This keeps the bit from being pulled through his mouth.

Longe the horse at the walk and trot, re-inforcing his response to voice cues until he responds instantly. You will not be cantering him in harness, so he won't need a cue for that. After he responds well, add the rest of the harness — the breeching and breastplate — and longe him until he is comfortable with it.

> **SAFETY TIP**
>
> It's wise to put splint boots on all four legs when longeing or driving. These boots, which cover the inside of the lower legs, will protect them from injury if the horse happens to strike a cannon bone with the opposite leg. Leave him unshod during early training to mini-mize the risk of injury if he hits himself. Once he is trained, your horse will be more coordinated and agile.

Long-Lining, or Driving in Harness

Next, introduce the horse to the feel of reins putting pressure on the bit and to the lines touching his hocks and hindquarters as he moves. Attach a line to each bit ring and run the lines through the keepers of the surcingle or through the leather loops (for the shafts) on the back pad. You can longe the horse this way or you can just follow him. Either way, it helps to enlist an assistant for first lessons. Have your helper lead the horse (if you are following, so the horse will get used to your being behind him) and keep him under control until the horse understands what to do.

If you are long-lining the horse in a circle with you in the center, he may get a little spooky when he feels the one line around his hindquarters; it helps to have someone at his head to help control him. Work at a walk until the horse is calm and relaxed. He must get used to the feel of the rein against his hindquarters so he won't kick at the trace or breeching when hooked to a cart. If he resents it, start by leaving this rein up over his back. Accustom him to the feel of the bit on his mouth and ask him to stop and

Drive in long lines, out in the open, to accustom the horse to the partial harness and closed bridle with blinkers.

turn. A helper comes in handy when first teaching your horse to turn. Don't keep him circling too long in one direction; change him often from one side to the other until he turns equally well both ways.

When a horse wears blinkers, he cannot see you behind him and must rely on your voice. Early training develops confidence and obedience in the horse and can make this transition easier. During first lessons in harness, walking behind him, do *not* use a whip if he's nervous. If he feels or hears a whip coming from an unknown direction, he'll probably become more anxious, and this will undo weeks of training. Let him have the reassurance of your voice. Keep talking to him so he knows you're there. Take care to prevent him from running into anything he might not be able to see. If you've ground-driven him sufficiently *before* adding harness and blinkers, it won't take long for him to adjust.

"Spook-Proof" Your Driving Horse

After you are able to walk behind your horse without a helper and can stop and turn him, drive him at a walk and trot to expand his experiences. Take him out of the pen to new places, to new sights, and up and down hills.

USING A WHIP

Your horse should already be used to a whip (see page 57). He must respect but not fear it, and should be accustomed to having it touch various parts of his body. A buggy whip will reinforce voice commands as you touch him on the sides or shoulder if he doesn't respond instantly. The whip is a major means of communication. Because you will always carry it while driving, make sure it is light-weight and comfortably balanced.

To become a dependable driving horse, he must learn to cope with potentially frightening situations. Ground-drive him up and down your lane, halt him at the main road, and let him watch cars go by. Once he is comfortable, take a helper along and ground-drive the horse on a quiet road before you try him with a cart. With time and experience, he'll become used to barking dogs, mailboxes, children playing, horses running along a pasture fence by the road, and the occasional vehicle.

Ask him to walk quietly past any spooky obstacle. A shying horse is more difficult to control in harness than under saddle, so spend time getting him accustomed to many things, including traffic. He should learn to trust your voice commands implicitly. He must be reliable and safe under all conditions. Having an assistant walk along when you are out on the road, to go to the horse's head to hold him, for example, or to lead him, can be a big help. The assistant should carry a short lead strap that can be quickly snapped on to the horse if necessary. Thus, it's always important to have a close-fitting halter under the bridle.

Selecting a Cart

There are many types of carts and carriages, both two wheeled and four wheeled. You may eventually get more than one type of vehicle. At first, however, you'll want one that is safe and easy for training the horse. Select a cart that is well made and that fits your horse as well as your budget.

Training Cart

For training, it's better to start with a two-wheeled cart. A four-wheeled vehicle is more comfortable to ride in but can tip over more

---

HITCHING TIP

Before you hitch up, make sure your horse has already felt a rein caught under his tail. This way, it won't be a totally new and frightening experience for him if it should occur, and you'll know how he'll react.

---

easily on tight turns and may jackknife if the horse backs up crookedly. A two-wheeled cart is more stable and will go wherever the horse goes.

If you want one all-around cart that will work for training as well as being a good, safe pleasure vehicle, most drivers recommend a Meadowbrook cart. This cart is solid and durable — it's made of oak — yet fairly light and portable because it has only two wheels. This type of cart is built by craftsmen all over the country and is commonly available. The Meadowbrook is a good training cart because it has a low center of gravity, which makes it more stable over rough terrain.

A cross-country cart — the type used in endurance portions of combined driving events — is also a good, sturdy everyday vehicle. A new Meadowbrook or cross-country cart costs about $2,000, but you may be able to find a good used one for much less. If you do buy a used cart, make sure it is in good condition and safe. If you have no experience, ask a knowledgeable person to help you evaluate it. If you want a four-wheeled vehicle, find one whose body is cut under, so the wheels fit

The Meadowbrook is a good, safe, all-around pleasure vehicle.

under the body when making a tight turn. Many of the older vehicles — the antiques — are rectangular boxes on springs and not very practical, especially for traveling across country.

Check for Soundness and Safety

Be sure the vehicle you get is solid and serviceable. Check the wood for dry rot, even if it's new. Make sure the shafts are not cracked, that the wheels match, that the hubs aren't cracked, and that the wood spokes are solid and firmly attached. Don't purchase a cart with wire spokes; these can't handle rough terrain. Also avoid air-filled rubber tires; a puncture, which will create a bang or a hissing sound, could badly spook your horse.

Most cart and carriage wheels have solid rubber tires, but some are metal bands. These may be fine on a dirt road, but they'll make a lot of noise on pavement and asphalt. If the cart has metal wheels, have them fitted with rubber by a wheelwright. If the cart does not already have brakes — and many two-wheeled carts don't — you can have these added.

The Cart Must Fit the Horse and Driver

The comfort and safety of both the horse and you depend on the fit of the vehicle to the horse. This is especially important in a two-wheeled cart because shafts are attached to the vehicle, putting it at the same angle as the shafts. A horse too tall or too short for the cart will tilt it up or down. The seat should be level when the horse is hitched to the cart, which means the shafts are level (horizontal) also. They should be positioned as close as possible to the points of the horse's shoulder.

You can use larger or smaller wheels to make the cart higher or lower, but a change of 4 inches in wheel diameter will change the height by only 2 inches. Some carts can be made higher by blocking up the springs, but this raises the center of gravity. It's best to measure your horse before you look for a cart. You can't adapt a cart made for a 12-hand pony to fit a 15-hand horse.

The cart should also fit you. If your feet don't touch the floorboards, there will be nothing to brace against, although a toe rail can be added. The other extreme is to have no leg room. Be sure the cart fits the driver as well as the horse.

Preparations for Pulling

Introduce the horse to pulling before you hitch him to a vehicle. One way to do this is to put traces through the tugs and breeching, then fasten an 8-foot length of rope, with a loop on the end, to each trace. The loops can be hooked to a singletree (the crossbar behind the horse when pulling a cart), which a helper can hold while following the horse. The traces should be slack at first. Longe the horse, asking him to walk around you in a circle as your helper follows the horse.

When the horse is freely going forward, the helper should begin to pull back slightly on the singletree — gently at first, as the horse gets used to pressure on the breastplate. As the horse learns to lean into the breastplate, more weight can be put on the traces. Then you can take the horse for walks, driven in long lines, with the helper pulling on the singletree.

Hitching Up

Before you hitch up, accustom the horse to the cart so he won't be frightened by the sight or sound of it. An easy way to accomplish this, after he's been led up to the cart a few times and allowed to check it out, is to have a helper pull the cart in front of the horse as you ground-drive him. This is less scary the first time the horse hears it moving than if it's behind him, squeaking and rattling. When the horse no longer worries about it, have your helper pull the cart along behind him. Don't hitch up your horse to a vehicle until he is at ease with how it sounds. If the cart has a singletree instead of shafts, get him used to feeling something between and on his legs before you hitch up.

Hitch in a Safe Place

Have the horse in a pen or enclosed area so he can't go far if he pulls away. While hitching, enlist your helper to hold him, then bring the cart up behind the horse, with shafts raised high so there's no danger of

> ### SAFETY TIP
>
> Before hitching up, make sure the cart is safe — no loose bolts, for example, no bad rubber on wheels. If it has brakes, check that they work evenly. Keep the harness in good repair and clean it frequently so you can check for weak spots and loose stitching.
>
> When hitching up the first time, station a helper in front of the horse. Even better, have two helpers — one on each side of the horse. If only one person is holding the horse from the side, the horse may try to move sideways or spin to the side, risking hitting someone with the shaft.

poking him. Stand beside the shafts, *not* between them, and slowly bring them down to gently touch the horse's rump. Don't try to hitch him the first time unless he is utterly relaxed. It may take several sessions of putting the cart behind him before he is ready. The person at his head should talk to him, distract him, and help keep the horse calm.

While hitching, let the horse face any activity that might be going on around the barnyard. This way, he can see it clearly and won't be trying to turn and face it to see it better. Create a regular routine for hitching so he will be comfortable with it. If you always put on the harness the same way and bring the cart up to him the same way, this helps the horse know what to expect and what is expected of him. He finds comfort in routine and won't be alarmed by something new. Be consistent, and require that your horse be consistent too. He will learn that when he is being hitched, his job is to stand still and stand squarely, even if he gets bumped or the harness moves around.

Facing a Wall or Fence

The horse that might be skittish should be hitched by a barn wall or solid fence. Put the cart 9 feet from the fence or wall. Then bring the horse, wearing his harness, with a longe rein snapped to each side of the bit. Let him sniff the cart until he is relaxed, then lead him in front of it to where he is facing the wall or fence. Hold him on his left side with your left hand, keeping your right hand free to guide the shaft through the tug, while your helper holds the horse on the other side, brings the cart up to him, and puts the shaft through the right tug.

The traces can be quickly and temporarily tied on with string, using a quick-release knot, while you hold the horse still, then the surcingle is buckled. Hitch up quickly and quietly, with every move planned in advance so the horse won't be frightened by a careless action.

Having the horse face a fence or wall helps keep him from moving during the hitch-up but makes it more difficult to move off because he must start by turning. To help the horse make the turn, pull the left shaft toward you with your right hand. Simultaneously give him the verbal command to walk while pulling him to you as a helper pushes the right shaft. Your horse is thus taken in a circle to the left to get away from the fence or wall so he can travel.

Make the first lesson short and stop him facing the fence or wall again, where you can carefully unhitch him. Hold the horse and reassure him while your helper unhitches the cart and pulls it away.

Unhitching

The procedure for unhitching is the reverse of hitching up. First undo the surcingle, then unbuckle the breeching straps, and finally unhook the traces. Before the cart is pushed back from the horse, make sure the ends of the reins are placed where they will not catch on anything, or he may get a jerk on his mouth.

It is better to push the cart back from the horse rather than lead him out of the shafts; he should learn always to stand still. If he is led forward, he may come to anticipate this and get into the bad habit of coming forward too soon. This increases the risk of having a shaft fall from the tugs and break, and in the process terrifying the horse. Never remove the horse's bridle while he is still hitched to the cart.

First Lessons Pulling a Cart

The first day you actually hitch up the horse, you won't do much driving. To be safe, don't even get into the cart. Stand by the cart with the lines in your hands and ask the horse to walk. Have a helper at his head to hold a lead rope. If the horse manages to walk a few calm steps, that's a good place to stop. If the horse is nervous, don't even ask him to walk. Let him merely stand there hitched to the cart.

Take several days, if needed, to work up to thirty or forty steps pulling the cart. For first lessons, have your helper at his head, to bring the horse under control if he panics. The cart will be noisy and the horse may become alarmed. Take your time. Even if things are going well, don't be tempted to rush the training. Build gradually on each success.

Driving Him from the Cart

After a few sessions of ground-driving him while walking beside the cart (with your helper on the opposite side of the horse), your horse may be ready for you to get into the cart. The helper should unsnap his lead rope from the right side and move to the left side and snap the lead to the halter while you walk beside the cart. When the horse is calm, get into the cart while it's moving. This is easier on the horse than having to start pulling the cart from a standstill. The unaccustomed extra weight may cause him to lurch or resist rather than ease into the breastplate.

After you and the horse have enough confidence that you can easily step into the cart while your helper is leading the horse, start practicing gradual turns and gait transitions — that is, from walk to trot, trot to walk, and walk to halt. If the cart has traces and a singletree rather than shafts, ask your helper to hold the lead rope the first time you make turns; the traces will bump the horse's hocks and he may try to get away from this strange sensation.

Soon your horse will be dependable enough to have both you and your helper in the cart. It's a good idea to take along a helper for a while: If you get into a tricky situation, the assistant can jump out and go to the horse's head. If all goes well, the horse will learn to depend on your voice commands and trust your signals, and in time you won't need the "insurance" of a helper.

Holding the Reins

Hold your reins properly and keep even rein pressure. Uneven pressure on the bit will give the horse the bad habit of *rubbernecking*, or turning his head to one side. Always have the reins even — with equal bit contact on both sides — before you ask the horse to start off. This way, he can move off straight and with the proper bit and harness engagement.

When getting in a cart, hold both reins in your right hand, with the left rein under your index finger and the right rein under your middle finger. This leaves your thumb and index finger free to hold on to the cart as you enter. The off-side (right) rein should be 3 inches longer than the near-side (left) rein, so if the horse steps forward while you are getting in the cart, a feel of the reins will keep him straight.

Get in quickly and quietly, then transfer the reins to your left hand. Hold the left rein over your index finger and the right rein under the middle finger so that two fingers are separating the reins. This lets you apply pressure to either rein by rounding your wrist up or down. There's more leeway for play of the wrist on the horse's mouth this way than you'd have with only one finger separating the reins.

Your thumb should point to the right, not upward, and should not press on the rein. The forefinger points outward and slightly to the rear. This keeps the near rein close to the knuckle; the horse can be easily moved across the road or to the left or right just by turning the back of your hand up or down. The middle, third, and little fingers press on the reins as they come through your hand, gripping them against your palm to keep them from slipping.

Keep your wrist well rounded and flexible and hold it about 3 inches in front of the center of your body, with your arm horizontal across your

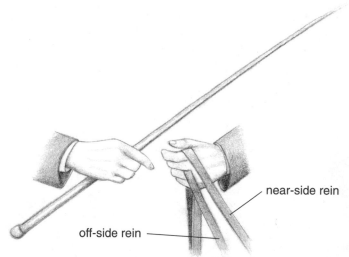

near-side rein

off-side rein

The proper way to hold the reins and whip while driving. The near-side rein lies over the index finger and the off-side rein goes under the middle finger.

body. A flexible wrist enables you to keep a steady touch on the horse's mouth without jerking. If your wrist is stiff and unyielding, the horse will pull against you.

Always keep the reins in your left hand so you can use the whip with your right hand without affecting the rein action. *Never* use the whip while the right hand is holding a rein. If you were to do this, it would put slack in that rein and the horse would immediately travel to the left.

Position in the Cart

Hold your elbows close to your sides. Your knees and ankles should be together and your feet planted firmly on the floor or footrest. It is not safe to drive without adequate leverage for your feet: If the horse starts to pull at the bit or suddenly stumbles, you could be pulled from your position unless you can brace your feet.

Using the Whip

Always use a whip when driving, and keep it in your hand, not in the whip socket of the cart. A whip takes the place of a rider's legs for giving signals, keeping the horse on the bit, and correcting him if needed. Apply the whip gently to his shoulders or sides. Do not use it on the hindquarters, as this might make him kick.

Overcoming Challenges

Tackle new things gradually. Work to develop good responses and good habits for the horse while being careful not to create any bad habits. If the horse balks at going through a puddle, for example, have your helper lead him through it. Avoid fighting with or forcing the horse. If he has a bad experience with the puddle, he will associate all puddles with conflict and then always try to avoid them.

Don't insist on immediately repeating a hard task or reconfronting a scary obstacle. Your horse may become tense and resentful and will develop negative associations. It's better to wait and repeat the difficult lesson the next day. It usually becomes easier for him to deal with if there wasn't a big to-do the first time.

Driving in Company

Once your horse is comfortable with the cart, take him out in company at every opportunity to introduce him to the sight and sounds of another horse and cart. The sound of another cart going next to him — which he can't see because of the blinkers — can be quite frightening to a horse. At first drive him behind the other rig so he can see it, then in front, and finally alongside it.

Gradual Conditioning

Just as in training a horse under saddle, a driving horse needs to be brought along slowly; again, you are training his muscles and body as well as his mind. If you work him too hard before he is in shape for it, he will be tired and stiff the next day and in no mood for another lesson.

Regular short lessons are always better than infrequent long ones. Work with your horse a little every day and not just on weekends. Even when a horse is fully trained, weekend workouts can be hard on him. If you ask him to do more than he is physically conditioned for, his muscles will get sore. You can't expect him to do as much as when he is being worked regularly.

Always do the first half mile going away from home at a walk, to warm up your horse. And always walk him the last half mile when coming home. This helps cool him down, and it also teaches him not to rush home.

When teaching him to drive in an arena or a training ring, remember that a horse expends more energy working in a circle than he does working on the straight or on a road or trail. It takes more muscle effort to make a circle, and the cart pulls harder as well. Half an hour of ring work may be roughly equal to an hour out on the road. Keep in tune with your horse and don't ask too much of him, especially in the beginning, or he will become sour about his work.

Also, always give him time to absorb one lesson before you advance to the next. For instance, when making gait transitions, make sure the horse knows the difference between a medium (working) trot and an extended trot before you ask him to go from an extended trot to a road trot (a fast trot that can be kept up for several miles).

If for some reason you are unable to work with the horse for a while, take the time to bring him back to condition slowly.

Be Consistent

For your horse to have good, dependable habits while driven, you must have good, consistent driving habits. In both riding and driving, your training goal is to develop good habits. Whether your horse develops good habits or bad ones depends on you.

Turning the Horse

To make a left turn, round your wrist so that the back of your hand leans toward your body, with the thumb down. This puts pressure on the left rein and releases the right rein enough to turn a light-mouthed

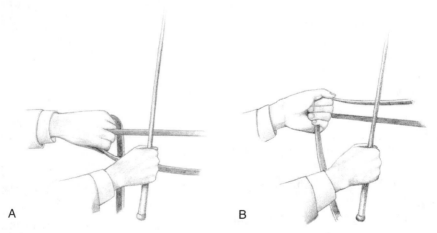

A   B

Proper hand position for a left turn **(A)** *and a right turn* **(B)**

Pay Attention to Body Language

Always pay attention to the horse and stay attuned to his mood. If things start to go wrong, you have just one or two strides — a second or two — to react and correct or reassure him. By observing his ears, head carriage, muscle tone (is he tense?) and body position, you can predict how he'll react to something and be better prepared to deal with it. Heed his tail position and the bend of his body. A second before he shies, for instance, a horse usually gives a clue as to which way he'll jump. If you know your horse, you will know whether to use a half-halt (a quick give-and-take on the reins), to pull on the left rein or the right, or to touch the horse with the whip. If you are always reading the horse, you can prevent a lot of problems. By observing the horse, you will also know if part of the harness is bothering him and can determine which part just by the way he holds his body.

---

horse, one that is very responsive to a slight signal with the bit. To make a right turn, turn your wrist so the little finger is upward toward your body. This puts slack in the left rein and increases the pull on the right rein. A common mistake is to swing the hands to the left or right in the direction of the turn — your hands should not move from their position in the center of your body.

A novice driver's usual reaction when turning clockwise is to grab the left (outside) rein to try to keep the horse turning smoothly instead of cutting the corner too much. This often results in poor corners; the horse will lean to the side of the bit that is held tighter and go into the turn with his head tipped to the outside instead of bending with the turn. With his head tipped to the outside, his shoulder is thrown to the inside and he makes a stiff corner. This can become a bad habit for the horse if you do it to him very often. Practice on the straightaway to keep even rein pressure and have the horse working well on the bit, not avoiding it or resisting it.

Just before you reach a turn, give a tiny half-halt with the bit to alert the horse that you're about to ask for something new. If you are making a turn to the right, keep the left rein steady while you very gently give and take with the right rein. This will help the horse make the turn with the proper bend to his body.

TURNING TIPS

If the horse is not sensitive to the rein, you may need your right hand to assist on a turn. Both reins should remain in your left hand, but for a right turn place your right hand a little in front of the left and over the right rein, to put more pressure on that rein. For a left turn, your right hand assists in a similar manner, over the left rein.

Going Up and Down Hills

When going uphill or downhill, keep in mind the way the horse pulls a cart and remember how to help him. When he is pulling uphill, he has his weight leaned into the collar or breastplate. Just before he gets to the top, in the last few strides, ask him to shorten his stride by checking briefly with the bit (half-halt). This does not mean slow down. Shortening his stride allows the momentum of the cart to catch up with him on top of the hill, so the breeching can contact his hindquarters before the cart starts down the hill on the other side.

As the horse heads over the hill, again check him briefly with the bit (another half-halt or two, depending on the situation) to ask him to shift his weight more to his hindquarters. This will enable him to hold back the cart a little better on the down slope. If you wait too long to help him make the transition from pulling to holding back, he may start down the hill with too much of his weight on his front legs. The cart coming behind him thus could push him down the hill.

You must give the horse his cues while he is at the top of the hill, just as he is starting down, *before* the cart starts down. Otherwise, your cues are too late. He will not be in the proper position to slow the cart, and the cart pushing into him may give him a very bad experience.

SAFETY RULES FOR DRIVING

- Always wear protective headgear and require your passengers to do so also.
- Wear gloves while driving.
- When driving alone, tell someone where you are going and when you plan to return.
- Shoe the horse appropriately for the conditions, especially when traveling on paved roads.
- Place a SLOW-MOVING VEHICLE sign (an orange triangle) on the back of the cart.
- Always carry a tool kit for any emergency cart or harness repairs. Your repair kit should include rein and trace splices, a wheel wrench, a hammer, a hoof pick, a hole punch, a halter, and a lead rope.
- Carry a flashlight if you might be out after dark and use reflective tape on harness and cart.
- Never allow anyone in the cart if the driver is not in it.
- Never leave a horse hitched without supervision, and never allow an inexperienced person or a child to watch your horse if you are not present.
- Don't take the horse for a drive with other drivers if he has not been out in the company of other horses before. Give him lessons first.
- Never ask a horse to do anything while hitched that he did not learn in his ground work.
- Never come up behind a harnessed horse without speaking to him; always let him know you are there.

9

FIRST MOUNTED LESSONS

Your young horse is well halter-trained, has been bridled and saddled, and is used to the bit. He knows about turning and stopping. Now the big day has arrived when you will ride him for the first time.

Get the Horse Accustomed to the Saddle

When your horse is saddled, get him used to the feel of it as you pull on a stirrup from each side, wiggle the saddle, and put a little weight across the saddle seat. Put on the saddle and take it off a number of times and from both sides. Lead him with it on, at both a walk and a trot, so he gets used to it when moving as well as standing still. At the trot, he may be alarmed if the saddle makes noises or the stirrups bounce and bump him in the ribs. He must overcome any nervousness about the saddle before you mount.

Traveling with the Saddle

If your horse is a little spooky about the saddle, spend some lessons longeing and ponying him with it on so he becomes familiar with its movement on his back and the flopping of the stirrups. It's best if he never learns he can buck with a saddle on (or a rider); if he is alarmed by the saddle, however, and tries to buck it off, you must help him get beyond this stage of insecurity *before* you get on him. Longe or pony him

> ## PLAN FOR SUCCESS
>
> First rides are important in setting the stage for the rest of your horse's training; make sure these lessons go well. Choose a time when the weather is right and when there isn't much going on to distract or upset your horse. Dogs should be tied up and no children should be playing in the area. Saddle and bridle the horse. Check that everything fits properly and comfortably. Tighten the cinch sufficiently so the saddle won't slip or turn when you get on.
>
> A young horse doesn't have much in the way of withers — he gets his height at the croup first and the withers catch up later — and doesn't hold a saddle as well as does an older horse. You don't want the saddle even slightly loose or it may turn if your horse spooks and wheels, forcing all of your weight into one stirrup. You also don't want the saddle pulled forward onto his neck if he tries to buck; putting down his head forcefully will pull on the saddle horn if the halter rope is tied to the horn. Last, the cinch should not be uncomfortably tight, but it should be tight enough to firmly hold the saddle during an anxious moment for the horse.
>
> For first mounted lessons, you may want to use a Western saddle. You can leave on a close-fitting halter under the bridle and tie the halter rope to the saddle horn, with slack enough that it doesn't interfere with normal movement of his head and neck but tight enough that he can't get his head down far enough to buck.

as many times as necessary until he is completely at ease with it. (See pages 149 and 158 for more on longeing and ponying.)

Conduct the first saddle lessons longeing and ponying in a pen. This way, if the horse gets too excited and pulls away from you, he won't be able to go far. After he is more at ease with the situation, you can pony him out across country carrying the saddle if you'd like to.

First Mounting

For first mountings and first rides, use a corral or pen. Some trainers mount a young horse in his stall the first time; he can't get away if something unexpected happens. You can undertake the lessons that follow working by yourself or with a helper, depending on your preference and on the horse. A horse may have been so calm and comfortable about everything you've done to this point that you're confident he'll stand quietly for this, too. Another may be more nervous, and you may want a helper to hold the insecure horse.

Longe or pony the horse before getting on him the first few times. This will take the edge off his high spirits and allow him to be more calm and settled in his mind. If he's been working a little, he'll be more willing to stand still and relax when you get on and off.

Stand Still for Mounting

Your horse should learn to stand quietly for mounting and dismounting. As you work around him, require that he stand. If he moves forward or fidgets, tell him *Whoa* and make him stand still. If you let him get away with moving, you'll have problems later. If he is in the habit of standing still at your command, you will be able to check him instantly with voice and rein cues if he starts to move when you are mounting for the first time.

Use a Mounting Block or Bale of Straw

If you are short and your horse is tall, use a mounting block of some kind. A bale of hay or straw works well for this; it is something the horse is not afraid of. Lead him up to it and ask him to stand. Let him get used to you standing on a bale of straw next to him. This also accustoms him to having you above him, as when mounted. Talk to him, rub him, keep him relaxed. With the

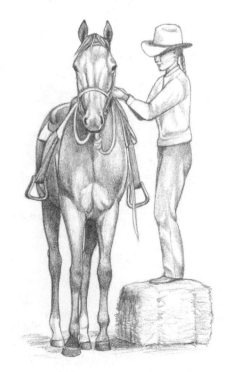

Use a bale of straw or a mounting block as an aid when you mount for the first time.

aid of a mounting block or straw bale, it is easy to lean over the saddle and start getting him used to having a little weight on it.

Have Someone Hold the Horse

It's handy to have a good helper as you accustom the horse to having weight in the saddle. From both sides you can fuss with the stirrups, put weight in them, lean over the horse. A person holding the horse should be calm and at ease. A horse can sense when someone is nervous and will become nervous also.

The helper stands on the same side you're on and holds the horse by the halter. You don't want the horse held by the bridle; this may hurt his mouth and startle him if he starts moving and must be controlled by your assistant.

Put Weight in the Stirrups

Before you get on, put weight in the left stirrup. Pull or push down on it with your hand, and if this doesn't bother the horse, put your foot in the stirrup and put your weight on it. If you are short, stand on a bale of straw.

Hold the left rein in your left hand, along with some mane, so you can check the horse or pull his head around in a circle if he starts to move. Use your right hand on the saddle horn as you step into the stirrup. Stand with some weight in the stirrup but don't swing into the saddle yet. Step back down and repeat this until you are sure he accepts the feel of your weight in one stirrup.

Standing with weight in the stirrup

When your horse is calm, stand with your weight in the stirrup and gently pat him on the neck and rump. This gets him used to your being up there, yet you are not in the saddle and can easily step back down if he starts to jump around. If he's nervous, just put your foot in the stirrup again and again, without putting weight on it, until he is calm.

Lean Over the Saddle

To get the horse accustomed to your weight, lean your upper body over the saddle. You can do this most easily from a mounting block or bale, but it can also be done from the ground. This is a safe position — you can readily slide back off if the horse becomes alarmed.

> ### SAFETY TIP
>
> When working with a green horse, *never* put your foot all the way into a stirrup during mounting lessons. That way, if he jumps, you can safely pull out your foot. If you're just getting up and down in the stirrup, use your toe. Don't put your feet clear into the stirrups until you are actually on him and plan to stay on him.

Hanging your body weight on a saddle, while a helper holds the horse, helps to determine whether the horse will stay calm for mounting.

Hold on to his mane and the reins with your left hand; pat him and grasp the saddle with your right. When your horse is relaxed, spring lightly up and across the saddle, keeping the horse checked with your left hand. If you are tall enough, or if the horse is short, simply lean over the saddle when standing on the straw bale. After a moment, slide back off and praise him.

If you are using a flat saddle (no cantle), move into mounted position from leaning over his back. After balancing over his back on the saddle — during either the first lesson or in subsequent lessons, depending on how soon he stays relaxed and no longer stiffens with nervousness — swing your leg over, taking care not to touch his hindquarters, and sit on him. Keep your body low at first. With some horses, your sudden towering above them is what's startling.

If you're using a Western saddle, it's not so easy to swing into mounted position from leaning across the saddle; and you're more apt to alarm the horse by trying. It's better to mount the traditional way (see page 31) after getting him used to having your weight across the saddle.

If you are working alone with a nervous horse, one way to get him used to your weight, with the least amount of risk if he is accustomed to being in a stall, is to conduct the lesson in a large box stall having no low beams, mangers, or feed racks. Place a straw bale about 4 feet from the wall on one side and lead him between the bale and the wall so that he is facing the far wall. He is thus contained by two walls and the bale. Stand on the bale, talking to him to keep him relaxed, then lean your weight over the saddle.

Getting on the First Time

When you get on, do it in a pen, so if the horse moves out, it will be in a safe place. Talk to your horse, rub him, get him relaxed and standing still as you move around him, tugging on stirrup leathers, putting weight in the stirrup (always working on either side). When he is calm, and after you've stepped up and down in the stirrup a few times, go ahead and mount. Do it smoothly and easily, taking care to keep your leg from touching his rump (which might startle him). Ease into the saddle; never drop your weight into it. Keep a short rein so you can check your horse if he starts to move. If he gets nervous as you begin to mount, step down and calm him. When he is reassured and standing, try again.

FIRST MOUNTING

A. *A helper holds the horse while the rider pats and reassures the horse in preparation for mounting.*

B. *Put weight in the stirrups a few times before mounting.*

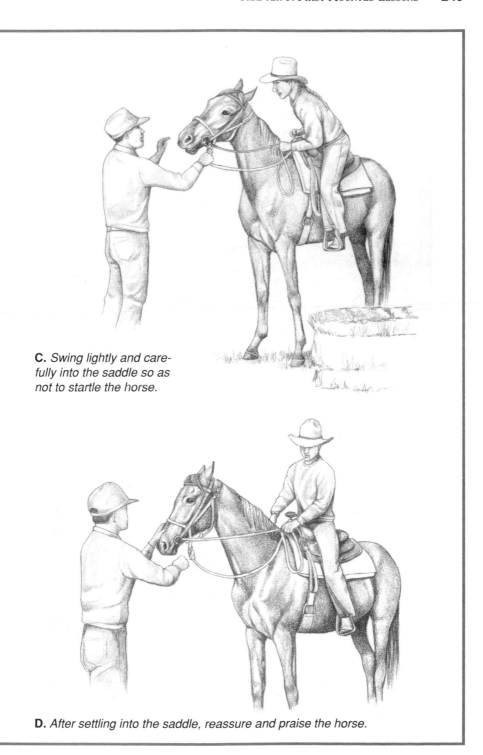

C. *Swing lightly and carefully into the saddle so as not to startle the horse.*

D. *After settling into the saddle, reassure and praise the horse.*

Repeat Lessons

When you're in the saddle, don't put your foot into the off stirrup just yet. Sit there a moment and talk quietly to your horse. Rub him and praise him; get him used to your being up there. Then dismount and mount again, repeating this several times so he learns that this procedure is nothing to be afraid of.

By getting on him and right back off, you let him realize there's nothing to fear. You have actually ridden him before he knows it; you've been on his back and nothing bad happened, which gives both of you more confidence. You may continue with the lesson, or this may be a good place to stop for the day. If your horse is insecure, do only a small step at a time; his security and confidence will increase with each success and you'll be less apt to run into trouble. When you get on him again the next day, he will be more at ease, ready to progress to the next steps.

Each time you get on, stay on a little longer. When your horse is comfortable with it, move around in the saddle, lean over his neck, rub his neck and his rump — get him used to your being up there. When he accepts this calmly, he's ready to move out.

Moving Out

If your horse starts to move when you first mount, either check him (halt and make him stand) to practice mounting and dismounting or let him move, as long as it is in a controlled manner. If he moves because he is startled, check him and have him stand quietly a moment. If he is moving in a calm manner, it's fine to let him continue; you can get him

---

BE SURE THE HORSE UNDERSTANDS

If the horse won't move when you ask him, enlist a helper to lead him a few steps as you give the cue to move out. The helper can snap a lead rope on to the halter, then unsnap it after the horse gets the idea. If the horse is still reluctant to move or won't keep going, have your helper longe him at a walk. This is something with which the horse is familiar; he will soon figure out that when you use leg pressure and clucking sounds, he should move.

used to repeated mounting and dismounting later. For the moment, let him move about, in whatever direction he wants to go, while he gets used to the extra weight of a person on his back.

Encouraging a Reluctant Horse

Many horses don't move when mounted the first time. They are bewildered and stand rooted to the ground. After you've mounted and dismounted a few times and are ready to move out, cluck to him and then squeeze a little with your legs. He doesn't know about leg pressure yet and must learn to make the association.

Never kick or use your heels, and don't put pressure back toward his flanks. Kicking generally startles or confuses a young horse, and he may jump forward and start to buck. Kicking is like slapping or hitting. It's a form of punishment rather than a cue. He won't understand what you want. Up to this point, you have been teaching him to stand still while you mount; now you have to teach him that it's all right to move.

Squeeze with the calves of your legs where they hang in the stirrups, just to the rear of the cinch. Lean forward slightly so your position will

The rider pulls the horse's head around and leans a little to get him off balance so he will take a step.

encourage him to move in order to restore balance. If you have used clucking noises as a cue to move forward or move faster during ground work when teaching him to lead at the trot, for example, the horse will associate the cue with the idea that he is to move forward. If he still doesn't move, turn his head a little, pulling it to the side and down, with one rein. This will usually put him off balance and he will have to take a step.

The first step your horse takes with an unaccustomed weight on his back may thoroughly alarm him because your weight changes his balance. Speak reassuringly to him. Allow him to take several steps but be prepared to halt him at any time if he becomes so frightened that he tries to run or buck.

Stay in Control If He Panics

If your horse explodes, stay calm and speak reassuringly as you try to keep your balance and seat. If the halter rope is tied to the saddle horn, the horse can't buck hard, and you can probably keep your seat. To control a buck or bolt, pull the horse's head strongly around to one side. Don't pull on both reins at once, which he'd be able to pull against; he can't run or buck if he is turning in a tight circle.

Get him halted and let him stand for a few moments. Speak calmly to him so he knows it's just you up there on his back, nothing to be afraid of. Rub him and allay his fears. After he is calm again, ask him to move

---

HOW TO PREVENT BUCKING

Usually, the green horse will never buck during first lessons because he has confidence in you. You have acquainted him with the use of reins, carrying a saddle, and other fundamentals. Generally, when a horse bucks, it is because of discomfort or confusion — too many new things being done to him all at once. By preparing for first rides gradually, and getting him accustomed to your commands and signals and the equipment (and making sure it fits properly), you should have done away with any confusion or irritation. Prepare for first rides with lots of ground work, so the horse is relaxed with what you are doing and never discovers he can buck with someone on his back.

out once more. Do not dismount and end the lesson until he is calm, or he may think that by bucking or bolting he can make you get off.

If he is still nervous and you are not confident about riding him, have a helper snap a lead onto his halter and lead him around the pen. This will keep the horse under control until he becomes more at ease with a rider on his back as he moves around. Once he gets over his initial fear or unfamiliarity with the extra weight and settles down, have the helper unsnap the lead and step back, allowing you to continue around the pen.

Stopping and Turning

After the horse has walked around the pen, stop him with *Whoa* and a pull on the reins. Use a gentle squeeze and release, one rein at a time in a seesaw action. This is more effective than a steady or hard pull. He should stop readily, as *Whoa* and the pull on the reins should be familiar to him from leading lessons with the bridle or from driving lessons. If he is nervous, this is a good place to stop — after a successful walk around the pen and a quiet halt.

If the horse is calm, however, you can continue the lesson. Practice moving out and stopping. Repeat this several times: Walk him forward and let him go at random for a few moments, then halt him and let him stand.

Use of Rein and Leg Signals

Once your horse is moving out at your signal and stopping well, begin guiding his direction as he walks. Use gentle leg pressure on the side away from which you want him to turn. He will tend to turn away from the leg pressure, especially if you did some ground work teaching him to turn on his quarters or forehand (see chapter 6) — he will know he is supposed to move away from pressure. As you use your leg, gently give and take on the direct rein to pull him around. Use your hands low on the reins.

If at times your horse does not want to turn, don't pull firmly or jerk on the rein; this will only hurt his mouth and create more resistance. He will want to pull away to avoid the pain. At this stage you don't need his response to be perfect. He is learning many things in addition to reining, and they all take time.

Use leg pressure when turning. Encourage your horse to turn right by using the left leg strongly behind the girth. Keep your hands low on the reins.

Using a Fence as an Aid

As you begin working on his turning, use the fence to advantage. As he comes toward it, use the reins and your legs to encourage your horse to turn one way or the other. If he begins to turn a certain way of his own accord, encourage the turn so he will begin to associate your signals with the turning. Help him along.

Short Rides around the Pen

For the first few days, take short rides, building step-by-step on previous lessons. By keeping rides brief and always ending on a positive note, you are setting the stage for a successful lesson the next day. The horse will have a good memory about the lesson and be in a positive frame of mind the next time.

Use the fence as an aid when turning. When the horse reaches the fence, he's better able to understand your cue because he knows he can't continue forward.

Don't Overdo First Lessons

Ten to 15 minutes of riding the first few times around the pen may be long enough. First rides are just more small steps in schooling. Ride your horse just long enough that he gets used to your being on him, but not so long that he starts to resent it. Remember, you are improving his muscle tone and refining his awareness of mounted cues in these first sessions.

Unsaddle your horse, talk to him, praise him. Goals are achieved by going gradually and not continuing beyond his attention span. If you keep lessons short, your horse will stay attentive and continue to learn; you don't want him to become tired, bored, resentful, or rebellious.

Frequent Short Lessons

Most horses make more progress with frequent short lessons rather than occasional long ones. Once you start riding the green horse, ride him daily, even twice a day. Just keep first lessons brief. And ride him regularly for the first weeks to keep him coming along well.

This is where some trainers overdo it. Let's say you've been working with a young horse for two or three years and are finally riding him, but now you're spending more time saddling and unsaddling than riding! Remember that you're laying an important foundation, and it's best to go slowly at this point. Also keep in mind that your horse is still young (especially if he's a two-year-old) and physically immature. Even if he is well grown, his bones and joints are not yet mature. Too much riding too soon can cause permanent damage. Give him lessons on mounting and dismounting and go for short rides around the corral only.

BE CAREFUL DISMOUNTING

When you end the lesson and dismount, do it carefully. Though you have practiced getting on and off, this is still a new experience for the horse and you don't want to startle him. Halt and have him stand quietly. Ease your left foot partway out of the stirrup so If he jumps while you dismount, your foot won't hang up. Put both reins in your left hand, with the left rein a little shorter, so if he moves, he will move toward you instead of away from you as you get off. Hold the mane in your left hand so you can use your right hand on the saddle to ease yourself off quickly. Tell him *Whoa* again. Shift your weight so he knows you're going to do something and your dismount won't startle him. Be careful not to touch his hindquarters with your right leg as you dismount. If he starts to move, check him, tell him *Whoa,* and start over. Make him stand as you get on and off a few more times before you put him away.

Stick to a Routine

Once you start riding a horse, ride him briefly every day. After the two of you are well started, you can give him a day off now and then, but it's best to have regular lessons at first. This keeps him in the mood for learning. If your horse has too much vacation between rides, he's more apt to goof off or test your control.

Avoid Boredom

Keep progressing and doing new things so your horse has new challenges to keep his interest. Though you are doing a lot of repetition to solidify lessons, don't drill too much on the same things. Some horses become sour or lazy if they get bored; others think up devious or prankish behavior. Some need to get out of the corral soon after the first lessons and see new things. To keep their minds fresh and willing, get out and across country as soon as you've ridden enough in the corral to establish the basics of stopping and turning for control of the horse (see chapter 10).

Continue His "Calmness" Lessons

You don't want your horse to become startled out on the trail if you move your arm to adjust your helmet. While making short rides around the pen, get him used to things that happen while he's being ridden. Halt him and move around in the saddle and get him used to weight shifts. Lean forward and back, patting his neck and his rump so he won't be jumpy if something bumps his rump.

When you first start riding a horse, minimize your actions and keep them slow and quiet. But after he is used to your being on his back, gradually familiarize him with more actions. Move your arms, and as he gets used to that, move them more. Then if you wave at a friend or grab your hat in a gust of wind, you won't scare your horse out of his wits.

First Trotting Lessons

After you've ridden him a number of times, you can tell whether your horse is ready to move on to faster gaits. This varies greatly from horse to horse, depending on physical and mental readiness. Your horse must be not only physically ready — able to balance himself with a rider on his back and handle himself at the walk, stop, and while turning — but mentally prepared as well. He must be calm and comfortable about everything. He must respond properly to cues and enjoy the lessons. If he is at all insecure or awkward, he is not ready.

The Signal to Trot

When teaching your horse to trot, or to make a transition from one gait to another, or to slow down or speed up a certain gait, give him a cue that is easy for him to understand. If you've been working with him at the walk, he should already know that leg pressure means to move forward and that increased, intermittent leg pressure means to speed up the

---

WHEN TO TROT

When the horse is ready to start trotting, you'll both feel comfortable about trying it. For some this will happen within a couple of days or a week after you start riding; for others it might take two months and many long walks. Each horse is different. Don't rush yours. If you try faster gaits before he is ready, it will be difficult for him physically and will possibly create a problem. Remember, your goal is to see how well you can train your horse; be patient and progress at a speed that's right for him.

---

walk. If you have used leg pressure correctly, your horse has already made the association that moving faster gives him release from pressure.

A press with your legs when he is standing encourages him to move forward and walk: He knows the pressure will stop when he starts moving. Leg pressure while walking taught him to move a little faster in order for the pressure to cease. A continuation of this logic will help him move into the trot. Rather than introduce a new command he might not understand, expand on cues your horse already knows. He knows that leg pressure accompanied by a slight forward shift of your body weight means he should restore balance by moving forward or going a little faster.

Don't Kick

It may be tempting to kick a little if your horse does not respond adequately to increased leg pressure, but *don't* do it. Kicking a horse in the sides to try to encourage a faster gait just confuses him and usually brings a response that is the opposite of what you want.

Imagine the situation from the horse's point of view. He's walking along briskly because you encouraged this with a squeeze of your legs; he has sped up his walk in response to a squeeze. Suddenly you kick him in the sides. He doesn't know what this means and thinks he has done something wrong. A typical reaction to that, for many young horses, is to stop. They perceive the kick as punishment, thinking they are being reprimanded for moving forward.

If the horse stops suddenly, you may be taken by surprise and momentarily stop kicking — which makes him think he did the right thing. He's rewarded for stopping; you stopped kicking. You have inadvertently taught him that a kick means stop! Then your natural reaction is to start kicking again in an attempt to make him move forward. He is again confused. He may tense up and just stand or he may try a different movement to get away from the kicking — he could back up, go sideways, or start bucking. If he does go forward, he will still be confused unless you immediately stop kicking. It will take several patient lessons to reestablish communication; avoid this mistake and *don't kick*.

Gradual Steps to First Trotting

The goal is to teach your horse to move into a trot at your cue. This is best accomplished by making every effort to help him understand what you are asking, and the way to do that is to expand on what he already knows. Walk him around the arena, then ask him to speed up the walk by applying light leg pressure. If you used clucking or kissing noises in earlier training, this cue can be helpful now. Your horse will understand that the leg pressure with clucking means to move forward and respond with more energy in his walk.

Even if it lasts only a stride or two, his moving more energetically in response to your cue should be rewarded. By your release of leg pressure and no more clucking, the horse knows he did the right thing. After he travels a bit farther, ask again — and again reward him if he responds by moving faster. Soon he will realize that he should walk faster when these cues are given and eventually will break into a trot after you have pushed him faster. When he does, immediately reward him with release of leg pressure.

Let him trot a moment, then allow him to drop back to a walk. If he doesn't slow down on his own, give a slight check with the reins and shift your weight back by sitting deeper into the saddle. If he trots exuberantly, keep control with a light feel on the bit (and halter, if needed) so he won't go on to a canter. Don't overdo the lessons. Work at the trot for short periods at first, as this gait can be physically taxing on a young horse until he learns to carry a rider. Spend a few days working on transitions from walk to trot and back again.

First Cantering Lessons

Wait to teach cantering until the horse is well trained at the walk and trot. Young horses get emotionally carried away at the canter, as this is the

---

USE BODY LANGUAGE

Use body language — that is, rhythm — with seat and legs to help cue your horse as to which gait he should be in. At first you may have to exaggerate it until he gets the idea. Soon, though, he will find that your rhythm is the most comfortable one to follow and will adjust his gait accordingly — he'll walk when your body rhythm says walk and trot when the action in your seat and legs pushes more forward and says trot. To slow him, slow your body action. To cue him to speed up, your body becomes more active as you push with your seat bones. Soon your horse will be able to "read" your signals through body cues and won't need much rein pressure.

---

gait they use when frolicking and playing. If you try it before your horse is mentally ready, he'll want to run faster or buck; his focus certainly won't be on you. Save this phase of your horse's training until he's farther along and more cooperative and you have more control. This will come after a number of longer rides out in the open. Once you have developed good communication and control at the walk and trot, you can think about the canter. This gait is usually not attempted in early mounted lessons and is discussed in detail later (see chapter 14).

Nip Problems in the Bud

Correct any problems as they occur, before they become bad habits. If your horse tries to travel too fast or won't stand still for mounting, now is the time to correct it. Most bad habits start because a horse is allowed to do something, instead of being corrected the first few times it happens.

The Horse Won't Stand Still for Mounting

This is a common habit and usually starts because the trainer is not consistent about making the horse stand each time he is mounted. This can become a dangerous habit if the horse takes off when the rider is only halfway onto his back. It often starts subtly: The horse begins moving as the rider gets into the saddle and the rider doesn't stop him. Soon the horse is moving when the rider barely has a foot in the stirrup.

Don't let this become a habit. Always make your horse stand while you mount, settle into the saddle, and put your right foot in the off stirrup. If he moves, make him stand. Sit there awhile, if necessary. Let him know he is not to move until you give the cue. If he has a problem with this, repeat earlier lessons — make him stand as you prepare to mount, while you have weight on one stirrup, and after you're on his back.

MOUNTING TIP

If the horse won't stand still as you mount, make sure you are not inadvertently causing him to move off by bumping him with your leg as it goes over his back or by poking his side with the toe of your boot when you get on.

Enlist the Aid of a Helper

If your horse still won't stand still, go back to the very beginning and have a helper hold him while you get on and off many times at the start of a ride. If the horse is not allowed to move out, he will accept the fact that he must stand still. Then you can progress to making him stand without your assistant. You don't want your horse to get into the habit of being held when you get on, so use the helper only temporarily; keep working on the problem until you resolve it.

Practice Sessions at the End of a Long Ride

Some overeager horses are more willing to stand still after they have become a little tired. Practice mounting and dismounting after a ride, when your horse is tired or more relaxed. He'll quickly realize there's no point in trying to move off; he's already had his ride and wants to stay home now. A few calm practices at the ends of rides will help him get in the habit of standing still when asked.

The Horse That Tries to Go Too Fast

Some young horses are lazy, and your challenge is to get them to move briskly. Others are overeager and go too fast: If walking, they want to be trotting; if trotting, they want to gallop. Your goal with the eager beaver is to gain control until your horse has learned to adjust his speed to your wishes.

PROCEED AT THE HORSE'S PACE

Work on control of slower gaits before you try anything faster. Your horse should be well under control at a walk before you try a trot, and he should be able to perform maneuvers at a trot before you try a gallop. If he can't handle something at the slower gait, he'll definitely have trouble at a faster gait; speed is where you'll get into problems. Do a lot of training at the walk and trot, and it will be easier to keep the horse under control at a gallop.

Use Side Reins

You don't want to make a horse hard-mouthed by constantly pulling on the mouth or even by continual give-and-take checking with the reins. One thing that works well on the overeager green horse is to snap side reins onto the halter, which is worn under the bridle. You will thus be holding a halter rein and a bit rein in each hand. You can hold the reins in such a way that part of the pressure comes on the halter when you check the horse. He already knows about halter pressure from when you taught him to lead (see pages 135–137). He can thus have the signal via the halter as well as the bit, so you won't have to use the bit strongly while he's still learning how to respond to it.

You want his mouth to remain light and responsive. To achieve this, never pull hard on it. (The exception is a safety risk if your horse is startled into bolting or bucking, in which case you must pull his head around sharply to the side. If you have halter reins, much of this pull can be on the halter, and you will not hurt his mouth.)

Do More Pen Work

Another solution for the overeager horse is to work him more in the pen until you have control over his gaits. When he starts to get tired or bored, he will slow down. There is not much to see in a pen, not much incentive to go fast. In contrast, a lazy horse will be easier to teach to walk faster if you take him out of the pen and go on long rides: He will be interested in going someplace and seeing new things.

Tailor Training to the Horse's Mood

Use good judgment and good timing so you can correct a problem in the easiest possible way, at the best possible time, using your horse's situation to advantage for lessons. Don't teach a stand-still lesson when he's fresh and exuberant; don't ask for an extended trot when he's tired. If your horse is tired, work on things that require standing still, such as mounting and having his feet picked up. When he's fresh and wants to go, work on things that require movement — gait transitions, trotting, changes of speed and direction. After he is more relaxed and has the edge off his exuberance, work on things that take less physical effort and more concentration; he won't be fighting you to go, go, go. Time your lessons for when your pupil will be most receptive; he'll be more likely to respond properly and less apt to fight or pick up bad habits by resisting.

The Skittish Horse That Tends to Buck

Some horses are just plain nervous, and it doesn't take much to startle them into bucking. If your horse is ready to explode with extra energy each time you ride, longe or pony him before you get on (see pages 149 and 158). Taking the edge off before each mounted lesson will help settle him down. This is far better than having to fight him to keep him calm. A "cold-backed" horse that bucks when first mounted or one that humps his back and is tense when saddled may also benefit from warm-up exercise. If the horse tends to buck, be especially sure the saddle fits; he may be reacting to discomfort or pain. (See page 172.)

Tie Halter Rope to Saddle Horn or Use Overcheck

If a horse thinks about bucking, keep tying the halter rope to the saddle horn for as long as needed during early training sessions. This technique keeps his head up so he can't get it down enough to buck hard. You may have to do this for the first week or even the first month; it all depends on your horse. If you prefer, you can use the overcheck, from the bitting harness, instead of tying the halter rope. Whatever method you choose, stick with it until you're sure your horse won't try to buck anymore.

The Baby-Sitter Horse

When making first rides across country with a nervous, skittish horse that explodes when he gets upset, have another horse and rider accompany you. This will give your horse more security. He'll focus on the other horse and will probably be able to keep his cool in situations that might otherwise cause him to bolt or buck.

The Overly Exuberant Horse

Some horses are energetic and constantly want to be on the go. They're not interested in standing still or relaxing. If this describes your horse, using side reins (see page 262) can help if he is difficult to control. Even if he responds well to control, he may be so full of energy that you must go several miles before he'll settle down for lessons. Longeing or ponying before a ride can help, but once you are started in training, you may have to travel miles and miles across country before you ask the horse to concentrate on further lessons.

Riding miles at a brisk walk will help settle him. Once he gets fit, so his young legs and joints aren't easily injured, ride at a trot. Trotting a mile or so will relax him to where he will start paying attention to you rather than charging ahead. Galloping is counterproductive; a green horse doesn't know how to handle himself with a rider. At a gallop he would become *more* exuberant; it would be more challenging to teach him to control his energy and rate of speed. You'd be pulling on him too much. Use a walk and trot to settle him, then work on gait transitions and exercises to keep his mind on his work and on you. Make productive training experiences out of his desire to be on the move.

Ride Daily

The enthusiastic, energetic horse will be more of a handful if you skip a training day. This type of horse often progresses best if he has no time off from lessons until he becomes more settled. He will accept the routine more quickly if you ride him every day. A young horse that tends to be silly or skittish will be even more so if he has much time off. As the old-timers would say, this type of horse "can't stand prosperity." The easy life is not for him; he'll do much better if he works regularly.

TEACH PATIENCE

If a horse anticipates — he wants to move before you are ready, for example, or makes gait transitions on his own — make a habit of having him wait for your cue. Even if you were going to ask for a faster gait, if he anticipates your command and moves to a trot before your cue, slow him to a walk, then ask him to trot. He must learn to wait for your direction, or you won't be the one in control.

10

FIRST RIDES
IN THE OPEN

When you take your horse outside the corral, do all you can to make sure those rides go well. A good experience sets the stage for future training, making it easier; a bad one will make future lessons more challenging. The horse will encounter many "firsts" — a first big, scary-looking rock, a first hill to climb, a first deer bounding out of the brush, a first bird flying up in front of him. How you help him get through these experiences will determine whether he learns to relax about the outside world or develops phobias that are difficult to resolve.

Safety Precautions

If you've done the proper ground work and have trained him consistently, by now your horse should be accustomed to many things, but having someone on his back while he travels will be a new experience. He is used to your encouragement and control while working on the ground but is not yet secure with "on-board" control when something unexpected happens. There may be situations he can't handle, spooky or confusing things, and he may react explosively out of fear.

Though you'll try to be in tune with his mood and in control of his actions, you can't always predict what will happen. Use safety measures to ensure that a spook or a buck doesn't turn into a wreck. Try to prevent getting bucked off during an early training ride — for your own safety and to avoid a bad experience for the horse.

Tie Halter Rope to Saddle Horn

If you are starting the green horse in a Western saddle — as many trainers do, even if they plan to ride him English — a simple precaution is to tie the halter rope to the saddle horn (see page 263). This keeps him from getting his head down far enough to buck hard. Use a rope of proper length to tie to the horn. It should leave him enough slack for head and neck movement for ordinary travel at the walk, with his head level with his withers, but not allow him to get his head much lower.

You don't want it too tight or it will inhibit his head as he makes balancing movements or will cause him problems when going up or downhill or through a gully. Too much restraint of his head can frighten him if he gets in a bad situation and will cause him to rear. You want his head to be comfortable during normal actions, but you also want to halt his ability to put his head between his knees. An overcheck on the bridle can serve the same purpose, but it will put pressure on the bit rather than the halter if he puts his head down.

Use Side Reins

Because your greatest help in keeping a horse from bucking is being able to pull his head around to the side, you may want reins attached to the halter as well as to the bit. The green horse is not accustomed to having his head pulled around by the bit with a rider; he may know the fundamentals of turning, from driving and other ground work, but still needs to learn about being cued from his back. If he panics and starts bucking, you don't want to hurt his mouth by yanking his head around with the bit. Having

Side reins snapped to the halter give the rider added control.

a rein snapped to each side of the halter gives you the leverage you need, as it spreads the pressure between bit and halter. (See page 262 for more on this topic.)

Reins on the halter can also be an advantage if your horse wants to go too fast. You can check him partially with the halter as well as with the bit, and not make him hard-mouthed by constant use of the bit to slow him down. Reins on the halter give a good measure of added control that's comfortable to the horse.

If you are not used to having four reins, make sure you have the proper tension or slack. Hold two reins in each hand — the halter rein and the bit rein on that side — separated by whichever fingers feel most comfortable and enable you to put slightly more pressure on either rein when necessary. This way, you can check him lightly with the bit alone, or use both together, or use the halter rein more strongly, depending on the situation.

Snug Cinch

Before you go, make sure the cinch is snug. This is especially important when you start riding outside the pen, where unexpected things might spook your horse (see Shying on page 279).

Enlist a Helper

Your first mounted ventures away from home should be careful, well-planned sessions. If your horse is quite insecure, have a helper walking with you; he should carry a lead that can be snapped to your horse's halter, under the bridle, in case of problems. A frightened young horse may disregard the rider if he hasn't been ridden much and encounters a situation he cannot cope with.

Should this happen, the ground person can snap on the lead rope and help keep the horse calm or get him safely past the obstacle, preventing a serious problem or a bad experience. You can acquaint the green horse with new things in easy stages that he can handle emotionally and prevent bad experiences that might make him reluctant to leave home. When the youngster is farther along in training, you won't need a ground assistant. At that point, a mounted companion on a calm horse can be a help to give the youngster confidence when encountering new and scary situations.

The Companion Horse

If the horse is nervous or skittish, your first rides out will be less risky if he has a companion. A baby-sitter horse can keep the insecure youngster calmer (see page 263). You may want to have a buddy accompany you until you and the green horse are more at ease with these excursions. But don't become dependent on a baby-sitter. You don't want your horse thinking that he can't leave home without another horse. Make some rides alone as soon as you can. Even if he needs a baby-sitter to help him overcome new obstacles, do some riding alone on paths he has already seen; this will give him confidence. He must ultimately depend on you, not the other horse, for guidance. When you ride with another horse, get your horse used to being in front, behind, and on either side, so that the whereabouts of his buddy won't matter. Don't let him think he can do things only a certain way.

Everything Is New and Different

Things you take for granted when riding an experienced horse may send your new pupil into orbit. Be prepared for this. You must be able to stay in control if your horse suddenly jumps and to ease him through each scary situation with reassurance and praise. Stay alert and don't be caught off guard or he may jump out from under you. Don't be angry if he suddenly leaps or bucks because something scares him; try to understand how your horse feels and why he does what he does.

PREPARE THE WAY

Your horse has never heard the sound of a tree branch on your hat; if you ride under trees and he hears a strange noise, he may think a cougar is attacking. The sound of sagebrush hitting stirrup bows, the feel of tall grass tickling his belly, even the sound of a sudden sneeze — any of these may cause him to explode in alarm. Expose him to some of these things ahead of time. While still in his safe home environment, make rubbing sounds on your straw hat and brush-crackling noises near his stirrups. Sneeze a few times. Your goal is to get him used to some of the things that might startle him in order to ensure an uneventful outing.

Travel along the edge of the brush at first, until you know how your horse will react to branches touching him.

Riding in Brush

If there are bushes or sagebrush where you'll be riding, the first time your stirrup drags through a bush, your horse may feel jumpy because he doesn't know what the noise is. Even if he has encountered bushes in his home pasture, this is different. He knows what the sound and feel of bushes against his own body is like, but he's never heard a bush hitting your leg or the stirrup.

The first time you ride through a patch of sagebrush where it will be hitting your stirrups, get him used to it gradually. Don't head straight into the brush; he might panic and start bucking when he hears it on both sides and crash deeper into the brush — and now the two of you are getting into an uncontrollable situation. Go alongside the brush for a moment, so if your horse gets scared, you can head away from it again and get out of it. Go in gradually.

Use the same cautious approach riding through trees. Go along the edge of the trees at first, until you know how the touch and sound of branches is going to affect your horse. If you head right into the trees and he blows up, you'll be dodging branches as you try to control him; if he starts bucking, he won't stay on the trail.

Allow Him Some Leeway

If he panics at something, try to see how your horse looks at it. His reaction is first to run and to ask questions later — this is the flight-or-fight response (see chapter 1). Danger he can run from is always less scary for him than danger he can't escape. If he spooks and tries to take off and you hold him right there, he might fight very hard or try to buck you off so he can run away from what is frightening him. But if you allow him to take a few steps away from perceived danger, he may not become panicky and probably will be more apt to listen to your reassurance that it's just a tree stump, or a grouse sneaking through the grass, not a bear.

Go with him a little when he's scared. Let him take a few strides away from the frightening object before you make him halt, and don't create more panic in him by using tight restraint. If you try to make him stay there, disregarding how he thinks or feels, he will be even more scared and may buck you off and then run. Try to understand your horse and to think the way he does.

Short and Sweet

If your horse is nervous about his first ride out of the pen, keep the ride short. Be in tune with his mind and judge how long the ride should be by how he's handling it. It's always better to ease into the big world gradually to avoid a panic situation you can't handle.

The object of training is to keep your horse relaxed and thinking, to acquaint him with new things in a nonconfrontational way. If he panics and loses his concentration, you've defeated the purpose of the ride. A bad experience on his first ride in the open is hard to overcome. You get only one chance to make a good first impression on his first trip out; don't ruin it by overdoing it and risking a problem.

If the horse is nervous, you may be nervous, too, wondering if he is ready for this. It's best to make a very short trip and to end it on a positive note rather than trying to do too much. If everything is going well, quit for the day, even if you've only been riding outside the pen for 15 minutes. If you can make a good, short ride and get home again with no problems, you've accomplished something and made an admirable first step. A very short, successful first outing will give your horse confidence and reinforce his trust in you. You'll both be more relaxed about the second one.

Build on Every Success

The next rides out will be easier for both of you because you made sure the first one would go well. The horse is less confused and more prepared to pay attention to you. He feels more stable under you. He's had time for the first ride to soak in. He's ready for the next one — he's interested in seeing new things. Now you can build on his success.

Keep His Mind Occupied

A horse is less likely to get into trouble on first rides if you keep his attention. If his mind is on you and what he's doing, he won't have the opportunity to worry about a buddy at home, for example, or about something in the distance that might be scary. He's also less likely to daydream and be startled inadvertently. When startled, the inexperienced horse usually jumps or bucks without thinking — and you may find yourself suddenly on the ground. If your horse is concentrating on something, there's a lot less chance for an unanticipated problem.

On the green horse, always stay in communication with him. Always keep his attention focused on you and his mind busy. You want him alert and interested. Maneuver him around rocks and bushes, ride off the trail into interesting terrain, make him go left and right or fluctuate his speed — all to keep his mind challenged and interested and focused on you. If you just head up a road or trail without asking your horse to do anything, he'll become inattentive and may find something to spook at or decide suddenly to buck. An idle mind is sometimes asking for trouble!

FOCUS, FOCUS, FOCUS

Do your best to keep the horse's undivided attention. When not doing exercises for maneuverability, keep him focused on walking straight lines; he should never be "on his own." You are always directing him. Whether traveling across country or in an arena, you are the pilot. Pick an imaginary spot ahead of you and try to have your horse walk as straight as possible toward it. If you are continually directing him with constant communication, he'll stay focused and become more responsive to your cues.

Keeping a Nervous Horse Calm

Sometimes a youngster will be eager and interested and not a bit worried — until you turn for home. Then he suddenly starts thinking about his buddies back in the barn or pasture and wants to hurry. As you try to keep him down to a controlled walk, he may become nervous at being restrained. On early rides, this can be the difficult part — getting safely home on the horse who wishes he were home *right now*.

In this situation, keep talking to your horse. Sing and hum to him. Hearing your voice helps him keep his focus and pay more attention to you. Keep yourself relaxed, even with constant control on the reins; relay your relaxation through the bit so your horse will settle down and relax too. Singing or humming helps you both relax.

Getting through a Tight Spot

If your horse is unsure about something along the trail, your attitude will make a big difference. Because he has confidence in you from earlier handling, he will stay calmer and listen to you, especially if you are merely steadying and reassuring him as you go past or through the obstacle rather than trying to force him. You are his security if he trusts you. If you try to force him, however, he loses confidence in you; your actions merely add to his fears.

HOME, SWEET HOME

Some horses are so interested in the big wide world that they want to keep exploring and seeing new things and are not at all interested in coming home until the novelty wears off after several rides. Others are more insecure and think constantly about home. The length of the ride and how much you expose a horse to at first will depend on the individual horse and his security level. If you take your horse too far, you may find it hard to get back home without a problem.

Overcoming Fear of Strange Obstacles

Most horses are a little afraid of new things; some are very nervous about strange obstacles. A bold young horse will quickly get over his hesitation, but a nervous or timid one may require a lot of patient training to overcome his fears.

Spooking at Rocks and Bushes

A horse that's never been out of a small pasture may be afraid of things he encounters on first rides out. He may hesitate to travel through bushes and timber. He may be afraid of large rocks and low shrubbery that tickles his legs or brushes his belly. Get him used to things gradually. You never, ever want a horse so frightened that he gets out of control. New experiences, such as traveling through bushes or past large rocks, are often more easily accomplished if he has an experienced horse to follow.

> **SAFETY TIP**
>
> A lead horse should go slowly when traveling through brush or trees followed by an inexperienced horse. If this isn't done, the young horse may be so fearful of being left behind that he rushes through the trees with no thought to his or your safety, increasing the possibility of injury.

Reluctance to Step Over Logs

A horse that has never seen a log across the trail may refuse to go over it. To him it represents an impossible barrier. This situation is avoided if you've practiced at home, first leading, then riding him over poles on the ground (see page 146). After he walks willingly over them, set them up higher to simulate a large log on the trail. If you haven't had an opportunity to prepare him this way, and his first experience with this type of obstacle is out on the trail, don't get into an argument.

If your horse is adamant in his refusal to step over it, dismount and lead him. Usually a horse will lead over or through an obstacle because he has more experience being led and trusts your guidance. Tap your horse's hindquarters with a stick if he balks. Just make sure to be out of the way in case he decides to jump it.

If you are riding with another person, stay on your horse and let the experienced horse go first. Usually the green horse won't want to be left behind. Though he hesitates and fusses, he will usually go over the obstacle so he can catch up with the other horse. Be prepared, however, in case he leaps over the obstacle rather than stepping calmly. If stepping over a log is a real problem for him, take some time later at home, with posts on the ground, to help your horse overcome this mental block.

Balking at Water Crossings and Bogs

Your horse must learn to cross rain puddles, streams, and ditches if you ride across country or on back-country roads and trails. With the inexperienced horse, ground work with water crossings is helpful (see page 147). If you have already led the horse through puddles or little streams, he won't be so fearful of these obstacles when you start riding. You can cross bigger streams, and he'll learn to handle them, especially if you are riding with other horses and he can see them crossing calmly. He usually will follow — perhaps with a bit of urging.

Tips for Water Crossings

One way to get the green horse over his fear of water is to find a shallow place in a stream or a puddle-covered dirt road after a rain and ride around in the shallow water. There's no place to go to avoid it. After going up and down the stream or around and around in the large puddle, the youngster learns that this is nothing to fear.

If your horse balks at a shallow stream, ride him across; don't get off and lead him, even if there are rocks to give you dry footing. If he is spooky about the water, he may jump onto the same rock with you or knock you down in his lunge across. It's safer to stay on him. If you have to lead a horse across a stream or up a steep bank, be prepared for him to jump or lunge.

If he knows how to pony — that is, to lead from another horse — one way to get a reluctant youngster to cross a stream the first time is to have another rider lead him as you ride the horse or as you follow along behind on foot with a switch. There may be times when you are confronted with a situation in which he must cross, and this method will generally do the trick. Once across, go back and forth again a few times so your horse realizes that he can do it without problems.

With swift-flowing water, ride across at an angle, facing slightly upstream. The inexperi-

> ### SAFETY TIP
>
> The green horse may rush a stream crossing at first or may try to jump it if it's narrow. Be prepared for this so you won't be unseated by a sudden lunge. Even a small ditch or stream that could easily be stepped over may be broad-jumped by the nervous green horse that has not yet gained confidence about crossing water. (See Narrow Stream or Ditch on page 276 for more on this topic.)

enced horse may be confused by the swift water and move downstream with the current. If you're facing upstream, it's easier to keep him traveling across — and you won't become dizzy if you face upstream and keep looking at the opposite bank. Don't watch the water on the downstream side.

DON'T LET HIM WALLOW

If a horse starts pawing in a stream or pond, as when you are acquainting him with water and letting him stand in it, quickly ask him to move forward. Some horses want to lie down and roll in water, especially if they've been sweating and are hot and itchy.

Lessons for Water Crossings

If there's a nearby stream for practice sessions, take some time to gradually acquaint your horse with crossings so his first experience won't be out on the trail, when crossing it is essential. Having the time to convince him it's his idea makes future crossings easier; he won't associate water with struggle and confrontation, as he would had he been forced. Lessons for water crossing will be more pleasant for both of you if you have already ridden the young horse enough to have him responsive to your signals and direction.

Ride along the edge of a shallow stream at a distance your horse feels comfortable with. If he's not being forced into something, he'll be more relaxed about it. If he's nervous, spend some time just being by the water or going back and forth, or coming and going to the stream. Keep riding back to the water. If he'll stop briefly by it, ride away again as a release and reward. Don't ask him to cross yet. Soon he'll discover that his reward for going to the water and staying calm is being ridden away again, and he'll become more relaxed about standing by the water. If he can drink from it, this too will relieve some anxiety. He may even decide to walk into the water on his own.

When you ask your horse to cross, encourage forward movement with your legs, but not so much that he becomes defensive. The instant he takes a step forward toward the water, release the leg pressure. If he gets nervous, ride him away from the creek and come back to the same spot again. Your goal is to keep getting one step closer; release all pressure when he does.

PLAN FOR SUCCESS

For water lessons, choose a shallow stream with ground-level entry and exit points rather than steep banks; good footing, with no big rocks or boggy banks; and enough width that the horse will not try to jump it.

Let him stand and relax and check it out, then ride away again before he tries to leave. Don't increase pressure if he's trying; talk reassuringly to him to make him feel safe and comfortable by the creek. It may take several tries, but your horse will eventually attempt to enter the water. Let him move forward without urging. If he goes in, let him stand in it for a moment if he wants, or go on across. Once he gets out, ride a few feet away, then come back to the creek. If you follow these positive steps, he will gain confidence and be more willing to do the lesson again.

Narrow Stream or Ditch

Most inexperienced horses will try to leap a narrow waterway rather than walk through it or step calmly over it. Try to avoid narrow streams and ditches until you've ridden your horse more, have better control over how he handles himself, and have crossed wider streams, during which he learned it's okay to get his feet wet. Leaping a stream or ditch can be hazardous if the footing is bad or if the young horse decides to turn the leap into a bucking spree on landing.

If you must cross a narrow stream, select the most level spot with the safest footing and approach it parallel rather than straight on. This makes it a wider obstacle and you'll have better luck at making your horse walk through it rather than leap it. If he gathers himself for a leap, steer him away from the water and approach again, encouraging him to go slowly and walk through it. If he does leap it, turn him around and go back over it. Cross several times in a more controlled manner until he crosses the water properly.

Balking at Dry Ditches and Gullies

The same methods for teaching him to cross water generally work for a horse that balks at a dry ditch or gully. Choose the easiest place to cross,

CROSSING WATER

A. *Let the horse check out the stream but keep him facing it.*

B. *Stay facing the stream until he tentatively puts in a foot.*

C. *When he thinks he can do it, encourage him across.*

where the bank is least steep and deep. Ride him across rather than lead him, in case he leaps, so he won't knock you down. If he's very reluctant, it helps to have another rider go first. Not wanting to be left behind, your horse will usually try to get across. If he still won't cross, have the other rider lead him while you encourage your horse from behind.

Once a youngster realizes he can do it, it's usually easier for him the next time — especially if you were patient and did not make him more fearful by using an angry voice or punishment. He can understand a light touch on his hindquarters with a willow, especially if he is being stubborn rather than truly afraid, but inflicting pain with a sharp spanking is never a good idea.

Rushing Gullies

When crossing a deep ditch or gully, the green horse often picks up speed going down into it, then lunges up the other side. To prevent this, halt him before you start down, then make him go down in a controlled manner. Halt again at the bottom if he wants to hurry up the other side. Turn him parallel to the steep bank, make him wait until he relaxes, then ask him to turn and walk quietly up the bank.

If he is following another horse, he may think he's being left behind and panic. Unless he is too insecure to try the gully on his own, have your horse go first. If he's simply too fearful, have the other horse go first, but slowly, and have him wait for you on the other side of the gully so your horse can see he's not being left behind.

CROSSING A BRIDGE

The green horse may hesitate to cross a bridge, especially if he's never seen one or the footing seems insecure or noisy. If he balks, either lead him across the first time or follow another horse. Feel your way with each situation; he may just need a little encouragement. Choose a safe method that is best for your horse's ability and confidence level. (See page 147 for more about crossing a bridge.)

Wildlife and Dogs

The unanticipated often poses the greatest risk when riding a green horse. A playful dog, a bird that flies too near, a deer bounding out from a bush — all will likely alarm an inexperienced horse. If he's not used to wildlife, its sudden appearance can really frighten him. Until a horse gets used to darting wildlife, you must depend on keeping control while reassuring him. Once your horse starts listening to you so you can help ease him through his fears, it will get easier.

It helps to ride with another person if you know you may encounter scary wildlife or barking dogs. Your horse will be less likely to try to whirl and bolt home if there is a stable horse to give him confidence. He sees that the other horse is not afraid, and he'll also be reluctant to leave the other horse. A calm, dependable horse can be a good anchor for the insecure one, helping him hold his ground and minimizing his desperate desire to flee.

FACING HIS FEAR

A horse will usually stay calmer if he can see what's spooking him, especially a moving object like a bounding deer or a dog coming up behind him. If you try to ride away from it, the horse will more likely try to run, and a dog may be inclined to chase you. Face the dog or ride toward it. This often prompts the dog to leave. Any time a frightening object is coming in your direction — a barking dog, a curious kitten, a noisy lawn mower — it's better to face it than have it coming from behind. If a plastic bag is blowing toward you, keep your horse facing it; circle around it. If you must ride away from a moving object, ride at an angle or zigzag, so the horse can keep it in his sight until he's safely away from it.

Shying

Some horses spook at things: They snort and leap sideways or whirl 180 degrees to bolt. The young horse is often spooky because he's inexperienced; his instinct is to flee from anything he perceives as dangerous. He may never have encountered a bird that flies up in his path or a plastic bag billowing and blowing beside the trail.

Some shying is to be expected when you take a green horse out on the trail, but your goal is to develop a calm, easygoing horse that will stay relaxed in the face of new encounters. Develop communication and trust. Use various lessons to acquaint him with new things gradually.

Teach him to accept these matter-of-factly or with curiosity instead of panic and leap-away avoidance. This is what training is — gradually getting your horse accustomed to different things so he can accept them without reverting to instinctive evasion.

Your horse may still be afraid of something or shy occasionally when he's feeling exuberant, with energy to burn, but if you've trained him through the basic scary stuff, he will be less likely to develop a habit of violent shying that may unnerve or even unseat an unprepared rider.

Sometimes It's Eyesight

Horses see things differently from the way we do. They have a broader field of vision and see a different picture with each eye. They can't focus as readily on close objects, however, and their straight-ahead view is more limited. In the wild, this type of vision makes it difficult for a predator to sneak up on the horse. He can see behind himself with just a slight turn of his head; he can see movement behind and on both sides and be prepared to flee from danger instantly.

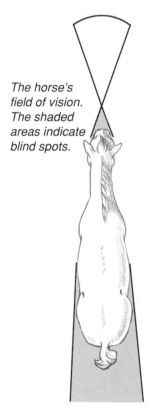

The horse's field of vision. The shaded areas indicate blind spots.

Although the horse has a wide view with his head up, there's a blind spot behind his body and beneath his nose. He can't see anything closer than about 3 feet in front of his forefeet unless he tips his head. With his head raised, a small log just in front of him may be difficult to focus on until he lowers his head to examine it closely. Anything that appears suddenly along the path beside him or in front of him will usually cause him to shy because he doesn't see it clearly.

He may react strongly to movement at the outer edges of his vision because his instinct tells him it could be a predator. Many horses spook at anything unusual along the path until they have a chance to examine it and satisfy themselves that it is harmless. Some horses have poor eyesight and are unable to decipher visual stimuli around them, and these may spook at familiar objects. In such cases, all you can do is strive for understanding and patience.

It's futile to succumb to frustration and anger. Being aware that the horse is going to spook at every big gray rock and preparing yourself for that is the best tactic. Simply reassure him and try to dispel his fears. If you stay calm, your horse will be calmer. If you ride nervously, anticipating a spook, the horse feels insecure and probably *will* spook.

Think Ahead and Be Prepared

When riding a green horse on first trips outside, remember that there are some things he hasn't seen before and he may be alarmed by objects an experienced horse takes in stride. It's up to you to recognize a shy potential before it happens, ease your horse's fear, and bolster his confidence. Give him time to check things out.

Learning how to handle a shying horse is an important lesson for you. You may never totally defuse his reactions of panic, especially if he has poor eyesight or a nervous nature, but you can tone them down and make the shying easier to deal with.

Many horses snort and spook at strange things, but if you give them a chance to look them over, they'll realize there's no danger and relax. The best you can do with a skittish horse is to exercise simple patience. Instead of trying to force him past the scary object, let your horse stand and snort and check it out at his own pace. He'll relax more quickly if you don't try to make him confront it at close range. When he's calm, he may want to go closer and smell it. When he finds it's no big deal, he'll continue on his way.

Gain His Trust

If you developed a good working relationship during early ground work, the horse looks to you for support. You must further that trust and communication while riding on his back. Reassure him in his panic; don't increase his fear by punishing him when he shies at something. When confronted with something scary, horses don't stop to think what it might be. To reassure a horse, you must relay the message that this "bogey" is harmless. Fighting him, punishing him for spooking, or trying to force him closer to a scary object will make him feel the scary object is truly bad, reinforcing his determination not to get any closer. A spooky horse can be made more skittish by an inconsiderate rider. Your goal, as always, is to instill trust.

You can't prepare a horse for every situation he might encounter, but you can increase his confidence level and develop his trust in you. If he listens to you, you can help him work through his fears. The dangerous horse is one who panics and pays no attention to his rider. If he has learned to trust you, he learns that spooking and bolting are not necessary to "save" himself in a frightening situation; he can rely on your calm and your direction.

Think Like a Horse

In a herd situation, horses rely on one another to warn of potential danger. The first one to see something suspicious will come to attention, body tensed, ready to flee. He focuses on the strange sight — head raised, ears pricked, nostrils flaring — and snorts. This alerts the rest of the herd and they, too, become ready to flee. But as soon as a member of the herd decides there's nothing to fear, he relaxes and drops his head to continue grazing and the rest of the herd takes the cue — they too relax and go about their business.

Be the Leader

The rider has the role of herd leader. At first sight of a spooky situation, you generally tense and focus in that direction, just as the horse does. This tension is conveyed to the horse through your seat, legs, and increased rein pressure. You must relax and reassure the horse that there's nothing to fear. If you stay tense, he will, too. After several episodes in which you relax, he'll usually start to believe you, and you can more readily encourage him to calm down in the face of something alarming. Your reaction — positive or negative — can greatly minimize or accentuate shying.

Riding the Shy

If you realize your horse is going to shy, don't stiffen your seat or shorten your reins to try to hold him steady or to prevent the shy with sheer force. You can't really stop an agile horse from shying. When a horse shies, do your best to stay in balance and keep light hold of the reins. You don't want to inadvertently punish a horse by accidentally jerking the reins.

With the green horse, however, you don't want the shy to turn into a whirl-around bolt for home. Keep an easy seat and calm rein, but be ready to halt a whirl. Any attempt to actively inhibit a shy may just induce your horse into more violent shying in his claustrophobic efforts to get away from what he perceives to be imminent danger. Your tension and apprehension (and possible punishment with bit and legs) are proof to the horse that this is a perilous situation.

Your ability to relax and ride through a shy without clamping your legs and hands, yet still keeping him under control, can cut down the incidence and intensity of the shying. And after-the-shy anger or fear on your part will increase your horse's tension, making the situation more frightening in his mind. If you react harshly to his shying, this initiates a vicious cycle. His reaction is to shy more desperately. Outbursts from you after a shy will reinforce his idea that scary things are dangerous. Get yourself under control. Understand your horse's reactions and never punish for what is in fact an instinctual response. Remember, too, to speak to him reassuringly when the situation has passed, and praise him when he's calm.

Some horses tend always to be spooky. It is their nature — a way of expressing high spirits. As you train a youngster and gain better communication, you can often tone down these explosive moves and mold them into something that can be more easily handled. If he is high-strung or timid, his insecurities can be greatly diminished and his trust in you greatly increased by consistent patience as you work together. After a while, you can minimize his cause for fear and the kind of insecure behavior that makes him a challenge to ride.

HE'LL FOLLOW YOUR LEAD

When riding through a spooky situation, stay as relaxed as possible. Go with the horse on as loose a rein as you can, looking straight ahead rather than at the scary thing. If it's a moving object, like a dog coming at him, he will probably have to face it. But a stationary scary thing can often be calmly ridden past. Pick a distant object to ride toward and keep your focus ahead; your horse may find it easier to follow your focus, gaining confidence from you. If you don't make an issue of the spook, he'll learn not to.

A Little Quiet Time

Sometimes a nervous horse needs extra lessons on relaxation and patience so he won't get into the frustrating habit of always wanting to rush home to the security of his pen, stall, or pasture buddies. The way you handle him during early rides can make the difference in whether or not he develops the rush-home habit.

Some Time to Relax

During rides away from home, find a place you can let him stand awhile. If it's a hot day, pick a place in the shade (a place that's comfortable for him) and ask your horse to stand and relax. Whether you dismount or stay mounted will depend on the horse and the situation and what you both feel most comfortable with. Let him know he can be comfortable out there and doesn't need to rush home. Make him stand quietly. Talk to him and rub him until he relaxes. If he is too nervous for that, it may take several sessions on several rides before you find a way to take his mind off his worry.

You may have to resort to letting your horse nibble a little grass somewhere out there, or have a rider along on a familiar horse so the young one will relax and stand, knowing he has the security of his friend. Whatever it takes, it pays to teach him early on that he doesn't always have to be traveling and working with a rider on his back. He doesn't have to rush home to get relief from being ridden. If the only time his work ends is when he's taken home, he becomes too eager to get there. He needs to know he can sometimes just stand and relax. The earlier you can impress this on him, the better, so he doesn't get the idea that he must get home in order to be comfortable.

Riding Along Roads

One of the most potentially hazardous situations when starting a green horse is riding along a road. Until your horse gets used to traffic, his unpredictable reactions can be dangerous. He may try to bolt or perhaps whirl into an oncoming car. How you handle this phase of training will determine whether he becomes safe to ride along a road or develops a permanent phobia.

If a frightening vehicle comes up behind you, turn your horse to face it, and keep him facing it as it goes by.

Get the Horse Accustomed to Vehicles

Acquaint your horse with cars and motor noises in safe and controlled conditions before you take him on a road. If you have a quiet lane, a driveway, or a back road close to home, ask a friend to drive a car slowly past you and the horse. If your horse becomes frightened, have the friend stop the vehicle, and speak to the horse while you ride past, so the horse recognizes something familiar. Take as much time as he needs to be reassured that the vehicle is nothing to fear. Let your horse approach and sniff it and check it out on his own terms so he can see for himself that this monster is not threatening. If the horse is skittish while being ridden, lead him past the car during the first lessons.

Ride with a Helper on an Experienced Horse

If you must travel a road with traffic on a horse not yet acquainted with this, go with another rider the first few times. (You may have to ride along a road to get to a place where you can ride across country, for example.) When meeting a vehicle, halt both horses off the side of the road. The green horse will gain courage and confidence from the calm one and be less apt to try to bolt or whirl away from the car. After he is more accustomed to cars going by, he can follow the experienced horse along the side of the road, not having to stop and face every car that comes along.

Handling Panic Situations

Sometimes you'll meet a vehicle that is more noisy, larger, or more frightening than anything the horse is used to. In this instance, try to find a place where you can get clear off the road before the vehicle meets you. This will give your horse more space. If a truck or scary vehicle — even a bicycle — is coming up behind you, turn the horse so he can see it better. Ride along the edge of the road, as far off as possible, toward the vehicle, and keep your horse's head facing it as it passes.

Be Prepared

If there is time, find a safe place to get off the road. If you often travel a busy road, make mental notes of places you can get off in an emergency. Do not, however, try to beat a vehicle to a safe place if it means increasing your speed to get there. This will create a potentially more dangerous situation. At a faster speed, the horse is more likely to panic and bolt or to start bucking.

Dismount If Necessary

If you can see there is no room to get safely away from a frightening vehicle and you know your horse is likely to panic, get off and hold him. This may be safer, with a horse that's well halter-trained, than trying to stay on. If the horse is so alarmed that he might pull away from you while being held, he needs more basic training before you take him into this kind of situation again.

Know your horse. You may prefer to stay mounted. If you don't have really good ground control, you may be unsafe on the ground. If he pulls away from you, he may bolt through a fence or into traffic. When riding out, always have a way to keep control if you dismount. Ride with a halter under the bridle; you'll have better control with a halter rope than with bridle reins.

Advance Preparation Is Best

Part of good training is knowing what your horse needs to be acquainted with — getting him used to these things, step-by-step, ahead of time. There are always things you can't anticipate, however. This is why it's important to build good communication and trust, so your horse gets into the habit of listening to you and relying on your judgment. If he has learned early on to take his cues from you — remaining calm and relaxed because you are — there will be less risk of serious problems out on the trail. Lessons in relaxation, using your voice and neck rubbing as prompts, for example, can teach him to relax on cue and be more steady and stable when a trying situation comes along. (See page 164 for more on relaxation.)

11

THE WALK

Most horses have three gaits, the walk; the trot; and the gallop, or lope. (The canter is a collected version of the gallop.) Some horses have additional gaits, such as the running walk, the amble, and the rack, and the Paso breeds have distinctive gaits that are quite different from all of these.

Knowing how the horse moves makes it easier to understand how and when to use certain cues to persuade him to do what you ask. This knowledge enables a trainer to use informed judgment in what he asks of the horse; the good trainer asks him to do only what is physically within his ability in a given situation.

How the Horse Moves

The horse has strong muscles in the upper part of each leg, but below knee and hock there are no muscles. Muscles in the upper leg extend downward as long tendons. Muscles in front of the leg pull it forward. Muscles at the back of the forearm bend the knee.

The Horse Is Front Heavy

The horse's center of gravity is immediately behind his withers. His front end, with head and neck, is heavier than his hindquarters. About 60 percent of his weight is carried by his front legs. A horse that is standing squarely on all four feet cannot lift a front foot until he shifts some weight off his front end. He does this by raising his head, by crouching a little on his hind legs, or by moving a hind leg forward to take more weight.

When you start to lead a horse that is standing squarely, he raises his head before he moves forward, shifting the weight so he can lift a front foot. If you don't have any slack in the rope when he raises his head, he'll pull you back a couple inches as you pull on his lead rope. To move in any direction, the horse must first compensate for the fact that he is front heavy.

Gaits

Some horses, due to selective breeding and training, have different or additional gaits, such as a running walk; rack (one form of which is called a *singlefoot*); a slow gait or stepping pace, which is a highly collected rack; amble; and fox-trot, which is often described as walking in front and trotting behind. Paso breeds have their own gaits and leg actions. Horses that just walk, trot, and gallop are often referred to as *straight gaited,* as opposed to *gaited,* which describes horses that do four or five gaits.

Some horses pace instead of trot; in the pace, the legs move forward together on the same side instead of diagonally. When pacing, the horse has a rolling motion from side to side, with less support, balance, and traction on slippery footing. Standardbreds trot or pace — and a few can do both — as fast as other horses gallop.

Variations on the Walk

The *fox-trot* is a slow, broken trot, faster than an ordinary walk. The horse's head nods as it does in the walk, and he brings each hind foot to the ground an instant before the diagonal forefoot strikes the ground.

The *flat-footed walk* is a natural gait of the Tennessee Walking Horse. Feet land in the same sequence as in a normal walk, but because of the loose ("free") action of the horse, he can move quite rapidly.

The *running walk*, a four-beat gait between a normal walk and a rack, is the fast walk of the Tennessee Walker. The hind feet overreach and land in advance of the tracks left by the front feet, overstepping by as few as a couple of inches to 18 inches or more, giving the horse a gliding motion. This gait resembles a pace, but the hind foot comes to the ground before the front foot on that side.

The *rack*, or singlefoot, is also called a *broken amble*. The rack is an exaggerated fast walk with four distinct beats. Each foot comes to the

---

LISTEN

When riding a horse, you can tell by listening to the hoofbeats how his legs are moving in sequence and thus what gait he is in.

---

ground separately and at equal intervals. At first glance it looks like a pace, but it is a four-beat gait rather than a two-beat; the hind foot lands just before the front leg on the same side. It is similar to the running walk, but the feet are lifted higher and with more up-and-down motion. At the rack, the hind feet do not overstep the front feet as much and may even not overstep at all.

Training at the Walk

All early training is done at the walk. The green horse needs to learn to respond to the rider, rather than doing whatever he wants. The walk is the best gait for teaching control with stopping, starting, and turning. He has a more relaxed frame of mind at the walk and is more receptive to the influence of his rider.

At faster gaits a horse tends to get more excited; these are his natural expressions of play or of fleeing a predator. Until he learns control, he is less apt to pay attention to a rider when trotting or galloping. The walk, then, is the gait to concentrate on for all first lessons, progressing to a trot only when your horse is very comfortable with what you've been teaching him at the walk.

First lessons focus on control and on developing the walk. When you ask the horse to move out and walk, you want him to move briskly and purposefully and not plod aimlessly. From the beginning, you try to channel his movements directionally and also teach him to maintain an even speed. He must be consistent in his gait rather than erratically speeding and slowing. This, too, is a subtle lesson in control.

Teaching a Faster Walk

Some horses are naturally fast walkers, while others are lazy. Some take long strides; others move with short, choppy steps. The horse's conformation — that is, how his body parts are put together — largely determines how he picks up his feet and moves them. No matter how a horse walks, however, he can be taught to move more swiftly. To teach him to walk faster, you must understand how he moves at the walk and how to use your hands and legs to give the proper signals. If you have to push him all day to make him move faster, you'll feel you've gone more miles than the horse.

Your hands check the horse and keep him from breaking into a trot, which is his natural inclination when urged faster, and your legs encourage him to move more energetically. If you check him with the bit, he will respond by walking faster instead of breaking into a trot. The best results are attained when you can synchronize your bit cues with the leg cues and use your legs alternately to stimulate the proper hind leg of the horse at the proper time.

Head and Neck Action

You control the horse's head and neck with your hands and the action of his feet with your legs. He uses his head and neck for balance, just as we swing our arms for balance when walking and running. The horse raises and lowers his head rhythmically to lengthen or shorten his neck to keep in balance as he travels. At the walk and at the canter or gallop, the head bobs at each step. His head drops each time a front foot comes to the ground, then rises to shift his weight each time he prepares to pick up a front foot.

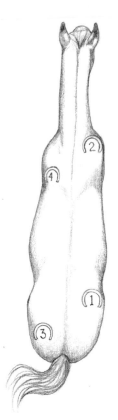

Leg Action

When a horse is walking, he moves each leg in a specific order, creating a four-beat gait. The sequence is right hind, right front, left hind, left front. As a hind foot comes forward, the front foot on that side prepares to take off and leaves the ground a split second before the hind foot lands, just out of the way of the approaching hind foot. As the speed of the walk increases, the horse picks up each foot more

Footfall order at the walk

swiftly. Because most of the horse's propelling power comes from his hind legs, with your legs you can encourage him to increase his stride length and walk faster by stimulating each hind leg at the proper time to make it push off more vigorously.

Using Your Legs

Your horse already knows he should move energetically in response to leg pressure, as this was one of the first things you taught him: how to move out or increase speed when you squeeze your legs against his sides (see page 251). How you use your legs, however, will make a difference in his ultimate response. The goal is longer strides, not merely faster ones. Short, fast, choppy strides will quickly tire a horse. When he pushes off with a longer swing of his leg, he covers the ground faster. If you stimulate him at the proper phase of his stride and not at random, you will get nice long strides.

As the horse's right leg and shoulder come forward, the rider squeezes with her left leg to stimulate the horse's left hind foot — which is preparing to leave the ground — to push off more strongly.

When you squeeze with your right leg, you stimulate the right hind leg. Squeezing with your left stimulates the left hind. By squeezing each side alternately, at the moment that leg pushes off from the ground, you are cueing your horse to move each leg more energetically as it prepares to leave the ground. This gets better results than intermittently squeezing together both legs. Once a leg is in the air, there is little your horse can do in response to a squeeze. The length of stride is determined by the amount of thrust given as the leg leaves the ground. Energy and the thrust of push-off determine the distance of a stride; once the leg is airborne, there is nothing to push against to influence its power.

The horse's left front leg comes forward an instant before the right hind; thus, to stimulate the right hind, use your right leg as the horse's left front foot comes to the ground. The simplest way to learn this timing is to watch the horse's shoulders. Execute your squeeze to their movement. As the left shoulder comes forward, the horse's left front is in the air. As the left foot lands, give a squeeze with your right leg. Then, as the right front foot lands, squeeze with your left.

It all happens so quickly that you will have the proper timing if you squeeze each side just as the opposite shoulder moves forward. Once you get the hang of it, you'll feel the rhythm with your legs and body and won't have to watch your horse's shoulders; squeezing your legs alternately at the proper time with the rhythm of the horse's walk will soon come naturally and won't take any concentration on your part.

SQUEEZE FOR SPEED

Kicking with your heels is not a good cue to use to make your horse walk faster. The horse responds better to the pressure of a squeeze. Each time he reacts with a faster step, release the squeeze to reward him with a decrease in pressure.

Squeeze

Squeeze where your legs hang in the stirrups, not back toward the belly. Exert the pressure through your upper calf muscles. Once your horse learns to increase his stride, all you have to do is swing with him, flexing at the hips and waist and rolling your weight in the saddle a little as you alternately squeeze your legs. You gently push him with your whole body, especially your legs and seat bones, as you move in rhythm with him.

This type of urging for a longer stride (faster walk) is easy on you and quite effective. Proper leg use, coupled with light hands and a good feel of your horse's mouth through the bit, will make the transition to a longer, quicker stride a simple matter for both you and your horse.

Using Your Hands

Good hands are as important in increasing the speed of the walk as are proper leg cues. Hands make the difference between a calm horse and a nervous one, between a responsive horse and a frustrated one. Proper use of the bit keeps your horse from continually breaking into a jog instead of lengthening his walk when you urge him.

Keep a light feel on the bit for communication and control. You never want to hurt the horse's mouth or make him resent the bit. He should not break into a trot when urged with your legs, nor should he prance or throw up his head because of too much restraint. Some horses want to walk slowly or to trot — no middle pace for them — and when you try to speed up the walk, they insist on breaking into a trot. It takes hand and leg coordination and some patience to teach a horse to extend the walk rather than to move into a trot.

You need constant communication with your horse's mouth, your hands giving and taking as his head and neck make balancing movements. A fixed hand — that is, one always in the same position — will bump the horse's mouth with the bit at every stride if the reins are held short or have so much slack that there is never direct communication with the mouth and you are unable to adequately check the horse. Keep your reins at the proper length to maintain a light touch on the bit at all times.

CHECK, THEN SQUEEZE

Learn to feel when the horse intends to break gait. Lightly finger the reins if the horse starts to break into a trot. Check him slightly just before he does, or you will be too late. Each time you check the horse, continue squeezing with your legs so he'll know he is to walk faster and not more slowly. If you squeeze immediately after checking him, he will respond by taking a longer stride rather than breaking into a trot. Soon you can increase the speed of his walk just by using body language to encourage him at the proper phase of his stride.

The Curb Bit

If the horse is past the snaffle stage and you are riding in a curb (see pages 188–193), you can keep a direct feel on his mouth even with slack in the reins. You will have adequate feel on a slack rein if the reins are heavy or when there's a snap between the rein and the bit. With a little weight in the reins, you always have a feel on the mouth without tightening them, and the rhythm and swing of the heavy reins as the horse walks help to maintain that feel so you can keep him from breaking into a trot. Reins should never be too slack, however; you need constant and instant control to keep the horse at exactly the gait and rate of speed you want. The reins are the right length when they have a nice curved line from bit to hand, without looking taut or having an excessive droop.

The Advantages of an Extended Walk

Many horses are never taught to extend the walk and therefore are not very alert as they go. They often daydream and thus are apt to stumble or even trip and fall clear down, unlike the horse that is "on the bit" and responsive to your hands and legs. The rider who just sits on a horse as a passenger rarely has a horse that walks swiftly.

Body Language at the Walk

When you are sitting balanced, with your legs relaxed, you can be responsive to your horse's movements as you sway from side to side with motion in your waist. You are moving like a hula dancer; your waist will be very supple. The thrust of the horse's hind legs moves you, from the hips down, from side to side in the saddle.

You get a rolling motion because as the horse takes a hind leg off the ground to move it forward, that side is unsupported for an instant and tends to drop a little. To be at one with your horse, let your back and buttocks sway from side to side with the horse's natural rolling movement. To follow his motion, let your seat bones move forward and down with each step, as if you were walking along with him. When you are moving "with" him this way, you can speed up your rolling motion and "push" with your seat bones, as part of the encouragement to walk faster, or slow it down to encourage him to stay in a walk rather than breaking into a trot.

STRIDE AT THE WALK

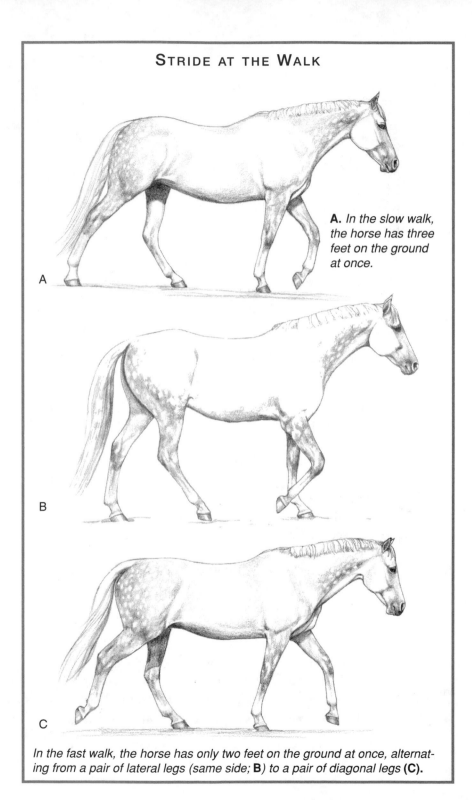

A. *In the slow walk, the horse has three feet on the ground at once.*

In the fast walk, the horse has only two feet on the ground at once, alternating from a pair of lateral legs (same side; **B**) *to a pair of diagonal legs* **(C)**.

You should be moving with your horse, attentive to everything he is doing, encouraging him with your body language. Then you'll have good control over his walk and can extend or slow it readily. He'll become more pleasant to ride, and more apt to respond willingly to your signals. He can walk swiftly on the way home instead of trying to break into a trot if he's overeager. By teaching him to respond to leg, rein, and body language cues to extend his walk, you have more control over his actions, whether you are encouraging him to walk faster on the way out or to do a relaxed fast walk on the way home without prancing and dancing.

Calming the Excitable Horse So He'll Walk

Some horses are excitable, with lots of nervous energy. Like a coiled spring, a horse that's poised and eager to go can be a challenge. He always wants to be going a little faster than you do. If he's walking, he wants to trot. If he's trotting, he wants to break into a gallop. He may be impatient if you are riding with other horses, wanting to be in front rather than left behind.

If you are heading home, he wants to be there right now and can't relax until he gets there. He'll start prancing and dancing because you won't let him run home. He may develop the bad habit of tossing his head and pulling at the bit if you try to hold him down to a walk, and it takes a lot of careful, patient work to relax him.

He may be insecure and not want to be left or to have other horses leave him. He may be a bit balky if he must leave home by himself with no horses for company. When you take him out, he may be in such a hurry to return to his stall, pen, or pasture that he's frustrating to ride. Trying to make him walk may result in an argument, with the horse becoming even more upset and tense.

The nervous, hyper individual can be trained to relax when you handle him with patience. The first step may be to give him more exercise and less grain, if the problem is accentuated by high-energy feed and confinement. The stall-bound young horse can go stir-crazy with pent-up energy. If training sessions and riding are his only outlets, you'll have your hands full. He will do much better if he can be turned out at pasture or in a large pen where he can run off some of that energy and be a little more mellow for his training sessions.

Patience, Not Punishment

When working with a nervous or overactive horse, make a conscious effort to keep your cool. You must remain patient and understanding. Remember that he is insecure. He needs reassurance, not threats. Never lose your temper. And don't punish him for his actions; that will only increase his anxiety and make him more jumpy. Punishment is appropriate *only* when a horse intentionally misbehaves — if he bites, kicks, or commits other aggressive actions. If he is misbehaving due to insecurity, punishment will make him more fearful, less able to relax.

CONSISTENCY IS KEY

The key to resolving the problem of the nervous horse is quiet, gentle firmness and consistency. If you are consistent in your actions and cues, the horse will learn what to expect from you and learn to trust you, which in turn will help him relax.

Be Calm

Your attitude while riding is crucial, just as important as your training methods, because the horse can easily sense your attitude. If he is tense and apprehensive, you must be completely calm. Gradually your relaxed attitude will be transmitted to him, and he'll be able to settle down and pay attention to what you're trying to teach him.

Being a herd animal, the horse takes his cues from other herd members. In a sense, you and he are a herd of two when working together, and you are the leader of this herd. When you maintain a calm frame of mind, your "herd mate" gains confidence from that and will lose some of his anxiety. But if you are tense, worried about the challenge he poses in the training situation, upset because he is trying to go too fast or not paying enough attention to you, your horse will have a difficult time relaxing.

If you worry about the fact that he is constantly testing the limits of your control, it will be even harder to calm him because he interprets your worry as a threat or a sign of danger and becomes even more agitated. When riding the tense or nervous horse — especially one that wants to prance and jig rather than settle down and walk calmly — the first step is for you to relax. Practice relaxing until your body becomes almost as limp as a rag doll. Transmit your serenity, not your tension.

Using the Reins

The way you use your reins is another factor in determining whether you can calm a nervous horse. The reaction of many riders when mounted on a jumpy animal is to try to keep him from getting out of control by tightening the reins or punishing him for his nervous actions, holding him tighter or even jerking his mouth. Remember, punishment is not warranted when the behavior is out of anxiety.

The horse's response to this constant restraint through the bit is to toss his head and fight the bit, and he becomes more fidgety instead of less. It is the horse's natural tendency to pull when pulled at, so pulling at the horse creates a tug-of-war that you won't win.

The answer to most problems is to become a better rider, to fine-tune your cues and improve communication through the bit, to develop better use of hands and body language. Communication is through your hands and bit, accentuated and helped by your legs, weight shifts, and seat. The whole attitude of your body — how you move, whether you are tense or relaxed, for example — is part of that communication.

Your hands are important because they control, encourage, steady, and direct your pupil. Calm, sure hands can help settle him if he's nervous, especially when coupled with patient horsemanship and lots of miles under saddle. This helps build a bridge of understanding.

EXAMINE YOUR METHODS

If a horse reacts adversely or does something wrong, it's usually the fault of the rider. Perhaps you asked him to do something beyond his ability and he doesn't know how to respond properly. Maybe you communicated your request unclearly, so he made the wrong response. If your horse performs poorly, resists cues, or won't do what you want, examine your horsemanship and rethink your tactics.

Relax to Relax the Horse

All too often, the rider of a nervous horse gets frustrated when trying to make the horse stop prancing: "It doesn't make sense to prance; the horse would get there faster if he'd just settle down and walk!" We must remember, however, that the horse reacts to emotions, not logic. All he can think about is that you are restraining him; thus, he becomes upset. Your tenseness and frustration increase his feelings of worry and panic; you inadvertently punish his mouth and make him even more fretful. It's very difficult in this situation to stay calm, yet this is exactly what you must do before the horse can relax.

You must communicate confidence, reliability, and calm so your horse can start to trust your judgment about the gait or speed he should go. Lean back a little and relax in the saddle. (Leaning forward is a form of urging, and staying tense is also a form of urging; your seat and legs are rigid.) If you lean back slightly while he is prancing, not only are you more relaxed, but also your body weight becomes more of an anchor; you are a little behind his center of gravity and he tends to slow down.

Try to communicate relaxation through your hands, reins, and bit. Never keep a steady pull or fixed hand on a nervous horse. This enables him to lean into the bit and pull harder. Don't pull at his mouth or he will become less responsive. Let your hands go with his head movements. A horse pays more attention to a bit that wiggles around in his mouth than to one that is constantly pulling. Give and take with the reins, using wrists and finger action. If you relax and are not pulling on the reins, eventually your horse will calm down.

GIVE AND TAKE

If your horse is trying to trot, keep giving and taking with the reins as you would with the head movements at a walk. When he's in that borderline space between a walk and a trot, your give-and-take rein action can often keep him down to a walk even though he'd rather trot. Keep that gentle mouth contact in walking rhythm.

Teaching Patience to the Prancer

Once you gain a constant but gentle feel of his mouth with give-and-take action instead of a pull, your horse will soon progress to the point where he is relaxed enough to walk freely. He can begin to be "on his honor" again, without constant restraint. When you reach this stage, your horse will begin to make noticeable progress.

The next hurdle to overcome is how to work through the times he still insists on going faster than you want him to. You must continue checking him with the bit but not enough to start a tug-of-war. How you check him determines whether or not he will react properly.

Start by giving him his head a little more. On a very nervous or overeager horse, you may be able to do it only a few seconds at a time or when the situation puts things in your favor. You may make the most progress after the horse has taken the edge off his exuberant energy with miles of riding. Perhaps then he will settle down and relax enough to stretch out his walk instead of trotting.

If his problem is trying to hurry home in a prance instead of a walk, you'll do best after a long ride when he's a little tired and more willing to relax. It may take many miles of steady, daily riding before you see a change in his attitude. Once you can let him have his head in a relaxed walk, even if just for a few strides (then checking him again to keep him from breaking into a trot), you'll be making progress. The goal is to create that cooperative moment amid the jumble of nervousness and discord. When your horse does relax for a moment, praise him and let him know this is what you want. When he learns that you are not going to pull on his mouth when he fulfills his part of the bargain, he'll do more calm walking and less prancing.

Reward Him with Your Trust

It takes a subtle, fine-tuned feel for how much freedom you can give your horse and how much restraint with the bit he will need to stay at the gait and speed you want. If you let him go more and more on his honor during the moments he is attempting to cooperate, however, the nervous charge-ahead attitude will diminish. Gradually, you will be able to enlarge that cooperative moment. If the horse walks a few strides in the middle of his usual prancing routine, praise him and keep trying for more walking steps the next time you ask him. If he regresses and prances, go back to a guaranteed success — a few walking steps — and start building on that again.

The key to relaxing the nervous horse so he will walk is less restraint, not more. Little by little, you can develop the rapport and confidence that will allow you to let the horse go more on his honor, without your constantly having to check him with the bit. Patience, understanding, tact, and a lot of miles of relaxed work in open country — with less concentration on fast work; skip those lessons for a while — will go a long way toward calming the high-strung horse. If you are patient, you can minimize or resolve the type of behavior that makes the nervous horse such a frustration and challenge to ride. The two of you can then become a real team: He will be willing to work with you instead of against you, and you will be trustworthy and won't let him down.

The walk is the most important gait to focus on at first, for it allows you to build a solid foundation for all of your future training of the horse. It is generally the horse's slowest gait and usually the most calm and relaxed. This makes it the ideal gait for starting to build rapport with, and to gain perfect control of, the green horse.

The walk also has the potential for development of very advanced and fine-tuned communication, because you can refine your bit cues, body, and leg cues to a bare minimum once the horse learns how to respond. You can allow or encourage him to do a very slow, medium, or fast relaxed walk, or a crisp and alert walk at any speed, or a highly collected walk at any speed. You can collect or extend his walk at will.

The communication you develop with the horse at a walk can be the basis for excellent understanding and trust, creating a very willing pupil for all further lessons. He feels at ease and secure in your calm, benevolent control.

12

THE TROT

In many horses, the trot is a neglected and poorly developed gait, hurriedly skipped over in training. Yet this gait is easiest for teaching many things, and any horse can benefit from work at the trot. A wise trainer spends a lot of time with the trot: It helps a horse develop strength, fitness, balance, and collection, all of which are essential for later training and for any sport that requires athletic agility. The trot is the best gait to combine speed, control, and endurance. By its very form, it makes some kinds of movement much easier for the horse. As a two-beat gait, the trot is regular and balanced. Because the horse's head and neck are steady, your job is simplified: You can keep your hands still and not interfere with his mouth. At the trot, the horse's muscular development and strength progress as a whole; it provides excellent all-around conditioning.

How the Horse Moves at the Trot

The trot is one of the three natural gaits of most horses (see pages 288–289) and is a two-beat, diagonal gait. The horse's legs move in diagonal pairs (left front and right hind; right front and left hind). At the fast trot there is a moment of suspension during which all four feet are in the air, but in the slow trot, one diagonal is striking the ground as the other diagonal is pushing off; there may or may not be a brief instant in which all four are on the ground.

The trot has constant, equal rhythm and balance; the horse has his best balance at the trot. He makes no balancing movements with his head, as he does at the walk or canter. At the trot, his head stays at a steady level unless he is lame — then he bobs his head as he takes weight more quickly off the injured foot. He has the best stability at the trot, as he is supported at both ends of his body and at both ends by diagonal legs at every stride.

First Trotting Lessons for the Green Horse

Early training is done at a walk until the horse is comfortable with the extra weight of the rider and is thoroughly at ease with your signals — responding well to moving out, stopping, turning, and extending his walk. All of these improve his balance (with weight on his back) and control, both of which are necessary before you begin trotting lessons. How soon you start trotting depends on the individual horse and how well he has progressed with his balance and control.

Cuing for the Trot

You will ask for the trot by squeezing lightly with your legs when your horse is briskly walking. Add clucking noises if he has been trained to this verbal cue. Your horse knows he is supposed to move faster in response to leg pressure and clucking. Squeeze with the calves of your legs, where they hang in the stirrups, not back toward his belly. Never kick (see page 258).

Your horse may increase the speed of his walk rather than begin trotting, but if you keep encouraging him and reward him by release of leg pressure each time he responds by moving faster eventually he will break into a trot. He will soon get the idea that you want him to trot when you give him these cues.

TROT TYPES

The speed and animation of a trot can vary greatly. There are four types of trot: the jog, which is a slow trot; the working trot; the extended trot; and the racing trot. There are variations of speed and animation within each category as well. Most horses never do a racing trot, and many don't know how to extend their trot while ridden. This usually takes training to develop.

TROT VERSUS PACE

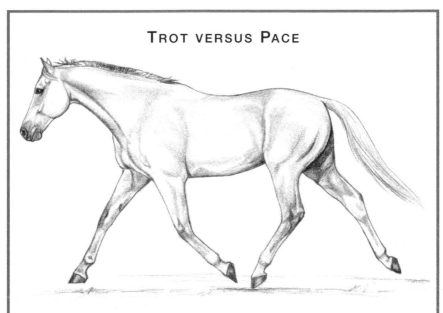

A. *At the trot, the diagonal legs move forward in unison, resulting in excellent balance and stability.*

B. *At the pace, the lateral legs work together. When pacing, the horse has a rolling motion from side to side, with less support, balance, and traction on slippery footing.*

Working the Green Horse at a Trot

Your first lessons at a trot will be in straight lines, heading across the arena or along a trail — wherever you do most of your riding. Work at the transition from walk to trot and from trot to walk as your horse becomes more proficient at understanding what you want. Do not undertake these first lessons until you sense that your horse is physically and mentally prepared to begin trotting.

The first times he trots, he will do a somewhat extended trot because he does not know how to collect himself. He may also be clumsy; he has never done this while carrying a rider. First trots should be brief, as your horse does not know how to balance with a rider at this gait and must develop his balance and muscles.

Out on the trail, the trot will become the best gait for covering ground quickly and with agility. It is the best gait for teaching balance and collection and for developing physical fitness. Give your horse as much (or more!) trotting work across country as in an arena. Once he can make transitions smoothly from walk to trot and back to a walk, start

The horse should have as much or more work across country as in an arena.

working on ways to improve his balance and dexterity at this gait. Training sessions now should be longer.

Arena work, or on any relatively flat ground with good footing when you are out riding, can help him learn balance and confidence, of course, as well as improving communication and control. Once your horse is comfortable carrying you at a trot, start doing large circles, figure eights, and serpentines. Doing circles and changes of direction at a trot help teach him to respond to directional cues — leg pressure and rein pressure to indicate the direction he must turn or circle — and how to adjust his stride and think about where he is putting his feet. All of this is necessary before he can develop good balance and learn how to collect himself.

Work on smooth transitions with a minimum of cues. Speed and slow him at the trot; perhaps begin some lessons in collecting and extending. (For more on transitions, see pages 319–320.) As you refine your cues, work on a smooth stop from a trot. The trot is the most versatile gait; the more you polish it, the better trained your horse will be.

The Jog, or Slow Trot

The *jog* is slow and unanimated. Some horses prefer to jog rather than walk fast and must be taught to speed up the walk or to extend it. The jog is a natural gait for many horses. The *jig* is also a slow trot, but in contrast to the lazy jog, the jig is a nervous prance. An excited horse may jig when he wants to go faster than his rider will allow. Jigging shows lack of training or inability of the rider to use hands and legs properly to encourage the horse to relax and walk.

In the jog, the horse's hind foot lands behind the track left by the front foot. There is little impulsion, or energy. The jog, because it is not quite two-beat, is not a perfectly diagonal gait. The hind foot may land a split second before the opposite forefoot. This can also occur in the racing trot, when a horse is traveling so fast that the unified movement of the diagonal becomes a little broken and disassociated, thus producing an irregular four-beat gait.

Using the Slow Trot

The slow trot is a good gait for training. It is useful when starting a horse, as it can be more controlled than a fast, extended trot. In the slow

This rider is trying to make a jigging horse walk by sitting back in the saddle, staying relaxed, and using gentle give-and-take with the reins.

trot, you can work on walk-trot transitions. (You'll do some transitions later from a fast trot, but they are easiest to accomplish first from a slow trot.) At the slow trot, work on communication with rein and leg aids until the green horse is comfortable at this gait and responds well, and thus will be more likely to respond at a faster speed.

You will eventually want to teach the horse various speeds at the trot. You'll want him to be able to shift instantly from a fast to a slow trot and vice versa, as well as from a collected to an extended trot and back again. But first you must perfect his slow trot so you have excellent control, balance, and unity.

Sitting the Slow Trot

When training a green horse, you will sit the slow trot. You will post the faster trots mainly because you can stay in better balance with your horse, which is very important, as he is learning his own balance while carrying a rider. When you are in balance, there's no danger of alarming, distracting, or hindering him by bouncing or banging the saddle. Because much of your work with the green horse will be at a slow and uncollected trot, you'll probably be sitting the trot much of the time or posting close to the saddle, whichever is more comfortable for you and the horse.

The slow trot is one of the best gaits in which to acquire and maintain a good seat on a horse. Most of your weight is in the saddle, your body is erect and slightly forward (never leaning back), and only part of your weight is in the stirrups. You will be using only light leg and stirrup pressure. Too much weight in the stirrups at the sitting trot makes a rider stiffen the knees, hips, and back and creates more bounce.

Don't slouch; your back should be straight and your shoulders square, but with your back muscles supple and relaxed. That way, your whole body is able to move with the horse. Then you can keep a good seat and never lose contact with the saddle.

Some horses have a rougher trot than others and can be challenging to sit. If you find yourself too far back in the saddle, you'll be jammed against the cantle with your feet braced forward against the stirrups, bouncing over the weakest and most tender part of your horse's back. This causes discomfort for the horse and hinders his ability to use his legs properly.

Too much weight over the loins makes the horse move less freely; he's forced to shorten his stride or develop an awkward gait. The rider is sitting directly over the area with the most up-and-down motion and bounce — from the thrust of the hind legs — and has to fight against the extra movement of the horse. The horse whose rider is bouncing and thumping on his back will stiffen in self-defense, tightening his back muscles and raising his head, which makes his trot even more uncomfortable for both of you.

Maintain Good Position and Relax

If you are tense as you try to stay in the middle of the horse and grip with your legs to decrease the bouncing, you will have trouble sitting the trot. Proper position and relaxation are essential. The flopping rag-doll effect back on the loins is not pretty, nor is it healthy for the horse. You must be over his center of gravity and moving with him — in the center of the saddle — relaxed.

Proper position at the slow trot — rolling your weight with constant saddle contact — enables you to have constant communication with the horse through seat and legs. Your hands remain stationary on the reins, rather than bobbing up and down, by flexing your wrists and elbows. You

The jog is a slow, unanimated trot. This is the gait for developing a good seat and balance.

can give proper signals through leg and weight-shift cues. You can't do a good job of training a green horse at the trot until you have a good, secure, well-balanced seat. If you are not "with" the horse, you'll inadvertently give him improper signals through your legs, body position, and hands. For steady, controlled communication through the bit, you must first have a secure and balanced seat at the trot, whether sitting or posting (see page 313).

Working on Better Horse-Rider Unity at the Slow Trot

If you have a little trouble keeping proper position while sitting the trot, slow-trot without putting your weight in your stirrups. This helps break the pattern of feet forward and seat too far back, which puts you "behind" the horse. Jog on a loose rein and sit heavily in the saddle with your weight on your seat bones. Let your legs hang in the stirrups without putting weight on them. Allow your legs to hang naturally, straight down from the hips, toes hanging down. This helps you sit more deeply in the saddle, where you should be, in a balanced position instead of back on the cantle. This will also help you relax and stay in the middle of the saddle, at the narrowest part of the horse's back — right behind the withers — where you'll feel the up-and-down motion the least. The farther back you sit, the more thrust you'll feel from his hind legs.

Keep hips and lower back relaxed and flexible; when you tighten your legs, your hips will tighten too. If you grip with your knees, you will interrupt the movement needed to stay in sync with your horse. Gripping your knees will tighten your muscles and cause tension, rigidity, and more bouncing. If you grip with your thighs, you can't sit deep in the saddle; if you grip with your calves, the horse may try to increase speed due to the leg pressure. You want a relaxed leg that dangles alongside the horse.

Keep your upper body vertical, and relax. Lean back only slightly, if at all, and let your hips and lower back swivel and move freely with the horse's motion. Don't sit too far back; you'll get more push from the hind legs and be bounced more. You want a relaxed, rolling movement with the horse.

Try First at the Walk

Once you relax your legs, work on relaxing your back. Practice at a brisk walk before trying a trot. At the walk, notice how your body moves with the horse and the way your back moves. The horse's motion moves your lower back and hips and is similar to the movement you should make while sitting a trot.

If he isn't walking fast enough, push the horse with your body, especially with your seat muscles, and urge him on. Swing with him. Use your legs alternately as you swing with his movements. As you stimulate each hind leg to push off harder, he will take longer strides and walk faster, and you will feel the extra thrust in your seat. This will help accustom you to the extra push he'll give at the trot. This kind of contact with the saddle and contact with the horse through your legs enables you to subtly control your horse's movements and speed. The way you move your body is as much a form of control and cuing as are your legs and reins. It's also the key to sitting the trot.

Keep practicing at the walk until your horse is walking strongly and you are automatically moving with him, coordinating hand and leg movements to keep him walking fast without breaking into a trot. Once you have the swing of your body and your hand and leg actions perfected, concentrate on letting your back relax as you sway with the horse.

Practice the Sitting Trot

Keep the horse at a very slow trot at first, fingering the reins if necessary, and concentrating on relaxation, so your back won't have to sway much faster than it did at the fast walk. Move in sync with the horse, relaxed but with a secure seat. The horse's back must be relaxed, too. If his back is flexible and relaxed, there will be some give to his muscles; he will be smooth to ride at the trot. Horses that are stiff and tense often have a very rough trot. When a horse raises his head above his withers, he tenses his back muscles. This makes his back unyielding; his trot is jarring. If he stiffens his back when he trots, go back to a walk until he relaxes, then try the trot again.

Rest your feet in the stirrup — on the balls of your feet — but don't put weight in the stirrups or push down your heels. If you try to put weight in the stirrups, you'll stiffen your legs and seat and lose that perfect contact with the saddle. Keep yourself loose, and your seat will stay more firmly in the saddle. If you stiffen your legs, you will lose saddle contact, and bounce. If you let your calf muscles relax, they will stretch and gravity will pull down your heels into the proper position.

If you start bouncing, you are putting weight in the stirrups, pressing down on them and pushing yourself up off the saddle. To keep from pushing down on the stirrups, practice letting your heels bounce down past your stirrups, with your toes just lightly touching the stirrups. Keep your ankles loose and flexible. This way you can keep proper weight distribution on your seat and continue to move with the horse.

Don't practice just in an arena. Go across country and trot at various speeds. Working on seat and balance in all kinds of conditions will give you and your horse the practice and confidence you need to have an excellent unity while sitting the trot.

KEEP LIMBER AND LOOSE

Most green horses need more work at the trot to relax and travel freely. If your horse trots with a stiff back, it may take patience to loosen him up enough to relax and carry his head at a normal angle. It will be good practice for you as well. Make your back muscles stretch and give as you swing with your horse, rolling your weight on your buttocks and thighs. Keep constant but not rigid contact with him.

Sitting the jog or slow trot

The Posting Trot

Posting is rising out of the saddle, with weight in your stirrups, to one beat of the trot, staying up long enough to miss the next. You are moving up and down in rhythm with each stride: up when one front leg is off the ground and down when it is on the ground. Rider and horse are working in unison. At a fast trot, it's easier to stay in balance with your horse by posting rather than sitting. You can keep your upper body forward over the withers so he can carry the load while expending the least amount of effort. When you are up, your weight is entirely on your stirrups instead of in the saddle.

There are two ways of posting. If you post *behind the trot,* you let yourself be lifted upward from the saddle by the thrust of the horse, simultaneously pushing your upper body forward to keep in balance with him, then let yourself down by flexing knees and ankles again on the next beat. The other method, posting *ahead of the trot,* is a little easier on both horse and rider. Lean your upper body forward, rolling your weight forward on your thighs and push your pelvis forward and then up. Your weight is on your stirrups and thighs. You can post very close to the saddle this way and enjoy a very smooth ride.

POSTING THE TROT

A

B

Posting the trot on the left diagonal. *The rider moves up and down with the action of the horse's left leg: up out of the saddle when the left leg is off the ground* **(A),** *down in the saddle when the left leg is on the ground* **(B).**

Posting ahead of the trot *(rising forward and up) is a way of urging the horse to trot faster.*

Posting behind the trot, *letting the motion of the horse push your body up (erect but not forward), acts as a drag on the horse's momentum, to help slow his trot if he's going too fast.*

POST LOW AND CLOSE

Posting the trot on a green horse should always be done well and "quietly" — that is, never come down hard in the saddle. It's best to post low and close to the saddle, staying in perfect balance, so you don't interfere with your horse's actions.

The Working (Medium) Trot

The *working trot*, or medium trot (sometimes called an ordinary trot), is a brisk, balanced gait. To a Western rider, this may be the fastest trot asked of a horse — moving out more freely and with more impulsion than in the jog. In some types of English riding and in long-distance sports, a working trot is the slowest trot the rider asks of a horse; the jog is rarely used.

When training a horse, you must know how to post so you can teach him more versatility. The slow trot is the only speed you comfortably sit. You must teach the young horse several speeds at the trot and how to make transitions from one speed to another. You'll be able to accomplish these goals more readily when you can post.

You can use posting as a means to help the horse trot faster or slower, taking advantage of your body position and balance. If the horse is sluggish in his trot and you are trying to encourage him to trot faster, you can post "ahead of the trot," pushing your body up and forward at each stride. Your more forward weight is a form of urging; your horse will tend to go faster to keep proper balance. Posting ahead of the trot, it is very easy to squeeze with your legs each time you roll your weight forward to rise because you can squeeze with your legs as a means to raise your body. Your legs and forward leaning are good encouragement for him to move faster. If he is trotting too fast, you can post "behind the trot," letting the thrust of the horse push your body up. This way your body is erect rather than forward and acts as an anchor to hinder his speed and make him stay slightly slower.

Your body weight and position send signals to the horse for speed and balance, especially when posting. Posting enables you to reinforce your leg and rein cues so you can minimize them and communicate with the horse with just a light touch of the bit or your legs, as when asking for more speed or less speed.

Don't Make the Horse One-Sided

When posting, be aware of the horse's diagonals (see box below and pages 321–322). Know which diagonal you are on. One reason for learning to distinguish between the diagonals when training a horse is so you can change them occasionally when trotting.

If you unknowingly always post on the same diagonal, it's more tiring to the horse and makes his muscle development one-sided. Then he'll feel awkward and unbalanced when you suddenly post on the other diagonal. Some horses get so used to a rider posting on one side that if you try to change, they will seem very rough or uneven in gait; a horse may even try to shift his strides so you are again posting on the preferred diagonal. Avoid this problem from the beginning by making a conscious effort to change diagonals often.

DIAGONALS

At the trot, diagonal legs of the horse move forward at the same time. To change diagonals when you are posting, stay up for two beats instead of one, then continue posting. You can also sit for two beats instead of one.

The Extended Trot

The *extended trot* is the lengthening of the horse's stride to the limit without changing the regularity, cadence, tempo, or rhythm. The legs are not moving faster; rather, they are taking longer strides and putting forth more thrust. This trot has tremendous impulsion. The strides are lengthened without speeding up the tempo of the hoofbeats, but the horse is being propelled farther and faster.

At a slow trot, the horse takes very short strides. He takes longer ones at a medium trot and very long strides at an extended trot. If you trot him a certain distance at a slow trot, then at a medium trot, and finally at an extended trot — and count the number of strides each time — you'll see that he uses the least number of strides at the extended trot.

A horse that doesn't know how to extend his trot merely speeds up his strides instead of lengthening them when asked to go faster; he just makes his legs move faster. A horse that knows how to extend will

Extended trot along a road

lengthen his stride when asked. His feet will hit the ground at the same tempo but will be going farther and faster; he will almost float above the ground because of the extra thrust.

Some horses extend more naturally than others. Some will do an extended trot when excited, but a good extended trot under saddle when asked is not learned quickly; it requires training and muscle development, as it takes a powerful thrust and suppleness of the horse's hindquarters and back. The horse lengthens his neck and his stride, and his hoofbeats may actually sound a little slower because he is propelling himself a much greater distance.

In the extended trot, the horse lengthens his frame and lowers his hindquarters. In so doing, he has more drive and thrust from the hindquarters and freer shoulder action, and can extend his legs farther. When some horses "shift gears" and go into the extended trot, the rider can feel the whole horse drop down. It is not as easy for a horse to break into a gallop from an extended trot as it is from a medium trot because of his lengthened frame.

SPEEDS AT THE TROT

The horse's trot can vary from a 4-mile-per-hour jog to a racing trot, which may be more than 30 miles per hour. The slow trot is 4 to 6 miles per hour with short strides. The ordinary working trot is about 8 miles per hour, with strides 9 to 10 feet long. Hind feet overstep front tracks several inches or even a foot. (The *overstep* is the distance the horse's body is traveling while in suspension — the thrust through the air with no feet on the ground.) There is a longer period of suspension between strides; the horse moves with more thrust. The extended trot is 12 to 20 miles per hour; the overstep is as much as several feet. At the racing trot, the horse may have a stride of 15 to 20 feet, with an overstep as great as 6 feet.

Walk-Trot Transitions

As you train your horse, you will work on improving his ability to make smooth transitions from one gait to another. Work first on transitions from the walk to a slow trot and back again. As the horse becomes more proficient, cue him to go from a walk to a fast trot and back to a walk again.

Use Weight-Shift Cues

The young horse is just learning about leg and rein cues. You can help him understand what is wanted if you add balance shifts and body language to reinforce those cues, especially when changing from the walk to the trot and from the trot to the walk. Leaning more forward and pushing with your seat bones will encourage your horse to move faster and break into the trot.

Likewise, to slow from a trot to a walk, finger the reins for some give-and-take pressure on the bit and shift your weight back to signal a slower speed. When trotting, both you and the horse have more energy and impulsion. If you return your body to walking speed by sitting back a little more and mentally shifting down to a walk, your horse will feel the transition you've made and slow down also — he'll want to keep the unity and balance that's most comfortable to him. If he's trotting and you sit down deep in the saddle in walking position, he needs to drop down to a walk to make things feel right. If you shift your weight back and relax, your horse will, too.

If he wants to keep trotting — as some young horses do when they feel good about traveling fast — give a little more cue with the bit as you shift your weight back. It may be sufficient to take the slack out of one rein with a quick touch of the bit or you may need to do several quick but gentle give-and-take actions. Either way, your horse will soon understand that your weight shift is the signal to slow down. With practice, you can use this weight-shift cue to decrease the speed of the trot or to have him drop from the trot to the walk.

Change of Speed and Stride Length

As you do more trotting with your horse and he gains better balance and control, you'll want to teach him to speed up the trot or to slow it on cue. You can do this with the use of your hands, legs, and weight shifts, just as you did in walk-trot transitions. Also, teach him to lengthen or shorten his stride, to extend and collect. This will help him become more agile and flexible for later training, whether to adjust his strides to approach a jump properly, perform a dressage test, collect himself to keep his feet and his balance while trotting downhill on a slippery trail, or to stop and turn quickly.

There is a difference between increasing speed and increasing stride length. The well-trained horse must be able to do both. After your horse learns to collect and extend, you can adjust his stride length without changing his speed. He can lengthen his stride, for instance, while still traveling the same speed, simply by taking fewer strides to go the same distance at the same speed. Changes of speed and change of stride length are brought about by synchronizing the use of your legs and hands, enabling the horse to respond to your cues.

Improving the Stop from a Trot

When the horse halts from a trot, he must balance himself and shift back his weight, so he won't be leaning forward and "lugging," or pulling, on the bit. The green horse must learn to shift his weight on cue. He may stop gracefully when running free in the pasture, but he has to learn to do it while carrying a rider.

Improving His Balance

Smooth stops can be accomplished only when the horse can balance himself, and he learns this gradually as he becomes more collected. As you work on walk-trot transitions and change of speed within the trot, this will also improve his balance and his stop.

Smooth starts and stops come naturally as you teach your horse to move forward with gentle leg pressure and lengthening the reins and to halt when he encounters bit pressure. Bit pressure should not be active. Once he knows that pressure ceases when he responds, your horse will halt when leg pressure gently pushes him onto the bit and he finds it unyielding. As always, rewarding the horse by instant release of pressure is the most effective way to teach him what you want him to do.

Trotting in Circles

Trotting in circles is good training because it teaches the horse to bend his body to the shape of the circle, which is not easy; his spine is fairly rigid, with little flexibility side to side. His loin muscles are not accustomed to this movement. Riding in circles, when done correctly, is an exercise that improves his flexibility and increases agility. Whenever he is trotting in a circle, his outside legs must travel farther than his inside legs, which therefore carry more weight. Pay attention to which diagonals you are posting on to make it easier for him to trot the circle smoothly.

Diagonals

When the horse trots, he alternates from one diagonal pair of legs to the other diagonal pair. He's on the left diagonal when his left front foot is on the ground and on the right diagonal when his right front is on the ground. When posting, you are posting on the left diagonal when you are sitting in the saddle as the left front foot is on the ground and rising in your stirrups when the right front comes to the ground. You are posting on the right diagonal when sitting while the right front is on the ground and up in your stirrups when the left front is on the ground. This can be a little confusing to figure out at first, as you can't see the horse's legs as you trot and because he is moving so swiftly.

It helps to watch the horse's shoulders. Work first at a slow trot until you catch on to the way his legs are moving as you are rising and sitting to post. When trotting in circles, you should post on the outside diagonal, for this is most natural to the horse.

If you are circling to the left, with the center of the circle to your left, you should post on the right diagonal. Watching the horse's shoulders as you trot, you should rise as his right shoulder moves forward and sit as the right shoulder moves back; you are moving with his right leg. When that leg is in the air moving forward, you are up out of the saddle; you are in the saddle when that foot is on the ground.

The reason you post on the outside diagonal when the horse is trotting in circles is because he has a "lead" at the trot. His outside legs have to travel farther than his inside legs — his body is curving to fit the circle. You tire him less when you post with his leading foot. When traveling in a circle to the right, his leading foot and outside diagonal will be the left.

Changing Diagonals

Change posting diagonals whenever you change direction. If you are making a circle to the left, posting on the outside diagonal, change diagonals when you start a new circle to the right so you will again be posting on the outside diagonal.

To get the feel of this, trot the horse in a large figure eight. When you come to the center of the eight after having made a circle to the left, change diagonals by either sitting an extra beat or standing for an extra beat, whichever is easier for you. You will then be posting on the opposite diagonal as you change direction and start a circle to the right. When practicing figure eights, make your circles large, well rounded, and of equal size, and have the horse go straight for a stride or two as you change and reverse the circle. If circles are too small when trotting, it will be harder on his legs; the inexperienced green horse may also accidentally strike himself, hitting one leg with the opposite leg.

GIVE HIM A BREAK

When trotting across country, change diagonals now and then. On a long ride, your horse will not get as tired if his muscles have been developed equally by posting on both diagonals.

When circling to the left, post on the right diagonal, sitting down in the saddle when the horse's right front foot is on the ground.

Circle Exercises

The trot is a good gait for working in circles to develop better balance. If the horse is bending his body to fit the circle, the outside loin muscles must extend as the inside ones contract. The fact that his legs are working in diagonal pairs requires quite an adjustment in the horse's stride to keep good balance and rhythm; remember, the inside legs travel a shorter distance and bear more weight than do their outside counterparts.

Trotting in circles is an excellent way to develop the horse physically for greater agility and versatility under saddle; it demands proper bending, adjustment of stride and rhythm, and increased impulsion on the inside leg, which carries more weight than the outside leg. This lesson is not an easy one for the horse. Always teach step-by-step. Trot him in large circles at first, and only when he has mastered them should you attempt smaller ones.

Attempting small circles before your horse is ready will result in awkward and clumsy movement and possible bad habits — he may hold his back and neck stiff, for example, because he hasn't yet learned to bend. It also risks leg injury; your horse may strike a leg because of lack of coordination. Save the small circles until he's had more trotting and has learned how to handle himself with balance and collection (see chapter 14). He must be collected to be able to keep his balance and bend his neck and body to correspond with the bend in the circle.

The Most Versatile Gait

A trot can be slow, medium, or fast, collected or extended, or just uncollected. A jog is a slow, uncollected trot. A collected trot can be fast, medium, or slow, depending on the rider's wishes. In the collected trot, the horse is more animated and vigorous, with more energy and spring to his step. He travels with head up and nose more vertical, his weight balanced back off his front end, moving with hindquarters well underneath himself to bear more of the weight. His weight is balanced back farther; he is not traveling heavily in front and thus becomes more agile and able to do intricate maneuvers with precision and dexterity.

The collected trot at medium speed is a good gait for tasks that require balance and agility, whether working in an arena doing complicated maneuvers or traveling through steep, rough terrain or slippery footing. The horse can maneuver with more surefootedness at the trot than at the canter or gallop.

The trot is the best gait for covering ground swiftly without tiring the horse. A medium-fast trot is easy on you and your horse and is much less tiring than a canter or gallop. The trot has more potential than any other gait for developing muscle and coordination. The more training you can give the horse at the trot, the more agile and versatile he will become.

13

Lightness and Collection

Collection refers to a horse's balance and animation. For the first few weeks of a young horse's training, don't worry about collection. Your main focus then should be on teaching the horse to move out, stop, turn, and respond to rein and leg signals, so he is always under control. At this point, he travels in an extended manner; that is, his hind feet are not well under him, his head and neck are fairly low, and he travels more heavily on his front end.

Most horses collect themselves naturally when running free, to balance themselves for various maneuvers. But when you add the weight of a rider, the green horse becomes clumsy and must learn his balance all over again. He travels heavily on his front legs until you teach him to collect himself and carry the weight properly.

Teaching a horse to move more lightly in front is one of the most important aspects of training. He can't perform well, with balance and agility, until he is collected with his weight farther back, thus freeing his front end for whatever new movement you request.

Training your horse to be light in front is not something that can be accomplished quickly. It is a gradual, progressive learning process. Begin at the walk and trot, and don't proceed beyond that until he's learned to collect himself at those gaits. After your horse is more fully trained, you can let him travel in an extended manner occasionally, but when there is work to be done or if a precise action is wanted, he must be collected or he won't be able to perform it well. He will also be clumsier and more likely to stumble.

What Is Collection?

When a horse is collected, his head is above the level (higher than his withers), his neck is slightly arched and flexed at the poll so his nose is down (more vertical), his chin is tucked in, and his face and forehead are almost perpendicular to the ground. His lower jaw is relaxed and responsive to the light hands of the rider. His hind feet are well under his body, and he is balanced and ready for movement in any direction — and like a coiled spring, he can do it very quickly.

The horse's back is slightly arched rather than swayed and hollow; thus, he is better able to carry the weight of saddle and rider. Because the horse's weight is shifted more toward the hindquarters, not only does he become more maneuverable, but stress and concussion on his front legs are reduced as well.

The collected horse looks shorter in length. Collection requires action of the back and hindquarters. When he takes more weight on his hindquarters, he crouches down slightly behind, which pulls somewhat on his back muscles, stretching them and making them more elastic. As he stretches his back muscles, he takes more weight off his front end. He also stretches and lengthens his neck, bending at the poll and bringing his lower jaw closer to his chest, decreasing the distance from the bit to your hand. The collected horse is on a rein shorter than that of the extended horse.

The opposite of collection is *extension* — that is, traveling with head and neck low, and heavy in front. The collected horse's weight is more on his hind legs. As you teach your horse collection, his hindquarters become stronger and develop the ability to carry more of the weight; his action becomes more elastic, elevated, and animated. The collected horse gives his rider a more comfortable and less tiring ride. He is more responsive to the rider. He is ready to move quickly and easily in any direction.

COLLECTION AIDS IN LOAD CARRYING

The lowered haunches and stretched back muscles make the back somewhat convex, allowing the horse to move his back more freely and carry a rider with greater ease. An arched structure can carry a load more easily than one that is straight or concave.

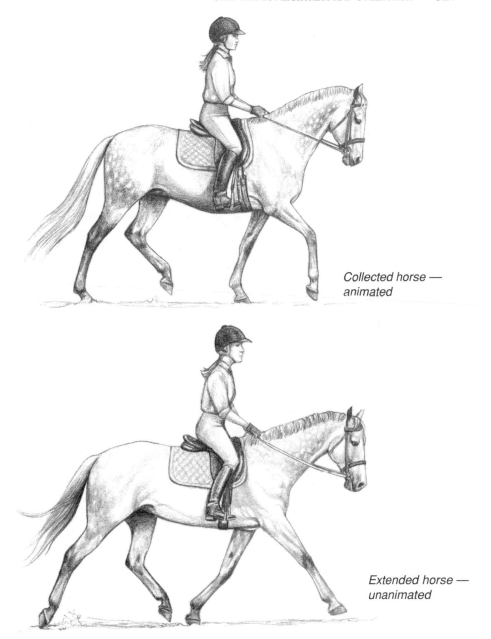

*Collected horse —
animated*

*Extended horse —
unanimated*

How the Horse Balances Himself

A horse keeps his balance by raising and lowering his head and changing the position of his hind feet. The lower his head, the harder it is for him to move freely because the heavier he is on his front end. He is not prepared for sudden movement in another direction. He must be able to get the weight off his front feet in order to free them for action. And to be light on his feet and prepared for a sudden change of direction, he must have his head up and more weight on his hindquarters.

A Green Horse versus a Collected Horse

A green horse with a rider moves in a manner much different from that of a well-trained horse. The trained horse is ready for action at any time. The stock horse, for instance, can leap into action in any direction you indicate. But the green horse is slow to respond if you ask him to head a cow. He is also awkward and clumsy.

He may understand the signal to turn and try to move as you indicate, but he's not collected. His legs aren't under him; he's not in the right position with balance for a well-controlled turn, fast start, or quick stop. He must learn collection before he can become well coordinated with you on his back. Do not ask your horse to perform fast maneuvers until he is far enough along in his training to do them successfully. If you try to make a horse do things beyond his ability, you may be tempted to kick him or to jerk him here and there and to a stop. None of these actions is instructive. Be patient. Proceed carefully and slowly, and don't blame your horse if he doesn't understand your requests.

Creating More Animation

The key to collection is to develop more controlled energy so the horse is moving with more up-and-down action. True collection begins from the ground up. You must enable the horse to engage his hindquarters more. Your goal is to create drive and energy from the horse's haunches.

Head Set

Proper head set alone — getting your horse to raise his head, for example, and to bend at the poll and tuck his chin — is not collection. Simply working on the horse's head, with your hands or with an artificial aid such as a martingale to bring his nose toward his chest, will not produce collection. Your horse will be artificially flexed at the poll, and his face may be up and down, but his head and neck will be stiff; he won't

---

DON'T NEGLECT COLLECTION

Some horses naturally travel with more balance and collection than others do, largely due to conformation and athletic ability. All horses, however, need some training when you start riding them so that they learn to be collected under saddle. It's easier to teach a horse that has natural agility and balance than one that doesn't, of course, but every horse can be taught better collection while carrying a rider.

---

have the balance and agility of a truly collected horse. Proper head set is a result of collection, not the cause. You want to enhance the horse's agility by encouraging him to bear more weight on his hindquarters.

First Lessons in Collection

Everything you teach a horse after his first basic lessons is aimed at making him more pleasant to ride and better able to perform the actions you want him to. You want him light in your hands, agile, and able to adjust his stride to any situation.

When playing out in the pasture, the green horse may handle himself with good balance, but his center of balance changes with the added weight of a rider. When you first start riding him, he travels heavy in front. If he raises his head, he still travels awkwardly because he hasn't learned how to tuck his chin, shift his weight, and arch his back. The uncollected green horse tends to hollow his back when he raises his head, which makes for more clumsiness. He needs some help from his rider to learn how to collect himself while carrying the extra weight.

Setting His Head

Setting his head when teaching your horse collection is done by proper use of the bit with very light hands. If you often fingered the reins lightly during his early training, your horse will now tend to lift his head due to the action of the snaffle. When he is farther along, he should perform well in any bit that fits him, but because a green horse moves with his head low, all early training should be done with a snaffle. You will be working on collection for quite a while; it takes time and patience to develop a well-collected horse.

Teaching the Horse to Give to the Bit

Your horse cannot progress in his training until he develops a soft, responsive mouth that gives readily to the bit. It may take several weeks of riding before he is ready to start collecting; first he must be responsive to the bit. If you try to push him into the bit with your legs (as you must do to get him to collect) and he has not yet learned to give to the bit, he won't understand what you want; he'll fight the bit or throw his head in the air. Be patient: It will take many lessons to get your horse used to the bit and responsive to it.

It helps if you did some ground work before riding, such as sessions in the bitting harness (see page 209), so he already knows about giving to the bit and not rooting with his nose when he feels a little restraint. Then when you start riding him, you merely have to keep working on fine-tuning his responsiveness.

Your first lessons at collection will be at the walk. Gently finger the reins, working ever so gradually at teaching him to give, to raise his head a little more, and to bring back his nose toward vertical rather than stuck out in front. You can't do as much with a snaffle in tucking his chin and relaxing his lower jaw as you can with a curb (that will come later; see page 192), but you can get a good start — you can get him to raise his head and shift some weight off his front.

Using Your Legs to Put the Horse "On the Bit"

Once your horse has learned to give to the bit, so that when you finger the reins he responds by giving his mouth rather than bracing against you, it's time to work with your legs. Begin by using a little more leg pressure. Using your legs — your horse already knows this cue to go forward or faster — makes him push his hind legs farther underneath himself, as he does when moving more quickly. But instead of allowing him to walk faster or break into a trot, keep fingering the reins lightly to keep him at the slower gait. This makes him put more energy (impulsion), and hence more animation and collection, into the slower gait.

It takes good judgment, practice, and lots of patience to know how much bit pressure and how much leg pressure to use, but as you proceed with these lessons, your horse will gradually raise his head and start to

THE JAW

When your horse becomes collected, he will travel with his nose down and his chin tucked in; but more significant, his lower jaw will be relaxed, not tense against the bit. If he is merely traveling with his chin tucked and his mouth behind the bit, not engaging it, you have not succeeded in collecting him. You have probably been too forceful with the bit and your horse has found a way to avoid it. When a horse is collected, his jaw is soft and pliable to the bit. He is actively "on the bit" and ready to respond to any movement of your hands.

flex his neck as he puts more energy into his walk. Slow, careful work in the snaffle, aided by judicious leg pressure, will get your horse traveling lighter in front.

Breaking at the Poll

Once he travels with his head up and starts bending his neck at the poll, just behind the ears, you'll really make some progress, especially if you can now start easing him into a curb rein. A Pelham is ideal for much of this training, as you can use four reins and keep the action of the snaffle to raise the head while adding the action of the curb to help set the head. (See page 194 for more about using a Pelham.) With judicious use of the curb rein, you can get the horse to flex better at the poll and to relax his lower jaw.

Using the Reins

Constant, gentle bit cues are the key to keeping the horse totally responsive; this way, you'll be able to collect and extend him at will. Although you use your legs to push him into the bit and keep him on the bit, it is your constant contact with the mouth, through gentle give-and-take actions with the reins, that directs his energy, letting your horse extend his gait or keeping him collected.

Practice at the Walk

First lessons in collection should be done at the walk, progressing to the trot only after the walk has been mastered. Practice collecting and extending the walk until your horse knows how to respond to your cues.

A collected walk is not necessarily a slow walk. You can have the horse collected at any walking speed. On long rides, you will often let him do an extended fast walk or a moderately collected fast walk to enable him to travel without tiring. There is a difference between speed and impulsion. A horse can walk fast in an extended manner with little impulsion or animation. He is more apt to trip and stumble, however, because he is so extended and heavy in front. Even when going miles and miles at the walk, the rider generally prefers to have the horse on the bit and somewhat collected, just because he picks up his feet better and is more alert and maneuverable.

Some horses are naturally fast and eager walkers and thus may be relatively easy to collect; you merely have to channel their drive into collection. It doesn't take as much effort to put these horses on the bit. A little leg pressure encourages an enthusiastic walker to keep up his drive and impulsion. Active and gentle use of the reins direct his energy into a little more animation — balancing weight back farther, tucking the chin, softening the jaw.

Other horses are lazy walkers; they don't learn to walk fast under saddle until they learn collection. If yours is a reluctant walker, collect him at the same time you are teaching him to speed up the walk. To get a collected walk, you will be urging the horse with your legs but fingering the bit enough to keep him from breaking into a trot. Once he learns how to collect, you can do a slow or a fast collected walk.

BOTTLING UP ENERGY

When you urge with your legs and gently restrain with the bit, your horse's energy in response to your legs is curbed somewhat by the bit and must come forth as a faster, more animated walk, rather than a trot. His energy is released as higher, springier action, not just forward motion. By fine-tuning how much bit and how much leg pressure to use, you can encourage him to do a slow collected walk, a medium collected walk, or a fast collected walk, and can slow or speed it at will, without your horse breaking gait into a trot. It helps to use your legs alternately, stimulating each hind leg in proper sequence, to speed his walk and get more impulsion (see pages 292–294 for more on leg cues).

Teaching a Collected Trot

Many horses can do a good trot naturally; others prefer to jog or gallop. They extend or collect only when prancing or playing. If you have pushed your horse into a good fast trot now and then during training sessions, it will be easy to encourage him to do a collected trot. Gently slow him from a fast trot with your reins, but keep using your legs sufficiently to keep him from slowing down to a walk. Soon he will learn to collect, slowing his speed but keeping the same impulsion, with the bottled-up, contained energy creating more springiness to his gait.

There is a big difference between a slow trot and a collected trot. Most horses, when allowed to trot slowly, merely plod along at a jog, barely picking up their feet and looking as if they might fall down. Their heads are low, and they travel heavy in front. On uneven ground they are apt to stumble. A collected trot is much more animated and the horse's weight is balanced farther back. His head is above the level, and he picks up his feet much better.

Collected trot

To get a more collected trot, keep fingering the reins to keep the horse slow, but continue to use your legs to keep him lightly on the bit and well gathered. With the horse "between the bit and your legs," you have much more control over all of his actions.

The Value of Teaching Several Speeds at the Trot

Teaching the horse to slow down or speed up his trot makes him much more agile and develops his ability to lengthen and shorten his stride. The slowing down also teaches him to respond to the increased feel of the bit (see pages 319–320 for more information).

Many horses carry too much weight on their front legs when ridden, which hinders their agility. Lessons in slowing down, especially to slow abruptly, as you'll do in later stages of training, will lighten the forehand, encouraging your horse to carry more weight on his hind legs. When slowing abruptly, his weight comes off his front legs and shifts back toward his hindquarters. He must know how to do this to do a good stop and to get weight off his front legs for a fast turn or a pivot on his hind legs.

Perfect Balance, Greater Agility

Once the horse starts collecting, he has much better balance and is not so awkward when performing the things you ask him to do. All later training will hinge on how well the horse can be collected — whether you want him to be a jumper or a cow horse or to run the barrels or perform dressage. In timed speed events, he must be able to collect immediately to do a smooth turn around poles and barrels, for example, to maintain his footing and balance for the precision work required between bursts of speed. He'll handle himself with more precision and agility when his actions are perfectly controlled and he is able to switch from collection to extension and back again.

Shifting from Walk to Trot

The best way to perfect a horse's ability to collect and extend is to develop several speeds at both the walk and the trot, along with his ability to shift fluidly from one gait to the other. When changing from a

A horse must be able to collect in order to balance himself and turn quickly around the obstacles in a barrel race.

walk to a trot, shorten your reins a little to maintain contact with the horse's mouth (since his head is steady at the trot and does not make balancing movements) and close your legs on the horse, squeezing slightly, to signal him to move into a trot. If you are posting, strive to keep your hands steady, even though your body is moving up and down in rhythm with the beat of the trot. The more you practice, the better able your horse will be to make swift, calm, and smooth transitions in gait, and from collection to extension and back.

Later, after he has learned collection at the canter (see page 346), you can add transitions from trot to canter and back to trot and eventually from walk to canter and back to walk. This will broaden his abilities and responsiveness to your cues, so that eventually you'll have perfect control over your horse's movements.

Refining Your Cues

Your hands control your horse's forehand — in slowing him, for example, in collecting him, and in indicating direction — and your legs control his hindquarters. Your legs give cues for turning, for pushing his hind legs farther underneath himself, and also urge him to move faster. Legs, body weight, and seat help develop more impulsion, or pushing power — and this is essential before the horse can become collected; you use your legs and lower back in pushing the horse to get his hind legs more fully engaged.

Collection at the Canter

The same principles of teaching collection at the walk and trot apply to lessons in collection at the canter, but it's easier for the horse to learn collection at the canter if he's already learned collection at the walk and trot. You will also have greater control of his actions; he'll be more responsive to leg and rein cues. With a young horse, the canter and gallop are the last gaits to work on because you want him to stay controlled. If you've built a foundation of control at the slower gaits, you've already trained his mind; he will be less apt to act foolishly at the faster gaits.

After you have done some cantering and galloping, and your horse knows when you want him to break into the faster gait rather than merely extend his trot, you can work on collection at the canter. If he already knows about leg pressure as a cue to continue moving more energetically and about bit cues for slowing, it's just a matter of practice and timing with leg and hand cues to gradually slow his canter and encourage him to keep the impulsion. His canter will be animated and springy but slow and controlled.

Once he understands the principle of a collected canter, you can vary the speed and still keep the collection. (The canter and gallop are discussed more fully in chapter 14.)

Don't Overdo It

Sometimes a horse will become "sour" from the rider's constant use of legs and bit. He will resist your cues or become grumpy about doing what you ask — protesting by putting his ears back, swishing his tail, or performing stiffly and awkwardly instead of willingly and smoothly. To avoid this, intersperse the lessons with time spent doing less concentrated things. Also give your horse some days every now and then just to extend himself and forget collection. If you have access to open country, take him out for long rides.

Traveling Miles and Lessons

The ranch horse has an advantage: His regular work in large pastures and on the range checking cattle gives him miles to travel. This takes up a greater percentage of his riding time than concentration on any one aspect of schooling. He is learning on the job and doesn't get bored or sour. He is introduced to a new maneuver after he's gone several miles and is warmed up and settled into his work, ready to pay attention to what is asked of him.

As he becomes more agile and balanced when carrying a rider, he begins to take part in more of the chores of moving and sorting cattle, and in so doing learns how to handle himself in all sorts of situations. After a summer of ranch work, the horse is well along in schooling and has become a useful cow horse. Except for more strenuous aspects of training and working cattle, which should wait until he is older and

Well-trained and collected, this cow horse is able to move in any direction instantly and with agility.

more physically mature, this horse has mastered most of the basic maneuvers that every well-trained horse should be able to do.

If you don't have this type of work for your horse, you can simulate it by varying his routine and doing more of his training and riding in open spaces. This keeps him fresh for precision work, even if what you want him to do is dressage or reining. If you do all of your training in a confined area, your horse will become bored and will probably resent the lessons. He will find ways to avoid doing what you ask. To keep him willing, make time for some miles between lessons. How much concentrated work and how much cross-country work you do will depend on what your horse needs to keep his body in shape and his mind fresh.

Cue Appropriately

Your horse will tell you if you are overusing the bit; he will fight it or find ways to avoid it. You walk a fine line when developing a good mouth; back off if you start to see signs of trouble. A horse that starts an evasive tactic or fights the bit is telling you that you have been too forceful. Your horse's good mouth is a sign of your good hands; his evasive mouth indicates that you need to improve.

The same is true of leg signals. Your horse should respond energetically and calmly to your legs, and you shouldn't have to use them hard to get his response. Whether teaching him to walk fast, collect, or turn properly, he should respond to a mere squeeze.

If your horse responds sluggishly when you use your legs softly, don't use them more and more forcefully. You're nagging when you do this, and it won't do much good. Instead, use a switch or whip emphatically in back of your leg. This calls attention to the soft signal you gave with your legs. Soon your horse will learn that if he doesn't respond to the leg squeeze, he'll feel the switch. He'll prefer to obey the soft signal instead.

In training a horse, exercise good judgment in everything you ask; don't command him to do something too difficult for his stage of training. But after giving a signal for something your horse knows how to do, be emphatic and demand obedience. He'll progress farther and faster if he respects your commands and if you are always consistent. Be reasonable in what you ask of your horse. Be patient and remain in control. This way, you will create a beneficial partnership.

14

THE GALLOP, THE CANTER, AND THE LOPE

The gallop is the fastest natural gait of most horses. The Standardbred's fastest gait is the racing trot, or pace. The gallop is a fully extended gait: The horse stretches his head and neck forward and takes long strides. In this four-beat gait, there is a brief moment of suspension in which all four feet are in the air. A slower version is called a canter or lope, which can be slow or collected (see page 346).

How the Horse Moves at the Gallop

Unlike at the trot, during which diagonal legs move in unison to create a two-beat gait, in the gallop each leg hits the ground in sequence. Even though two feet may be on the ground at the same time, they hit in a different order. At the fast gallop, the horse never has more than two feet on the ground at once.

Leads

Whenever your horse gallops or canters, one front leg and one hind leg come farther forward than their opposite leg. He is on either the left lead or the right lead, which means that during the phase of his stride when he is putting his feet down onto the ground, he is reaching farther forward, or leading, with either the left or the right feet.

If your horse is on the left lead, his left hind and left front come farther forward. His feet will land in a definite sequence and then all will be off the ground in an instant of suspension. If he's on the left lead, the first foot to land after that moment of suspension will be the right hind. Then the left hind will reach farther and lead. Next to strike the ground will be the right front foot, then the left front reaches forward and leads. When the left front is picked up, there will be another brief moment of suspension and the sequence is repeated.

The Gallop versus the Canter

At the gallop your horse is extended and traveling fast. His head and neck are stretched forward. His hind feet hit the ground in sequence but the first one to hit is off the ground again by the time a front hits. Leg sequence is nonleading hind, leading hind, nonleading front, leading front, then the period of suspension.

The canter or lope is slower. Though these terms may be used interchangeably, horsemen often consider a lope uncollected (a slow version of the gallop) and the canter collected. A loping horse and a cantering horse tend to have the same cadence — the rhythm in which the hooves

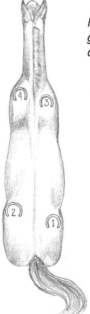

Footfall order at the gallop with the horse on the left lead

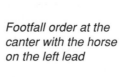

Footfall order at the canter with the horse on the left lead

land — though there will be some variation, depending on the slowness of the gait and whether it is collected or uncollected.

In the collected canter, your horse is not so extended; his head is up, his neck more arched. His weight is farther back on his hindquarters. He's more prepared for a quick change of direction than he is at an extended gallop because he is more balanced for a turn.

When he canters, he has one or three feet on the ground except during the phase of his stride in which all four are off the ground. The canter is usually a three-beat gait. If your horse is cantering on the left lead, his right hind comes down first, then his left hind comes farther forward and leads, landing at the same time as the right front leg for the second beat of the cadence. Then his left front reaches forward and comes to the ground for the third beat.

When the left foreleg is lifted again, there is a brief moment of suspension, as in the gallop, in which all four feet are off the ground. At a medium-fast canter, the right hind is lifted before the left foreleg comes to the ground, making a brief instant in which your horse has two feet on the ground instead of one or three, as in the slow canter. If the canter becomes quite animated, that is with a lot of impulsion and up-and-down movement, it may result in a four-beat cadence like the gallop, even though it's much slower. In this instance, the paired front leg and opposite hind leg do not land together; the hind foot lands slightly sooner than the front.

How to Ride the Gallop or Canter

As your horse starts to gallop or canter, bring your weight sharply forward and your knees and calves into closer grip with the saddle. Your back should be hollow; maintain it that way by pressing your buttocks to the rear. This also helps you keep your heels down. Your thighs, knees, and calves should be in close contact with the saddle, firm but not rigid. Your arms (if riding with a snaffle and two hands on the reins) should be extended and moving with the balancing gestures of the horse's head and neck. With a curb bit and one hand on the reins, that hand is forward and moving with your horse's head and neck, thus staying in perfect bit contact.

At the canter or gallop, you should stay up out of the saddle, floating close to it, with your weight in the stirrups. The horse will have more

BODY POSITION

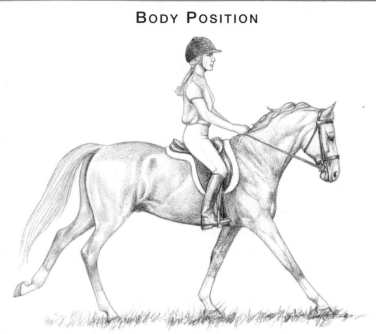

Rider's position at the canter. Her weight is forward; back is hollow; head is up; heels are down; thighs, knees, and calves are in close contact with the saddle. Her arms are extended, moving with the horse's head and neck.

At the extended gallop, the rider's weight is brought more sharply forward at the waist.

freedom of movement when your weight is balanced through your knees and stirrups rather than on his back. At the canter, the saddle should pat you, not bang you. No matter how fast the horse is traveling, the stirrups will hang straight if you are properly balanced.

At the extended gallop, your weight is brought more sharply forward at the waist as your hands shorten the reins: Leaning forward has shortened the distance between hand and bit. Apply your legs with greater force to start the horse into the faster gait, after which merely rest them alongside him, ready to urge him on if he slackens his speed. Keep your back hollow and your heels down; your knees are your main contact and pivot points, with your upper body leaning forward and your buttocks raised slightly from the saddle and pushed to the rear. Use your shoulders to maintain your balance as you move in sync with your horse.

Teaching the Green Horse to Gallop and Lope

After your horse is well started and has good control at the walk and the trot, add some galloping to his training sessions. You must determine when he is ready for this step. This will depend somewhat on his age and his attitude. With a very young horse (such as a two-year-old you are working with in preparation for later lessons), it's best to postpone fast work until he is at least three years old.

With a skittish or exuberant individual who may use galloping as an excuse to bolt or buck, wait until you have established more control at slower gaits so he has confidence in your authority. With a mellow individual, you may be able to progress to the canter quickly. You may reach a point during a lesson when you suddenly realize he's ready. It may be while working in an arena or pasture, or traveling across country. You just know that his attitude and the time are right. It's best to try the canter when riding alone, because if another horse is traveling fast, your horse may become more excited. Keep the first cantering lesson brief.

If you are riding across country when you decide to try a short gallop, choose an area free of obstacles, relatively flat with good footing, or slightly uphill. (A horse is much less apt to buck while galloping uphill and more likely to buck while galloping downhill.) Don't choose a spot that has a ditch or depression; if your horse jumps it, he may go right into a buck. Young horses often like to play and buck just from high spirits, and when they start galloping, they tend to slip right into play mode.

For a first gallop, choose ground that is relatively level or slightly uphill with good footing.

Cuing for the Lope

When training the green horse, put him into the lope or canter from a trot; it will be easiest for him. Later you can teach him to go into a canter from a walk or a standstill, but that takes more practice. He will be less confused if the first few lessons are from a trot.

Your horse should have good control at the walk and the trot and be responsive to your signals through the reins. He has begun to develop a good mouth and is able to understand your cues through the bit. He is ready for the canter after you have smooth transitions from walk to trot and from trot to walk and when he knows how to respond to the bit and to leg pressure without becoming confused or resistant. He knows how to speed up and slow down on cue.

To encourage your horse to go from a trot to a canter, squeeze harder with your legs and lean forward a little to signal him to go faster. His tendency will be to move faster to restore balance. Don't lean too far forward, however, for that will have the opposite effect: He may slow down or stop. If he tries to trot faster in response to your cue, gently finger the reins to check him with the bit. This tends to control his energy, and because he cannot trot faster, he will break into a canter or a lope.

DON'T RUSH HIM

If the green horse is hesitant to break into a canter or gallop, don't rush him. Keep asking for more speed at the trot and immediately release your cues when he increases his speed, so he knows that he has responded properly. Let him travel a few strides at the increased speed before you apply leg pressure again. After applying your cue several times, the horse will break into a gallop. He may gallop only a few strides and then drop back to a trot, but that's fine. Take as many lessons as needed to work on sustaining the canter or gallop. You don't want to push him too quickly, or he may try to bolt or buck. After several sessions of short gallops, the horse will figure out what he is being asked to do and will respond more quickly to your cue. You can then increase the distance he travels as he becomes more confident.

If your horse trots faster, continue squeezing with your legs and gently checking with the bit until he breaks into the canter. Then immediately cease checking with the bit. Follow his head and neck movements with your hands, keeping him on the bit with light contact but no longer checking him. Do not lose contact with his mouth or you will have less control over his actions if he tries to become playful and throw in a few jumps or bucks.

He should stay in good control, contained between your legs as you encourage him to keep cantering and your hands as you maintain good communication through the bit. Soon your horse will learn that your leg pressure and the bit signal are his cues to canter. If you synchronize your legs and hands for a smooth transition, he will shift gaits fairly readily, without leaping forward into a gallop or merely trotting faster.

If your horse is sensitive and skittish, be careful about giving too much pressure with your cues or he may try to buck. It may take several sessions to work him up to a canter when you begin this phase of training. Take all the time he needs, so the first lessons will go smoothly and successfully.

Encourage him to trot faster, then let him drift back down to a slower trot. Do this a number of times, slowly working your horse up to the point when he decides to break into the next gait. Eventually, he will shift into a lope on his own. It will be his idea — you are not forcing him — yet he learns that this is what you are asking and will make the connection. Soon he will understand to shift into the lope or canter when you ask him for it.

If your horse gallops too fast, spend a few lessons on slowing down. Use your hands lightly, taking and giving, pulling and slacking. He'll usually respond by wanting to drop to a trot, so use your legs to keep him cantering. By using your legs and hands, you help him learn what speed to go, keeping the canter controlled.

Don't Worry about Leads

When you start cantering, pay little attention to leads. Allow your horse to lead with his right or left foot. Just make sure he is never cross-cantering or *disunited,* that is his fronts and hinds are on opposite leads (see page 353). If he starts off that way, ease him back to a trot and start over.

When first teaching your horse to canter, you will not be making small circles where correct lead is essential for proper balance. During first lessons, it's best to work in straight lines. Conduct lessons in a large enough pen or arena where you have plenty of room for a straight canter, or in an open field or on a nice trail or track with good footing, so you can allow the horse to move out freely.

If you try to concentrate on leads from the start, some horses become confused; this is simply too much for them to think about all at once. The young horse first must learn new balance — how to handle himself with a rider. Work on leads later, after you are both at ease with this gait and the horse is moving well and completely under control.

Teaching a Collected Canter

A canter is a slow and collected gallop. Some horses can gallop slowly but have not been taught collection and are still traveling in an extended manner — just slowly. With their heads low, they look as if they're about to fall down at every stride. Neck and body are strung out; they are traveling heavy on their front feet. They lack animation. The uncollected horse is not in a position to change direction suddenly or gracefully. To do a good job of changing leads while cantering, a horse must be collected, so this will be one of the next things you teach him. (See chapter 13 for more on collection.)

Urge your horse with your legs to keep him cantering, and finger the reins to keep him from going faster. Direct his energy into more animated action instead of more forward speed. Your hand(s) should be moving with his head and neck as you communicate with him through the bit with gentle give and take. Your hands are moving with him, yet you are letting him know he is supposed to stay slow as you restrain him gently with your fingers on the reins.

Use your weight to advantage. Sit more upright than at the gallop; if you lean forward too much, your horse will take this as a signal to go faster. Using your weight farther back keeps him collected and traveling slowly. His head will be raised instead of extended, and his hind legs will push farther underneath his body. He will have more spring to his movements, as his propelling power is directed up and down as well as forward. Checking him lightly with the reins while urging gently with your legs tends to contain his energy somewhat. This encourages him to move more springily, and to travel lighter on his front end.

Speed Control While Keeping Animation

With practice, your horse will master a slow, collected canter. You'll be able to put him into a canter from a walk, trot, or standstill by leaning forward and squeezing with your legs. As your horse takes his first cantering stride, lean back slightly so your upper body is only a little forward, and continue using your legs and hands to keep him from slowing down or speeding up.

Working with Leads

A horse is well along in training when he has mastered collection at the canter. Now he can take either lead from a walk, trot, or standstill and can change leads smoothly while cantering. The gallop and canter bring out your skill and showcase your horse's training. Both gaits demand good response from a horse.

If you are familiar with the way a horse moves at these gaits, you can more readily understand how and when to use certain signals to encourage yours to perform as you want him to. Your judgment in what you ask of the young horse will be more informed.

Proper Lead Changing

When a horse changes leads correctly, he does it smoothly, always changing with his hind feet first — as they are the feet that come to the ground first after the period of suspension. If he were to change leads in front first, as he would if improperly cued by the rider, or changed in front and not behind, he would have a rough, unbalanced gait and could hit a front and hind leg together. You must give the proper cues so your horse will be able to change leads smoothly and correctly.

Why the Horse Has a Leading Leg

At the gallop and at the canter, the horse moves his legs in sequence rather than in pairs; thus, one front and one hind will always come to the ground ahead of the others. For balance and agility, then, your horse must be on the same lead in front as he is behind; when traveling free, without a rider to confuse him, this is generally the case.

If a horse is galloping in a straight line, he will normally change leads periodically to keep from tiring. When he is galloping in a circle or making a change of direction, he normally leads with his inside legs. This helps balance him and gives better traction on a tight turn in bad footing. A horse that is circling to the left should be on the left lead; if he is circling or turning to the right, he should be on the right lead.

How and When the Horse Changes Lead

When cantering or galloping, the diagonal set of legs that are on the ground together undergo more stress and are more subject to tiring. In the right lead, the right hind and left front get the stress; the left hind and right front are stressed in the left lead. For this reason, the horse often switches leads when running free. If you are riding him and doing a lot of speed work, ask him to change leads periodically.

Many horses do so automatically, without cues, but some get confused and change leads only in front or in back. You may need to give your horse cues for changing so he will do it properly. Then the stress will be on the opposite diagonal pair of legs for a while and the other legs get a rest.

When a horse changes leads, he does it during the instant all four feet are off the ground. If he's been cantering on the right lead, for

LEADS

Right lead: *The leading hind foot is still on the ground and the right front has just come to the ground farther forward.*

Left lead: *The right hind foot lands first, then the left hind comes to the ground and leads. The right front strikes the ground next, then the left front comes forward to lead.*

instance, he changes at the moment of suspension and the foot that then comes to the ground first is the right hind followed by the left hind, which reaches farther forward. He is now on the left lead.

Determining Which Lead He's On

There are several ways to tell what lead, gait, or diagonal a horse is on without leaning over to watch his feet; leaning over throws him off balance. You determine by feel, by watching his shoulders, and by listening to the cadence of the hoofbeats.

If he is in the left lead, a glance at his shoulders will show that his left shoulder is always moving forward about 4 inches in advance of the right one. The left shoulder, like the left leg, is moving farther forward than the right at each stride. If you are traveling in a circle, his inside leg (the left one, in this case) will be in motion; his outside leg will be comparatively still.

Control of His Leads

To put your horse into a canter or gallop on a certain lead, or to have him change leads while cantering, you will need to use proper leg and rein cues. Your horse will more readily respond to your prompts if you first put him in a position that leaves no room for confusion or hesitation. By correct use of your legs, shifting your weight properly, and a gentle but firm hand on the reins, you can put him into the proper position or take advantage of his being in the correct position. It is easier to teach your horse his leads after he learns collection.

Signaling for a Lead

There are several ways to put a horse into a certain lead. One may work better for you, or with a certain horse, than another. It's always good to know different methods; this way, you can tailor your lessons to suit your horse.

Use Leg Pressure

No matter which method you use for weight shifts and handling the reins, squeeze harder with your outside leg, behind the girth, when you

Putting the horse into the canter on the left lead from a trot. While sitting or post-ing on the outside diagonal, the right in this case, the rider gives the signal to canter on the left lead by squeezing with his right leg as the horse's right front foot hits the ground.

give the signal to canter. (The outside leg is the one opposite the side on which you want your horse to lead.) If you are making a circle to the right — that is, the center of the circle is to your right — and you want him to pick up the canter on the right lead, squeeze hard with your left leg.

This cue stimulates your horse to give his greatest push with his left (outside) hind leg and he will lead with his right (inside) leg. Because you have trained him to move away from leg pressure, squeezing with your outside leg helps signal for a turn to the right — the direction in which he'll be going as he starts to canter on the right lead.

Cue When Turning

An effective way to make him take a certain lead is to give him the signal to canter when going around a turn or making a circle. This takes advantage of his natural tendency to lead with his inside legs.

Let's say you want your horse to pick up the canter on the left lead. The best way to accomplish this on a green horse is to ask for the canter while he is making a circle to the left. This can be most easily accomplished the first few times if he is trotting.

Be sure to make him take the right lead as often as the left so he won't become a one-lead horse. And give him the signal to canter on a certain lead at various places when you're working so he won't learn to anticipate your signals at the same place or at the same turn.

To pick up the canter from a trot, give the signal when your horse's outside front leg is on the ground. This means that the inside hind leg is on the ground also and the outside hind is moving forward, ready to strike the ground. If you press with your outside leg at this phase of his stride, he will be in the correct position to begin the canter on the proper lead. He will be able to push off more strongly with his outside leg, the one you are stimulating with your squeeze (the right one, in this case, if you are asking for a left lead), which is now coming to the ground, thus making the transition to the left lead.

If you are posting on the outside diagonal, as you settle down into the saddle give the cue to canter, pressing with your outside leg while your horse's outside front foot — the right in this case — is striking the ground. If sitting the trot, press your outside leg hard as the horse's inside shoulder moves forward. In some respects, it is better to try this technique from a sitting trot, as you are more in contact with your horse and thus able to use your hands and legs more harmoniously. If it is difficult for you to tell which phase of his stride the horse is in while sitting the trot, however, it is better to use a posting trot and give your cues on the outside diagonal.

Turn the Horse's Head

Another method of asking for the canter on the left lead is to turn your horse's head very slightly to the outside of the circle — in this case to the right — as he is circling to the left. Turning his head to the right and leaning back a little at the same time as you push hard with your outside leg will usually result in him picking up the canter on the left lead. Turning his head to the right hinders his right shoulder action and frees his left one to move farther forward and thus lead. Pressing with your right (outside) leg pushes his haunches slightly toward the left, toward the center of the circle, and this helps him lead with his left hind foot as well as stimulating his right hind to push off hardest.

As you do this, lean slightly to the left. If you tilt his muzzle slightly to the right, which would be his natural tendency when taking a left lead, be careful not to turn his head too far or you'll put him off balance and he'll take the right lead instead. He'll push his left shoulder forward to catch his balance, and the result will be a right lead because he'll be at the wrong phase of his stride to take the left lead at that moment.

Prevent Cross-Cantering

When cantering or galloping, your horse will be on the left or right lead or perhaps disunited, which means he is on one lead in front and the other lead behind. Cross-cantering (another term for *disunited*) produces a rolling motion of the horse's body that is uncomfortable for the rider and dangerous because the horse is unbalanced. When his feet are not moving in proper sequence, he is more likely to strike a front foot with a hind foot and throw himself. If your horse ever starts the gallop or canter disunited in his leads, stop him and start him again.

The Flying Lead Change

When he takes the proper lead readily and on cue, you can teach your horse the *flying lead change* — changing leads at the canter. He must make the change while he is at that phase of his stride during which all four feet are in the air.

SOME HORSES ARE ONE-SIDED

Many horses naturally take the left lead when ridden; some will not take the right lead at all until they are taught to do so. These horses are, in effect, "right-handed." When a horse goes into a canter or gallop, he raises his front and gives a push with his hind legs. Most of the pushing power comes from the hind leg on the side opposite the leg he will use to lead. Because he is "right-handed," therefore, his main driving power comes from his right hind leg and he naturally goes into the left lead. At the same time that he's giving this push with his right hind leg, his muzzle will be slightly tilted upward and to the left.

Many horses will change leads automatically when working cattle, following the turns and dodges of the cattle, without training and without cues, because they know what they are doing and can anticipate the direction they need to go. But they must be taught the changes of leads for arena work and other types of riding in which they do not always know when the rider will request a change of direction.

The Figure Eight

This is a useful movement for teaching the flying lead change. Each loop of the figure eight should be a circle at least 30 feet in diameter. Canter the horse in a large circle on the proper lead, then give him the signal for the new lead as you change direction and begin making the other loop of the eight.

To make it easier for the horse, drop to a trot for a few strides in the center of the eight before you make the change, and straighten the horse. You don't want his body crooked or even a little bent from the last circle. Give him the signal to canter on the opposite lead after you've put him in position to begin the next circle. This way he'll pick up the next lead readily, especially if he already knows how to take his leads from the trot. A few trotting strides will give him time to get into the proper position to go into the new lead. The more you practice, the fewer trotting strides your horse will need, and soon you'll know he is ready to try the flying change.

Canter one circle of the figure eight. Just as you begin the next circle, check your horse with the reins for an instant, then immediately give the cue for the new lead. He should make the change of leads in midair.

You may find it helpful to keep your horse's head bent slightly in the direction of the old circle for an instant before you shift to the new direction. Squeeze hard with your legs as he makes a stride, exerting more pressure with your outside leg behind the girth. This signals for the next lead. Use your legs when you are farthest down in the saddle, at the phase of the stride when your weight is pressed down hardest in the stirrups and your body is lowest. Be careful not to lean too much in indicating the shift of direction; you don't want to put too much weight over what will become the leading shoulder. Don't prolong the lesson; flying changes will tire your horse if sessions are too long.

Changing Leads on the Straightaway

After learning how to respond to your cues, your horse will make the flying change with minimal prompts. Then he should be able to change leads at your signal, whether he is making a circle or cantering in a straight line. Continue to practice doing changes on the straightaway.

BE SUBTLE BUT URGE WITH FEELING

The method you use to put the horse into a canter on a certain lead, whether you turn his head to the outside or to the inside, will probably depend on the results you get. Some horses respond better to the first method, some to the second. Whichever one you use, be subtle and lean only slightly. Concentrate on using your outside leg energetically — a little behind the girth — to push the haunches toward the center of the circle and to stimulate the outside hind leg.

Picking Up the Canter from a Walk

A common way to ask a horse to pick up the canter on a specific lead is to lean forward and to the inside — to the left if asking for a left lead. Leaning to the left and forward, simultaneously using the reins and your legs to indicate left direction, helps urge your horse to turn to the left. He will then lead more readily with his inside legs (the left ones, in this case), his natural tendency when turning or making a circle. If you use this method, don't lean excessively; throwing your weight over the horse's shoulder will put him off balance.

Picking Up the Canter from a Standstill

As your horse becomes more proficient at taking the correct lead at your signal, you can teach him to start the canter on the proper lead from a standstill. The usual way to ask for a canter departure on a certain lead is with the help of diagonal aids, conforming with the horse's natural tendency to flex his body a certain way when he goes into a certain lead. When asking for the canter on the right lead, flex his head in that direction, use your left leg to stimulate his left hind, and push his hindquarters a little to the right. This puts him in the position he takes naturally when starting the canter in the right lead.

Using your cues this way makes it impossible for your horse to start off straight, however, which you may want him to do later. Eventually, refine your cues so that just a slight touch of the rein will signal the lead to take. Then your leg is not needed quite so much as a cue, and you can use whichever leg is needed to help him canter straight.

When he is taking leads nicely from a trot in a circle, you can gradually work your horse into taking either lead on a straightaway as well as on a turn, and from a walk or a standstill as well as from the trot. At first you may have to push his hindquarters a little to one side — to the left when taking a left lead; to the right when taking a right lead — but with practice he will be able to depart into the canter when his body is straight. *Canter departure* is a term often used to describe how well or poorly a horse executes this.

Refine Your Cues

As you and your horse work on leads and learn to communicate better, start to refine your cues. You may need merely to shift your weight a bit, finger a rein, or move your leg slightly, and your horse will know what you want him to do. When your cues are done consistently, they can become almost imperceptible.

When giving cues to canter, push with your inside pelvic bone as you shift your weight to the inside to increase the horse's impulsion on that side. You don't want to lean excessively; that will just put him off balance and impede his leading shoulder. You don't want to hinder that shoulder; you want to free it so it can move forward easily and let that front leg become the leading leg. As your horse becomes familiar with your cues, he will come to associate the canter on a certain lead just by your position in the saddle.

After your horse has reached this stage in his schooling, he is more useful for any kind of fast work and is ready to train for athletic events. One of the main purposes in training is to have your horse always under control, working willingly and responding readily to your signals. He can't do a good job until he is always where you want him, moving with good balance, with his feet under him properly. And this can be accomplished only after you have control of his cantering and his leads.

Your horse needs a lot of work and training at all gaits in order to develop dexterity and balance while carrying a rider and to learn to adjust the length of his stride to any given situation. As his trainer, you learn to give the proper signals at the right time to get the best response from your horse. Now you are working as a team.

Don't Overdo It

If at any time your horse becomes confused or won't take a certain lead, slow down and go back a step or two. Go back to putting him into a certain lead from the trot or on a turn. After a while, he'll associate your signals with the correct lead, regardless of where or when you give them — especially if you give the cues at the proper phase of his stride to use his natural inclinations to advantage.

None of the sessions should last so long that your horse starts to resent them. He'll become sour and refuse to respond or start resisting the cues. As with all lessons, when your horse responds properly and takes a lead successfully, end the cantering session and do something else. Intersperse short cantering lessons with lots of work outside the arena, and your horse will remain cooperative and enthusiastic.

15

FURTHER SCHOOLING

After your horse has learned balance and control at all gaits, there are many other lessons the two of you can work on. The more versatile and flexible your horse becomes, the more he will be able to do, and he'll be more pleasant to ride as well.

Fine-Tuning Your Communication

Every time you ride your horse, you are training him: Your goal is to continue improving communication with him as you strive to refine your cues and his responses. Whether out on the trail, in the show ring, heading over a jump, or chasing a cow, good communication between rider and horse is essential.

Work as a Team

The well-trained horse gives readily to the bit and to leg cues and is able to move left and right, to flex, and to extend. You and your horse have spent countless hours perfecting your teamwork. Your lessons and exercises have trained your horse to be more responsive to leg, rein, and weight-shift cues; at this point, you can signal him with the lightest possible prompts to get the desired result. Now, everything after first mounted lessons are geared toward that goal.

Teaching a Smooth Stop

Your horse must stop immediately and smoothly when asked. This is something you work at all along — teaching him to respond to *Whoa* from the time he is a youngster and then under saddle. Stopping quickly and smoothly from fast gaits takes practice. Your horse learns to stop through repetition and reward; you learn to cue at the proper phase of his stride.

How the Horse Stops

When he is asked to stop, the momentum of the horse's body is still moving him forward. He uses his front heels as brakes while he straightens his front legs. He shifts his weight to the rear, placing his hind feet on the ground in front of the normal standing position. He does this by raising his head and flexing his hocks and stifles — that is, by crouching in the rear to shift his weight to his hind feet.

When forced to stop unexpectedly or pulling up from a gallop, your horse must shift his weight to his hind feet before he can regain the use of his front feet. If he doesn't put his hind legs well under to shift back

To come to a sudden stop, the horse shifts his weight to his hind feet. To do this, he raises his head, crouches down on his hindquarters, and slides his legs under him.

the weight, he'll bounce to a stiff-legged stop on his locked front legs — uncomfortable for both the horse and rider.

From a Walk

To ask your horse to stop, lean back slightly, tell him *Whoa*, and give a light touch of the bit. If he stops, immediately release the tension on the reins and the feel on his mouth. This release of pressure is his reward. If he doesn't stop, increase checking with the bit until he does. Vibrate the bit with a series of give-and-take actions rather than a steady pull; your horse will pay more attention to it. Practice until he responds instantly and without resistance.

A horse that doesn't stop well usually has a rider who never gives him slack. Because the reins are always tight and the rider is pulling all the time, the horse pays no attention to the signal. Don't be that kind of rider!

From a Trot

Once you are satisfied with the stop you get from a walk, try stopping your horse from a slow trot. When he stops well from a slow trot, try a faster one. Lean back a little as you check with the reins. Your shifting weight helps him change his forward motion to a stop because of the balance, and it also becomes a cue.

TIPS FOR STOPPING LESSONS

If at any time your horse refuses to stop smoothly, work on the stop at a slower gait until he once again performs it well. Try to avoid stopping on rocky or uneven ground. This can hurt his feet; and he'll probably be reluctant to try this again.

From a Canter

The timing of your signal is important when you want your horse to stop from a canter or a gallop. If you ask him to stop during the wrong phase of his stride, his hind legs won't be in position to stop properly; the result will be a series of jarring hops on his front legs, with all his weight on the front. You must ask him to stop as his hind legs are coming forward under him, not when they are extended backward completing their drive.

To get the proper timing, canter the horse slowly. Feel his rhythm and watch the movement of his shoulders to determine when the lead foot strikes the ground. Time yourself to this rhythm. Ask your horse to stop when his lead foot hits the ground; this is when the hind legs are starting to come forward.

To cue him to stop at this moment, sit up straighter in the saddle and lean back a bit as you check him slightly with the reins, raising his head a little without pulling. By raising his head in this way, you are helping him raise his forehand and clear the ground with his front feet as he makes the stop.

Sometimes it helps to squeeze your legs a little as you lean back, encouraging your horse to drive his hind legs underneath himself. But if you do this, check him adequately at the same time, so he will stop instead of increasing speed.

From Canter to Walk to Halt

Going from a canter to a walk in one stride can be included in lessons for teaching your horse a collected canter (see page 346). As you alternate the canter and the walk, always keep your horse on a loose rein so he stays relaxed and learns to canter slowly. The first transitions from canter to walk will be gradual; never force him to slow down quickly. As lessons progress, however, your horse will begin to understand that you want him to drop to the lower gait, and soon he will require fewer trotting strides to accomplish this. He'll eventually be able to go directly from the canter to a walk.

There is one phase of his cantering stride during which he can change directly from a canter to a walk, and that is the moment when he has both front feet and one hind foot on the ground. This part of his stride corresponds roughly with the stage in his walk at which he also has three feet on the ground. At this moment of the canter, he can shift gears and go directly into the walk.

Give the signal to your horse with enough time for him to make the transition at the correct part of his stride. At the instant he's ready to make the transition — when his leading front leg comes to the ground and one hind leg is still on the ground — cease your checking signal with the reins and use your cues for walking, yielding slightly on the reins and

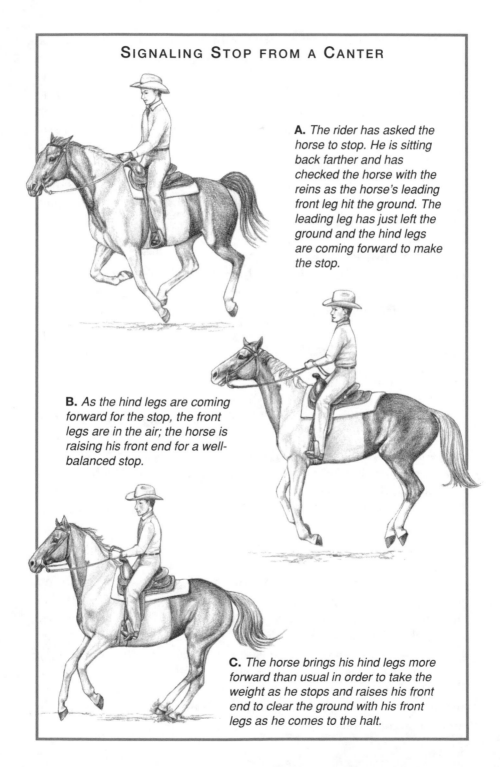

SIGNALING STOP FROM A CANTER

A. *The rider has asked the horse to stop. He is sitting back farther and has checked the horse with the reins as the horse's leading front leg hit the ground. The leading leg has just left the ground and the hind legs are coming forward to make the stop.*

B. *As the hind legs are coming forward for the stop, the front legs are in the air; the horse is raising his front end for a well-balanced stop.*

C. *The horse brings his hind legs more forward than usual in order to take the weight as he stops and raises his front end to clear the ground with his front legs as he comes to the halt.*

squeezing with your legs as your horse lengthens and begins his first walking stride.

Transitions from canter to halt are not difficult once your horse can do a smooth change from the canter to the walk. Making a correct halt with your horse standing squarely is more of a challenge. He must be checked from the canter but allowed to round off his stride so he stands squarely on all four legs; use a half-halt to put him into the first walking stride, then check him more fully to bring him to a smooth, square stop. (A *half-halt* involves briefly checking the horse with the bit but allowing him to continue on. It is generally used to signal the horse to slow down or to prepare for a transition, such as a change of gait or to change leads.)

Stop from a Hard Gallop

After your horse can do a good stop from an easy canter, gradually increase his speed so that he learns to do a smooth stop from a fast gallop. To stop well balanced, your horse must use his hind legs to stop, taking the weight on them under his body but not too far underneath. He should use his hocks, stifles, and hips to cushion the stop, keeping his front legs off the ground until he has come to a halt.

SLIDING STOPS

A stop in which the horse locks his hind legs and slides on them is a requirement in a reining class in the show ring but is impractical anywhere else — even in reining competition if the horse must do a fast turn after he stops. If his hind legs are locked underneath him in a slide, he must regain his balance before he can turn. A sliding stop is hard on the horse's legs, especially on rough ground.

The Fast Stop

The purpose of a fast stop is for the horse to stop as quickly as possible without causing himself or the rider discomfort. He must stop in a balanced position so that if you ask for a quick turn afterward, he can execute it immediately. He should always stop on his hind feet, rather than bouncing to a jarring stop on his front legs. A well-trained horse, if properly balanced, can stop from racing speed in less than 5 feet.

Teaching the Horse to Neck-Rein

As soon as you start putting your horse into a curb or Pelham bit (see pages 188–193 and 193–194, respectively), you will teach him to neck-rein. Because the action of any curb is to the rear, direction cannot be adequately given by direct-reining to the side. With neck-reining, you hold the reins in one hand. To turn, use a loose, indirect rein applied across the horse's neck to encourage him to move away from rein pressure. (The indirect rein is the rein opposite the direction you are turning.)

The First Lessons

Start teaching the neck rein while your horse is still in the snaffle so he will understand what you want him to do by the time you put him into a curb. Though you are still using both hands on the reins, begin using the indirect rein across his neck, following up this cue with the direct rein to guide his head around and reinforce the turn. This helps him make the connection, and as lessons progress, it gives your horse a chance to respond to the neck rein without needing the follow-up cue.

When first starting to neck-rein, change directions only slightly. As your horse begins to understand the signals for turning, rein him at a more acute angle, until you can readily turn him 90 degrees.

Practice neck-reining from a walk at first, then from the trot, and finally from the canter. If at any time your horse refuses to turn well, drop back to a slower gait. If he has

Start neck-reining the green horse while he's still in the snaffle. Use the neck rein, then follow up with a little help from the direct rein.

trouble figuring out the turn, use two hands again while using a snaffle or a Pelham with snaffle reins and curb reins and help him with the direct rein until he catches on.

Use Your Legs

Always use your legs and body when turning; your horse will get the idea more quickly. His natural tendency is to move away from leg pressure and in the direction in which you lean. Turning right, shift your weight slightly to the right and press with your left leg. Usually gentle pressure is enough, but you may need more if for some reason your horse doesn't respond to the soft cue.

You'll get a better response using your legs as an aid in turning rather than turning him with reins alone. Leg pressure encourages your horse to turn with his whole body. Horses that are just reined around on a turn may get "rubbernecked" — that is, they turn their heads while continuing straight ahead. Proper leg pressure will prevent this. It also helps the horse bend his body correctly on the turn.

Teaching the Horse to Turn on the Center

Turning on his center is what a horse does naturally, but he often has to be taught to do it under saddle, as his balance is altered when he is carrying a rider. This is a good movement to learn: It's the fastest way a horse can turn, in the least amount of space. When doing fast work later, in a sport like polo or for cow cutting, for example, this skill will come in handy. The turn on his center requires less effort and motion than a pivot on his hind legs.

Perfecting a Natural Movement

When a horse turns on his own, without human interference, he moves his forehand in one direction and swings his hindquarters the other way. The pivot point is the center of his body. To teach your horse to do this under saddle, use your legs and body language, with rein cues, to indicate the direction you want him to turn. Teach this lesson while he is moving rather than standing still; this way, he'll see the purpose in the change of direction.

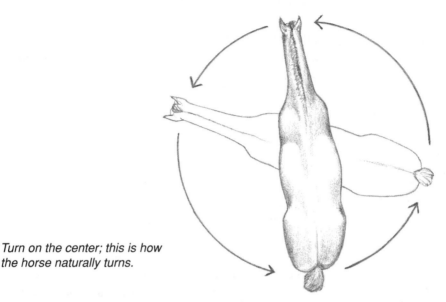

*Turn on the center; this is how
the horse naturally turns.*

It is easier to teach a horse to turn on his center if he already knows how to collect. While doing a collected walk, you have precise control of the horse's total movement. You can direct his front end one way while, with a bit of leg pressure, you swing his hindquarters the other way. You are spinning his body with your legs as the pivot point.

While your horse is gathered and collected, moving with impulsion, signal for a tight, controlled turn with your reins, simultaneously using your legs to spin him on his center. Much of your directional cue comes from your legs. Rather than squeezing just with the outside leg, as you'd normally do in a turn, to hold hindquarters in place, use both legs, squeezing steadily, to direct the hindquarters to rotate in the opposite direction from the front end. Your legs are turning the horse in place. Once he gets the idea, this becomes natural for him and greatly increases his maneuverability when you need a quick and precise change of direction.

Teaching the Horse to Turn on the Forehand

Turning on the forehand teaches a horse to move away from leg pressure, preparing him for advanced movements such as lateral work, and also helps overcome resistance to the bit. It is handy if you need your horse to be more maneuverable for opening and closing a gate from horseback, for example, or to reposition him for other movements. Maneuverability is always an asset.

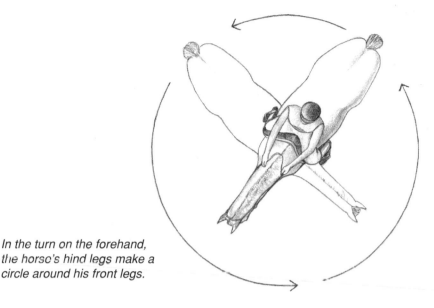

In the turn on the forehand, the horse's hind legs make a circle around his front legs.

In its most basic form, the turn on the forehand is a lesson in which your horse learns to obey your leg and move his hindquarters around to the side. Ideally, this movement is a small circle on two tracks. Your horse's hind legs make a large circle around his front legs (which make a very small circle), with the inside front leg almost marking time in place. He learns this movement easily, even though it is not a natural movement; remember, when turning on his own, he turns on his center.

Cuing for the Turn on the Forehand

The turn on the forehand is an excellent exercise for developing the muscles of the loins, because the hind legs take large sideways steps. Developing and maintaining the elasticity of the loins enhances the athleticism of the horse.

If you have already taught your horse to turn on his forehand during ground work (see page 177), it will be very easy to accomplish this while mounted. Stop your horse squarely. To move his hindquarters to the right, use pressure with your left leg. With a rein in each hand, pull back the left rein slightly, toward the withers (that is, pulling toward the horse's right hip). His head and neck should be bent slightly to the left. When the movement is executed properly, it produces good flexion of the neck and jaw, and his jaw is relaxed, not stiff. Your hands hold the front of your horse relatively still while your outside leg pushes the hindquarters over, with your horse retaining the feeling of forward movement.

If he tries to move forward, your hands check him lightly to keep him in place. If he tries to move backward, your legs keep him thinking "forward," so he can round his back and free up his hindquarters for movement. He can't make sideways steps properly with his hind legs if impulsion is to the rear rather than forward.

Ask for one step at a time. Your leg and rein cues keep his body in position — increase pressure to thwart improper direction; release pressure when your horse yields or takes the step with his hind legs in the proper direction. After your horse completes a 180-degree turn, walk him forward.

Reining and Pivoting: Smooth Turn on the Hindquarters

The turn on the hindquarters (pivot) involves moving the front legs in a circle around the hind legs. Your hands control the forehand and your legs control the haunches. Your legs aid in turning the whole horse, not just the neck. They help keep the horse collected so you can teach him to shift his weight back and turn on his hind legs. You want your horse to turn smoothly, and he can do this only if he is balanced and collected. Otherwise, he'll turn awkwardly on his front legs instead of easily on his hind legs. You may want him to pivot on his hind legs to spin around and change directions — necessary for reining, cow cutting, and any sport that requires quick maneuverability.

From a Halt

For your horse to pivot on his hind legs, his front end must swing around while his hindquarters stay in place. If you taught the turn on the haunches during ground work, now it's time for the lesson under saddle. Use your legs to hold your horse's hindquarters in place while you rein his front end around. If the horse is in a snaffle, you will use your direct rein. If he's in a curb, you will be neck-reining. Don't expect him to turn on a large angle at first; ask for one step at a time. It helps to halt parallel to a fence. With a fence on one side, your horse will more readily turn

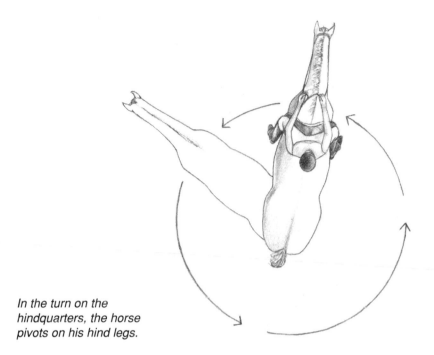

In the turn on the hindquarters, the horse pivots on his hind legs.

away from it. He'll understand your signal to turn and will see that he can't turn into the fence.

When he is standing still, the horse carries most of his weight on his front legs. He must learn collection and how to shift weight to the hind legs to lighten the forehand so he can move his front legs more freely and easily. (See page 326.)

Use your legs to urge him to make a step and your hands to rein him to the left, for example, instead of forward. Check just enough with the bit so your horse will move his front legs but not his back legs. He may want to move his hindquarters to the side instead of his front legs; this is easier for him to do, especially if you taught him to turn on the forehand. To thwart hindquarter movement, press with your outside leg — this will keep your horse from moving into your leg.

Use your inside leg very gently, just to steady the hindquarters, and press harder with your outside leg behind the girth to keep his hindquarters from swinging out as your horse turns. He also must learn to position his inside front leg properly, leaving room for his outside front leg to cross over it. The horse will catch on faster if you taught him to pivot on his haunches in earlier training from the ground.

When your horse has learned to turn correctly, he should be able to pivot 180 degrees in four distinct steps. Throughout the turn he will keep his body perfectly straight from poll to croup, with no bend to his head and neck. (Your hands keep his head, neck, and shoulders straight.) If he tries to pivot on his front legs instead of his hind legs, move your outside leg back farther and give a good squeeze or tap him with it to keep his hindquarters from swinging out.

From a Walk

After your horse has learned to pivot well, you can turn him this way from a walk, then a trot, and finally from a canter. As always, do it gradually. If you try to progress too quickly, your horse will become confused, you'll end up giving cues more harshly than you should, and he will react by fighting the bit or resisting. Always work step-by-step, and never rush your horse.

USE SLIGHT PRESSURE FIRST

When applying leg pressure for any purpose — to signal the horse to move, for example, or to move away from that leg — first use the slightest pressure from your upper calf. If your horse doesn't respond, follow up with pressure from your middle calf. If he needs more persuasion, use a heavier squeeze with the lower calf. If he doesn't respond to that, it's time to tap him with the leg or touch him with a spur. (See pages 58–59 for more on using spurs.) By giving him several chances to do the right thing before he has to feel the tap or the spur, he soon learns that a quicker response saves him from greater pressure.

From a Trot, Canter, or Gallop

Progress to faster gaits only when your horse is ready, after he performs the pivot very well from a walk and can work in collection. Work him along a fence so he'll catch on more quickly. Always halt him before you give the signal to turn, and gather him back onto his hind feet before you turn him. If he doesn't have his hind feet well under his body, he can't pick up his front end and won't be able to pivot. It helps to lean a little in the direction of the turn. The more sudden and fast the desired turn, the more you should lean and the more decisive your hand and leg action should be.

It will take time for your horse to learn to pivot quickly and well. Never rush him or jerk on his mouth. Before he can master the pivot from a fast gait, he must be able to shift his weight back and come to a smooth stop. He can't do it if he is traveling heavy in front, so again he must first be collected.

Teaching the Backup

Save the backup under saddle until your horse is well started in training, rather than teaching it during early mounted lessons. When a horse is asked to do a new exercise that's difficult for him to grasp, he'll often try to do something he has already learned, often the backup.

If you save backing lessons until he has learned other maneuvers and has good response to cues, he won't be so apt to back up while being introduced to things he doesn't understand.

Some young horses use the backup as an evasive tactic when they are reluctant to do what the rider asks. You don't want an argument with a green horse out on the trail or road, only to have him rush backward blindly over a steep bank or into an oncoming car. By saving the backup until later in training, you won't have this problem.

The Proper Backup

Backing a horse looks simple enough when it's done well, but it can be a little tricky for the green horse. When a horse backs calmly and well, his legs work together in diagonal pairs. Watching it, it looks like the horse is trotting backward slowly. His head and neck are somewhat extended and his mouth is closed. His movement is smooth, relaxed, and straight.

Many horses back poorly because they have not been properly taught the movement. They move one leg at a time instead of in pairs, or have their jaw set or their mouth open. They may back crookedly instead of straight, or may tuck in the chin, drop the bit, and rush backward.

The horse uses mainly his front legs to propel himself backward, not his hind legs. Many trainers use the same cues for backing as they do for stopping, but this isn't correct. In the stop, most of the horse's weight is on the hindquarters, whereas in backing up most is on the front legs.

In a proper backup, the horse's diagonal legs move in unison.
Position him near a fence to help teach the reluctant horse to back up.

Use of Reins and Legs

To teach your horse to back, halt him from a walk and wait until he is relaxed. Sit in a balanced position, then put weight in your stirrups. Collect him and flex him slightly, getting his attention through the light use of your legs and from your fingers on the reins. His head should be up a little and vertical, neither overflexed nor lowered. Now squeeze with your legs and lean back slightly to indicate the direction your horse should go. Most of your weight is on your knees and thighs, and your back is somewhat braced.

Your horse will be impelled to move because of the leg pressure and your braced back, but his movement will be backward because your hands are checking forward movement. Don't pull on the reins. Keep your hands fixed to resist any forward impulse, perhaps vibrating the bit a little in his mouth by opening and closing your fingers. As he begins the first step backward, slacken the reins immediately and relax the pressure of your legs — these are his rewards for the proper response. Remember

---

SMOOTH BACKING

As your horse learns what is expected of him in backing, he'll soon be able to back up several steps and then step forward again without halting. If you apply leg pressure and release the rein pressure, he can go backward and forward in smooth transitions.

---

always to reward by relaxing the cues so your horse knows he did the right thing and can relax.

As soon as he completes the step backward, ask your horse for another, repeating the process. Your hand is fixed in place and your legs push him into the bit. Because he can't go forward, he goes backward. After he has done three or four steps backward (asked for one at a time), walk him quietly forward. Do not prolong the first lessons. If your horse responds properly, it's time to stop and do something else.

Things to Avoid

In planning your lessons on the backup, consider some of the don'ts.

- **Don't halt a cue before your horse responds.** If you stop squeezing before he begins to take a backward step, you lose the benefit of the leg cue pushing him into the bit. Instead of coming into the bit, he may try to go forward; you have to synchronize bit and leg cues so he is impelled to move, finds that the bit is halting forward movement, and thus goes backward instead.
- **Don't allow incorrect movements to continue.** If your horse backs crookedly or swings his hindquarters to one side, use your leg behind the girth on the side that is "bulging out" to move his hindquarters back into line. Slightly increase the tension of the rein on that side, too.
- **Don't let your horse get out of your control.** If he starts to rush backward, stop him between each step with leg pressure and release of rein tension. Don't overdo this, however, or he'll think it's a signal to go forward. You must carefully synchronize your hands and legs.
- **Don't try to back your horse by jerking on his mouth, pulling steadily, or pulling on reins without the leg aids.** This confuses

him. Instead of being pushed into the bit, he will fight it by throwing his head in the air or bracing his neck. To teach him to backup smoothly and willingly, you must reward him by relaxing the cues as he obeys.

Dealing with a Horse That Refuses

The green horse is usually easy to teach — especially if you taught him to back at halter and he knows the voice command — but sometimes you encounter a horse that has had bad experiences with the backup and just won't do it. He doesn't pay attention to your cues because he has been hurt in the mouth or is so confused that he is "locking up" and avoiding all movement. You must show him what to do by making it easy for him to choose the proper response.

For this lesson, use a fence or a wall to block all forward motion. With light contact on the reins (one in each hand), ride your horse toward the barrier, using your legs if necessary to keep him headed straight. When his head reaches the barrier, stop him and let him stand and relax for a few seconds, with no leg or bit contact. He realizes he cannot go any farther forward.

Once he's relaxed, with his head at normal level, squeeze with your legs to push him forward and take up any slack in the reins. He can see he can't go forward, and may try to duck his head left or right. Use your legs and hands to keep him pointed straight at the barrier while continuing to push him forward. Your horse knows he must move, to get away from leg and rein pressure, and will take a backward step.

The instant he responds by going backward, release rein and leg pressure. Then ask for another step — with fixed hands to keep the bit steady and leg pressure to drive him into the bit again. If he doesn't respond well, it helps to vibrate the bit using a quick give-and-take with your fingers so there's no solid pull against which he can brace his jaw. As soon as he takes another step back, release all pressure and praise him. He'll realize he did the right thing, and now the lesson is over.

Repeat the lesson later, still using the barrier. If your horse resists your cues and locks up his body or braces his jaw, you can always push him forward to the barrier to reinforce the cues. Each time he responds properly, release all pressure. Do a backup lesson each time you ride. After several sessions of the correct response when asking him for only one or two steps, start asking for a few more steps, still one at a time.

Soon your horse will realize how he can be rewarded with instant release of pressure and won't need the barrier in front of him as a crutch.

Using a Helper

If you are working with a spoiled horse, go back to ground work and teach him to back up at halter (see page 138). Do several sessions if needed, until he moves back willingly at halter. Then repeat the procedure while mounted. Enlist a helper to give the signals you gave at halter as you cue your horse while mounted.

Lateral Work

Training your horse to move to the side in either direction with light rein and leg cues increases his maneuverability. It makes it easier to handle difficult obstacles out on the trail, to open and close a gate from horseback, and to control a dodging cow without having to turn away from her.

Being able to move your horse to the side and on "two tracks" — in these movements, his hind feet follow a different path from that of his front — gives control over his actions. It also strengthens his muscles and makes them more supple. When done correctly, the horse's legs, front and back, cross over each other. This stretches the loin muscles and aids his ability to collect himself. Turns on the forehand, turns on the hindquarters, and lateral work are some of the basic early stages of dressage training and generally make a horse more maneuverable for all kinds of athletic work (see pages 384–388 for more on dressage).

TYPES OF LATERAL WORK

There are several forms of lateral work. In a *two track,* the horse is going sideways and forward at the same time. In the *sidepass,* the horse travels sideways with little forward movement. In the *leg yield,* his body is relatively straight with only a slight bend in the head and neck, curved away from the direction he is going. The *shoulder-in* is more demanding and requires his whole body to be flexed as he moves sideways. The *half-pass* is more difficult; it requires the horse to move ahead as well as to the side and to arc his body in the direction he is traveling.

Teaching Him to Move Sideways

When you begin teaching your horse to sidestep and to move on two tracks, he will be clumsy at first; moving his legs forward and sideways at the same time is not natural for him. Once he has learned the turn on the forehand, however, he has an idea about bending his hindquarters away from leg pressure, and then it's a simple matter to move into lateral work. Teach him to two-track forward and sideways first — it's less confining — and then teach the sidepass.

For the sidepass, you use the same leg and rein cues for sideways movement as for the two-track or any other lateral move, except that you check the horse's forward motion more completely. Thus, he travels sideways only, rather than moving forward and sideways.

INCREASE YOUR HORSE'S FITNESS

All of your training in these lessons is aimed at getting your horse to shift more of his weight to his hindquarters, which makes him not only more pleasant to ride but also less prone to leg injuries because there is less jarring and pounding. Your horse is also improving his athletic ability.

Leg Yields

When he performs the leg-yield maneuver, your horse's head is bent slightly away from the direction he is traveling, but his body is relatively straight. He is moving sideways at a 45-degree angle to the alignment of his body. This maneuver teaches collection, because he reaches well underneath his body with his outside hind leg and begins to round his back, and this helps him lighten in front.

In the leg-yield position, your horse's head and neck are bent away from the direction he's traveling, just enough that you can glimpse his eye. You are pressing his hindquarters slightly to the side with your outside leg; his outside hind leg and the inside foreleg are in line with each other when viewed from directly behind the horse. In a sense, he is actually walking on "three tracks." The outer hind leg takes more of the weight.

Leg yields can be done at a walk, a trot, and a canter, but start your horse at the walk. At the walk, he'll be going sideways more than forward, at the trot and the canter, he'll be going forward as much as sideways.

To teach leg yields at the walk, ride a little way out from the arena fence. (Your horse's natural inclination is to go down the fence if you have been schooling him along it, so he wants to return to it.) Now as he walks, use your inside leg strongly. The horse will tend to bend around your leg and curve his body away from the direction of travel. Use a little pressure on the inside rein, if needed, to curve his head. He will bend in a gentle arc as he moves sideways and forward toward the fence. Return to the fence gradually, letting him travel sideways several steps.

If his hind legs lag behind the rest of his body, your horse is traveling on four tracks instead of three, and he's not getting the gymnastic benefit of having the outside hind reaching underneath himself. (When the hind leg reaches under, taking more of the weight, it can be a foundation for more collected work.) If his haunches drift too far to the side, use your leg with more pressure on that side or tap behind your leg with a whip until he learns to respond to lighter pressure in order to avoid the stronger cue.

You can also work your horse along the fence. Apply the outside rein to get the proper bend; at the same time, press your outside leg slightly

Leg yield: To teach the horse to move on "three tracks," use the right leg strongly behind the girth to move the horse to the left. The horse's right legs cross over his left legs.

back of the girth to move the haunches. This makes him face the fence a little. If he has already learned to move his hindquarters over in response to leg pressure with turns on the forehand, he should soon figure out what he should do.

After moving sideways a few steps, let him travel straight for a while, then try again. Practice in the open as well as along a fence. Go side to side from one track to the other while traveling along a jeep road, or practice out on a trail, sidestepping a tree or a badger hole. Practice at the walk in a relaxed manner. Eventually you'll be able to position your horse for any movement.

Maintaining forward impulsion is crucial. If your horse slows down, get him moving forward again and start over. If he breaks gait when you ask him to move sideways, slow down and begin again. He should soon learn that pressure from one leg means to move sideways and equal pressure from both legs means go forward and straight. If you are consistent, your horse will understand the difference.

Shoulder-In

From doing the leg yield, it's an easy step to the shoulder-in, a movement that demands more bending and leg engagement. It brings his hind legs more under him, which develops more carrying power in the hindquarters and leaves the front quarters free, and with better balance, for more athletic movements. One way to accomplish the shoulder-in is to ride your horse in a circle, bending his neck and body to conform to the circle. At some point begin pushing him sideways, still keeping him bent in the position he was traveling on the circle. Keep his body bent at a 30-degree angle with the line of his progression.

Doing the shoulder-in, his inner hind leg will carry more load as it moves sideways under his body. This requires hock flexion. If your horse is clumsy and his hock action is poor, and thus he loses impulsion and rhythm, continue forward in another circle. When hock action and impulsion are good, again try the shoulder-in.

After he accomplishes the maneuver while traveling in straight lines, have him do it while circling. When first starting this lesson, use a large circle for your departure, with the shoulder-in on just a slight angle; it will be easier for your horse. For the shoulder-in on a circle, his hind legs must take longer strides than the front — more so as the circle becomes smaller. The smallest circle is the turn on the forehand.

Shoulder-in on a circle

Half-Pass

In the half-pass, the horse's body is not flexed. He moves sideways and forward. In contrast to the leg yield, in which his neck is bent slightly away from the direction of travel, in the half-pass his head is bent slightly toward the direction of travel. This means he has to reach farther underneath himself to bring his outside front and hind legs toward his center.

This movement gives you precise control over your horse's hindquarters and shoulders, paving the way for easier *flying lead changes*, which are lead changes made mid-stride (see page 353). Use leg and rein cues as needed to keep your horse's body straight as you cue him sideways. To help keep him straight and balanced, concentrate on staying balanced and centered yourself, with your weight evenly distributed on your seat bones and in the stirrups.

The half-pass can be done at the trot; in this gait the horse has the energy, impulsion, and diagonal leg movement that makes it possible. Trot him energetically in a circle. Use light outside leg pressure, applied slightly behind the girth, to move his hip inward and maintain impulsion, and exert inside leg pressure at the girth to keep his body following the circle. Use inside rein pressure as needed to keep your horse looking in the direction of travel.

As you start the half-pass, increase outside leg pressure and use rein contact to keep his head bent. If you see more than the corner of his eye, you've tipped his head too far. Keep forward movement as your horse starts to go sideways. As he moves away from your outside leg, you should feel his back rise beneath your seat as his body compresses and collects. His hindquarters should lower; his shoulders and poll should rise. Ask for at least three lateral/forward steps, then reward him by letting him continue trotting straight.

Repeat the exercise several times, going right as well as left, and gradually ask for more half-pass steps as your horse becomes better at it. After he can consistently perform this in both directions at the trot, graduate to a slow canter.

Opening a Gate from Horseback

Being able to open and close a gate from horseback is handy, and once your horse has learned to move away from leg pressure and to back up, he can be positioned wherever you need him for working a gate.

First Lessons

Walk your horse along a fence and halt next to a post. Don't worry about a gate at first. If you try to make him stand next to a gate, all he may think about is moving away. Let him stand parallel to the fence, relaxed and on a loose rein. Ride along the fence in the other direction and halt again, next to another post (the fence is now on his opposite side). Again, let him stand. This gets him used to positioning himself as

---

LEAD BEFORE YOU RIDE

Before you try to open gates mounted, accustom your horse to leading through a gate. Never hurry through a gate. Lead him slowly through it. Have him stop occasionally in the opening; get him relaxed about it. Let him come around and stand next to the gate when you close it, so he gets used to that, too. You want him to remain calm as you lead him through a gate; it will then be easier to teach him to work a gate with you on his back.

he would next to a gate. If he doesn't want to stand still, don't force him. Make a circle and come back to the fence. Do this as many times as is necessary. If he has to walk quickly in the circles, he'll soon realize that standing still along the fence is a lot less strenuous.

Another exercise is to ride your horse up to a fence head on, with his nose almost touching it, then ask him to move his hindquarters one step to the right (or left), then his front legs one step to the right (or left). Let him stand, and praise him. Repeat the maneuver, walking him sideways down the fence a few paces, stepping first with the hindquarters, then with the front. Practice it the other way, too, then travel along the fence a few steps in a true sidepass.

The Gate

Ride your horse up to a gate, then stop alongside it as you did along the fence. Allow him to stand on a loose rein, relaxed, then move away. Ride him to the gate several times without trying to open it, until he is comfortable standing alongside it. Then reach down and unlatch the gate. Keep a hand on the gate and the reins in your other hand to make any adjustments in your horse's position. If he moves out of position, correct it, even if you must let go of the gate. When he is relaxed, ask him to move his body properly to allow you to open the gate, walk through it, swing around, and move properly for closing it again. Position him alongside it so you can reach down and latch it.

After you latch the gate, pause briefly. Turn your horse toward the gate for a moment, then ride away. This way, he won't develop the habit of rushing off before you're ready. Likewise, practice doing gates from both directions and from both sides of your horse, so he pays attention to your cues. He should not anticipate what you will ask. You want him to know how to respond to whatever you ask, and to move his body whichever way you direct him.

Again, always strive to create a situation in which your horse can succeed. Take your time in early sessions, making sure he is comfortable about every movement. If he gets out of position, take time to work on that. It doesn't matter if you get the gate opened (or closed) properly at first. The main thing is to keep your horse calm and thinking, and responding to your cues.

If you worry too much about unlatching or closing the gate, and your horse moves wrong and you struggle with him, he'll associate gates with

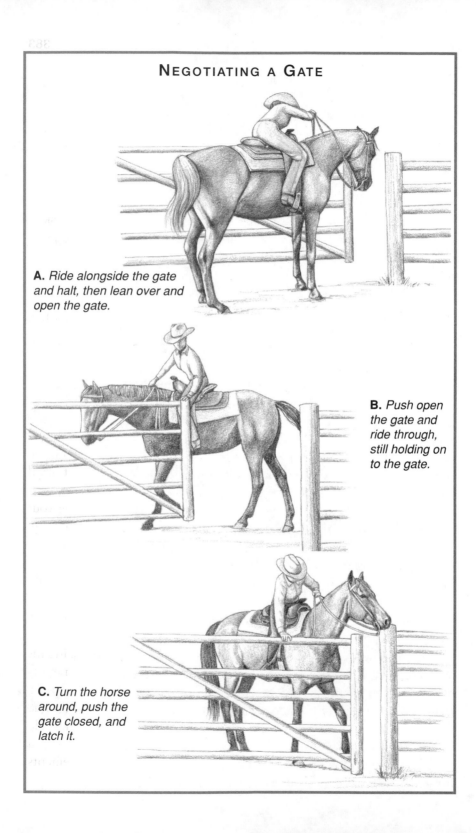

NEGOTIATING A GATE

A. Ride alongside the gate and halt, then lean over and open the gate.

B. Push open the gate and ride through, still holding on to the gate.

C. Turn the horse around, push the gate closed, and latch it.

ACQUAINT HIM WITH THE GATE

Make sure your horse is comfortable with a gate before you open it while mounted. Some horses do fine the first time, especially those that were prepared with preliminary lessons. Others need more sessions. If yours needs more lessons to be at ease with the gate, open it halfway and spend time walking through it. Go about 50 feet, turn around, and walk through it again. When he can do this quietly, ask him to stop in the gateway. Pet and praise him, then walk on through.

---

conflict. If he doesn't understand, forget the gate; ride away from it and do something else, then work on early lessons again until you feel he's ready to try again. Soon your horse will put it all together.

Using a Helper

If your horse is skittish about a gate moving toward him, have someone swing it for you. Concentrate on training the horse for proper movements without having to hang on to the gate. Station your helper on the other side of the gate and ride parallel to it so that the gate opens toward your horse. Halt by the gate and let him stand and relax. Have your assistant open it toward you while you sidepass away from it, keeping your horse's body parallel to it. Walk him slowly through and around the gate, as if you were hanging on to it yourself, and halt when you are again parallel with it.

Your assistant should then move around to the other side of the gate and hold it open. Sidepass again as he pushes the gate closed. Back the horse along the closed gate until you are even with the latch, so you can reach down to fasten it. Let the horse stand quietly, then walk away. Your horse will get used to the gate being opened and shut while moving his body in proper relationship to it, and you can soon do the maneuver without a helper.

When teaching any new maneuver, be patient and take things one step at a time. This keeps your horse from anticipating your cues. He needs to wait for your signal. Never get into a battle at the gate. If you have a problem, take him somewhere else and practice basic movements until he gets them right.

Work on gate lessons only a few times during a ride, and not just when going through a gate to an arena or back to the barn, or your horse will associate the gate with going back to the barn. He'll be in a hurry going through when he thinks he's returning to the barn. If you are ending your ride, come back through the gate, away from the barn; dismount in the arena or training area; then lead him back to the barn. This way, he won't start anticipating and rushing gate lessons.

Beginning Dressage

Although you may never intend to compete in formal tests, dressage can still be of value. Any schooling you give your horse that makes him more agile, more responsive to cues, and more maneuverable under saddle is beneficial in whatever role you plan to use your horse.

Rather than using gimmicks and quick fixes, dressage training is a progressive method of strengthening and suppling a horse to enable him to perform certain tasks. Circles, leads, reining spins, and other maneuvers become much easier for him to perform properly; not only has he been taught the cues, but his body has also been strengthened and made more supple so he can accomplish them easily.

What Is Dressage?

The concept of dressage means different things to different people: It can encompass basic training, harmony between horse and rider, perfection of the gaits, development of a horse's physical and mental ability, and horse ballet. This term is often misunderstood to mean a type of riding that can be performed only in a certain way and one that is just for English riders.

The term comes from the French word *dresser*, which means "to train" — a kind of training that goes beyond simply breaking a horse and making him willing to carry a person on his back. *Dressage* is the art of improving a horse beyond this stage, making him more agile, willing, easier to control, more pleasant to ride, more graceful, and better balanced. It involves a type of consistent horsemanship that is necessary for developing perfect obedience and perfect lightness and agility.

Dressage teaches a horse more fully to understand the aids and become more responsive. Dressage is therefore beneficial for any horse

— it will help him to become more well rounded in his education and less apt to become spoiled or one-sided. A little dressage makes for a better-trained horse. A broader experience of dressage trains a horse, but more than that it develops him physically and mentally so he is truly "one" with his rider, able to understand whatever the rider asks of him, and physically competent to perform it.

THE GOALS OF DRESSAGE

The horse must be willing and able to move forward freely in all gaits and throughout all movements. He must also be able to remain straight through the body when moving on straight lines and be evenly bent from poll to tail along curved lines. He must have equal flexibility and development on both sides. He must also be relaxed, submissive, and attentive in both body and mind to what the rider is asking. Dressage principles are a logical progression from easy to difficult movements; more is asked of the horse as he becomes mentally and physically ready and able to respond.

Teaching Suppleness and Control

A good trainer keeps a horse continually improving athletically and mentally, developing the horse gradually, correcting any problems as they arise. In dressage, each new exercise can be done correctly only when the foundation has been properly laid by previous lessons. For instance, you cannot collect a horse until he first has impulsion. Gymnastic exercises are designed to develop muscles, poise, dexterity, and balance and are the basis of dressage training.

One of the goals of training a horse is to strengthen his muscles, especially the back and neck, so that he can perform easily while carrying the weight of a rider. This is accomplished while the horse is learning to obey signals, stop, turn, and handle himself at various gaits. He should be developed gradually, first at the walk with exercises and traveling across country. He is encouraged to extend his head and neck and push his hindquarters well under himself, reaching confidently with his hind feet. The next step is to develop a strong trot with a slow cadence and long strides.

After strengthening the horse and developing strong impulsion, the next goal is to make him more agile and flexible. This requires exercises

Lessons in making the horse more supple: The horse's neck and body are bending to fit the circle.

to help him bend his spine (shoulder-in, for example; see page 378). Suppleness and good balance are demonstrated by a horse that can canter a small circle smoothly and easily. He bends from poll to croup, the bend in his spine matching the bend of the circle. When he is bending properly, his hind feet follow exactly the path of his forefeet. Horses do not travel this way naturally; they must be taught these exercises to develop greater agility and dexterity.

For best athletic performance, the horse must be taught to follow his head and to bend his body around curves. When running at liberty in a pasture, a horse seldom follows his head; he inclines his body slightly away from the direction he is traveling. While cantering in a circle to the left, for example, his body is slightly to the right. He keeps his balance fairly well — until he gets into an unusual situation, such as slippery ground, an unexpected change of direction demanded by the rider, or a rider's request to canter in a small circle. Good training teaches a horse to collect and handle himself with more balance and control than he has when running free.

Teaching Response to Subtle Cues

Training enables you to be in tune with your horse so you can ask him to respond to very refined cues. He has better lateral suppleness for circles and better longitudinal flexibility for work in straight lines. Dressage exercises enable him to respond immediately to a subtle request such as to halt from a trot or canter in just one stride or to make a smooth change of gait. Speeding up, slowing, and stopping all require perfect balance. He comes to a square stop with no resistance or head tossing.

Your horse must be able to shift his center of gravity forward or backward as you request. One purpose of dressage is to make him willing and completely obedient. It teaches him to move in any direction, at any gait, at any speed, laterally or on a diagonal, with his forehand active and his hindquarters passive or vice versa. He is taught pivots and the half-pass, and can also change easily from an extended gait to a collected one and back.

Three Stages

Dressage training is categorized into light or basic, intermediate, and heavy. *Basic dressage* consists of teaching the horse to be obedient at all gaits on straight lines and in circles. *Intermediate dressage* takes him through schooling exercises that include shoulder-in, pivots, and half-pass; flying change of leads; and the beginning of the *passage*. (This is a trot with high action, in which the horse seems to travel in slow motion: He moves very slowly forward in perfectly cadenced steps, with each diagonal pair of legs suspended in the air.) *Heavy dressage* includes highly collected movements such as the *piaffe* (a trot in place), more work on the passage, flying change of lead at every stride, and the *pirouette* (a turn on the haunches at the canter, the horse cantering in place as he turns). When executed properly, these movements represent the highest degree of collection, flexibility, and suppleness.

Classical dressage included "airs above the ground" such as the *levade, courbette,* and *capriole* — specialized military maneuvers that were used in the 1500s when well-trained horses were ridden in battle against foot soldiers. Modern dressage is quite different from its classical roots. Horses of the old school were smaller, with high knee action. Traditional movements had the greatest degree of collection, and the highlights of these displays were movements on two tracks, in which the horse was presented with head and neck bent. Today, the horses used in dressage are larger, faster, and more powerful and move with a lower, longer stride. Extended gaits are emphasized as much as highly collected movements, with continuous transitions being made from one to the other.

Beginning Jumping

Some horses have natural jumping ability, but most horses can be taught to handle low to medium jumps. Many exercises used in teaching a horse to jump — working over rails on the ground, cavalletti, and low jumps, for example — teach better control, coordination, dexterity, and obedience. They help a horse become more flexible and relaxed, as well as developing the muscles for jumping.

CAVALLETTI TRAINING

Cavalletti consists of six or eight poles laid on the ground about 4½ feet apart (space them appropriately for your horse's stride length), with a pair of jump standards (uprights to hold the poles) at each end of the last pole. The horse is trotted over the poles, putting one foot down between each pole. The last pole can be raised onto standards to make a low jump. As the pole is raised to make the jump higher, the distance between the last ground pole and the jump is increased accordingly, thus providing the horse with enough distance to make the jump successfully.

Work on the Flat

Before your horse can learn to jump, he must learn how to space his strides and prepare his approach. This is easily accomplished by riding over rails, or *cavalletti*, on the ground. Practice walking and trotting him over the poles in both directions. With a green horse, walk over a single pole first until he is confident, keeping him in a straight line before and after the pole, and progress to the trot after he does well at the walk.

Then use multiple rails 4 to 5 feet apart, depending on your horse's stride. The average distance for most horses when trotting is to have them 4½ feet apart. Three poles is enough to start with, then add more, to help your horse regulate and adjust his stride. Post the trot (see pages

313–315) when riding over them, or stand in the stirrups so that you are not bouncing on his back. He needs complete freedom of movement to figure out his stride. At the sitting trot, you are more apt to interfere with his mouth and alter his rhythm.

After your horse is comfortable about trotting over several evenly spaced poles, return to single poles and canter him over those. This helps both of you to judge the distance and take-off position. Once he canters nicely over single poles, space them so that he takes a certain number of strides between each one.

Work Over Small Jumps

When your horse can trot and canter over poles on the ground, raise them to make small jumps, about a foot high. Trot him over those, then raise the poles after he is at ease with the small jumps. Stay well forward, in jumping position, as he goes over the jump, taking care not to hinder his mouth with the bit or bang down in the saddle. Your horse needs complete freedom of head and neck. If you don't interfere with his mouth or land hard in the saddle, he will learn to use his head and neck for proper balance and keep his back rounded for smoother jumping.

If your horse has trouble regulating his strides while trotting over small jumps, put three poles on the ground ahead of each small jump, spacing them about 4½ feet apart, with about 9 feet between the last pole and the jump. You can also use just one pole in front of the jump (again, 9 feet ahead of it), if that works better for your horse. If he is nervous or overeager and tends to anticipate the jump too much, one pole is better than three; multiple poles may make him want to rush even more.

If you are cantering over the jumps and using a pole in front, space the pole 10 to 12 feet ahead of the jump. Placing poles ahead of the jump encourages your horse to get there at the right speed and with the proper spacing of his stride to make a smooth jump. As he gains confidence, set up a miniature jump course so he can go from doing single jumps to doing several.

Ways to Avoid Souring the Horse

As you train your horse, you'll make the best progress by going at his speed and not drilling excessively during lessons. Repetition is important

Trotting over poles to help regulate the horse's stride, with the last pole 9 feet from a small jump

in training, but keep him fresh and eager to learn; always give him some "time-outs." If every time you ride you are doing training exercises, your horse will begin to resent the lessons. Be creative and vary the lesson plan enough to make your rides pleasant and interesting for your partner.

Constructive Loafing

Every now and then go for a ride during which your horse does more "lollygagging" than working. Go out several miles for a pleasant hike across country or make an occasional rest stop with some grazing — use these occasions just to enjoy each other's company. An occasional "do-nothing" ride can help keep your horse from getting sour and burning out. This gives him a mental and physical break from intense training sessions and makes him feel happier about the whole program.

PROPER JUMPING POSITION

Your position while jumping should be similar to the position for the gallop — seat out of the saddle, leaning forward over the neck, with proper leg position to give grip and balance. If your legs go back too far, your body will come too far forward when the horse lands and you may fall off. If your legs go too far forward, your upper body will go too far back and you'll be "behind" the horse on the jump, likely to catch him in the mouth as well as bang down on his back. This will discourage him from jumping. Keep your weight in your stirrups, relaxing your knees and ankles. Use the strength in your lower legs for part of your grip and balance.

Good jump position

When training a young horse, stay attuned to his frame of mind and gear lessons accordingly. Intersperse some "at-ease" sessions whenever you feel your horse needs them. In order for him to give you his maximum effort and cooperation in his lessons and in future work, he needs to learn early on that being ridden is not just a constant grind of hard work.

As he progresses in training, he becomes more able to work for longer periods, with more demands on his span of attention and concentration. Even as your horse matures, still plan some restful sessions or he may grow tired of the predictable routine.

Work and Play

One way to keep your horse's attention is to integrate several short but intense lessons during a long and relaxed across country ride. Take advantage of a level open area along the trail for a quick practice of a figure eight and lead change, or a couple of rollbacks (pivots on the hind legs from a gallop), circles, and gait transitions, or whatever he needs work on — then continue on your ride. This way, your horse will come to realize that work can be a spur-of-the-moment thing, anywhere, anytime, not simply constant daily drilling in the same old place.

A long aimless ride can also calm an overeager, nervous, or keyed-up horse and help him relax. If you have been trying too hard to reach a certain goal in lessons and thus are drilling your horse too much, the intensity and tension will be counterproductive. Back off and give your horse a break. He may just need time off to improve his attitude about his work.

Often he will come back to his sessions with more willingness and progress more steadily because the "vacation" gave him the space and time he needed for some lessons to sink in. If you don't give the horse a break now and then, you'll probably end up in a confrontation: He will have mentally "quit" on you while you keep making demands. This situation will only make him more resistant and resentful and destroy all your progress. Don't let things get to this point. Think about your horse. Keep him happy, and he'll be more willing to try and learn and give his best.

THE SKY'S THE LIMIT!

If you commit yourself to a goal, there's practically no limit to what you and your horse can accomplish together. The key ingredients for success are time and patience. A good horseman always has time for whatever the horse needs, never rushing him, and recognizes the value of a strong foundation built on mutual understanding.

If a certain goal is worth pursuing, it is worth your time and commitment. Recognize the importance of keeping the horse's future progress and well-being as your highest priority. Put the welfare of the horse first, above your own feelings or any perceived schedule demands. A good trainer remains alert to his pupil's attitude, willingness, and abilities, progressing at the horse's pace and no faster.

Like a good teacher, a good trainer makes his pupil strive for more and stimulates him to advance today beyond what he learned yesterday or last week but never gives him more than he can assimilate. You want to keep the horse willing and eager to learn, challenging his mind as well as developing his physical capabilities and stamina for athletic accomplishment. But if you ask too much or if the work becomes too hard, he will lose interest in trying.

The secret to creating a top performer, a perfect teammate in your chosen riding discipline or sport, therefore, is knowing when to push and when to be patient, interspersing work with relaxation. A happy, interested horse will go a lot further in his schooling, both mentally and physically, than a horse that is rarely challenged or one that is confused or pushed too hard.

16

TRAILER TRAINING

Almost every horse owner at some time must transport a horse — to a show, to a trail ride, to a veterinarian, to be bred, to a new home. If a horse is not easy to load, the frustrated horseman may resort to force. Rather than attempting to pull the horse into the trailer or making it so uncomfortable outside the trailer that the horse finally decides to "escape" into it to get away from the whip, train your horse to load *before* you must take him anywhere; then he will load willingly and without fear.

There are many different trailer-training methods that work well. The important thing is how you go about the task. Even the best methods can fail if you don't have good communication with your horse and if you don't stay calm. Most unsuccessful attempts at loading are the result of two extremes in human behavior. One is to become impatient, angry, and too forceful; the other is being too passive and unable to ask the horse for enough response to your requests. Be patiently persistent.

Training the young or inexperienced horse to load is generally easier than retraining a spoiled horse that has become stubborn or fearful due to bad experiences, so start training early.

Load the Foal with His Mother

If possible, trailer a foal with his mother. He'll gain confidence from her, and if his early trailering experiences are good, he will be off to a good start and receptive to further lessons.

If Mama willingly enters the trailer, her foal will want to follow her. The foal takes his cues from Mama, and if she is comfortable and confi-

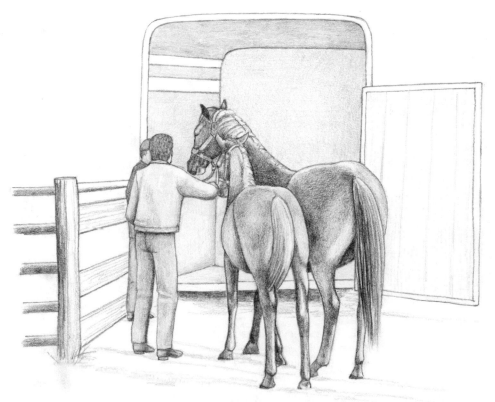

Using a buddy horse, especially Mama, can put the weanling at ease and make his first experience with the trailer and loading less frightening.

dent, he will be less afraid. Load her first and lead him right after her. If he has been trained to tie (see pages 93–96), an older foal can be tied by his dam in a two-horse trailer for a short trip.

The safest way to trailer a foal, however, is in a four-horse or stock trailer. Either tie the mare and leave the foal loose or leave them both loose. A small foal traveling with his mother will nurse and lie down when he gets tired, so he's best left loose. If the foal is loose in the stall next to the mare in a two-horse trailer, make sure the back is closed so he can't jump out.

Putting Him in the First Time

If your foal balks when you try to load him, boost him in. The easiest way is for one person to lead him, to keep him aimed straight, while two others lock hands behind him and gently move him into the trailer.

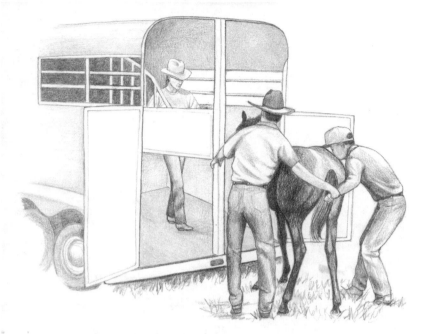

With a reluctant youngster that you must load at his first session, two people can lock arms behind him to encourage him to step in while a third person keeps him aimed into the trailer, with no pressure on the halter.

The Weanling or Yearling

If your horse is a weanling or older before his first trailering lessons, plan a few training sessions before you have to take him anywhere. Then you can let the youngster accept the trailer at his own pace and on his own terms, when everyone is calm and there's no urgency.

Using Feed

At this age, if he is accustomed to eating grain, you can usually lure your foal into the trailer with food. Be patient and take several days, if necessary, to overcome his apprehension. Some bold youngsters walk right in the first time, with some encouragement, but others need more sessions to decide that the trailer is a safe place. Feed your foal in a tub inside the door, where he can reach it from outside, then move it farther inside each day until he must put his front feet in to reach it. Then place the tub far enough inside that the foal must step all the way in. Because he is not being forced, walking into the trailer becomes his own idea.

DON'T CONFINE OR RESTRAIN HIM AT FIRST

The first few times a horse goes completely inside, don't fasten the rump chain or bar (in a two-horse trailer) or tie him. Leave him free to back out if he wants. This way, he won't feel threatened or panicky. He can come to terms with this new experience at his own speed. Once he is relaxed enough to stay in and finish eating his feed in the manger, you can start fastening the rump chain and tying his head.

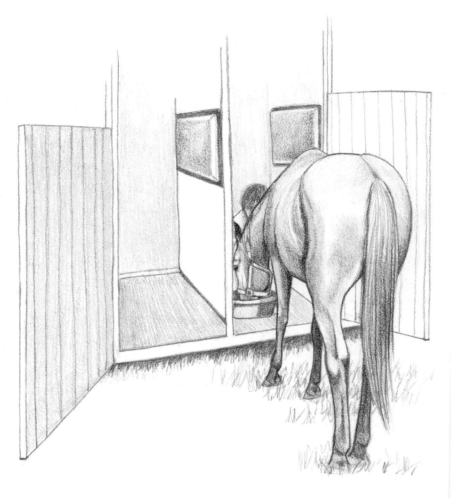

Feed grain in a tub in the trailer for a few days and gradually move the tub farther in to encourage a horse to step into the trailer on his own and without apprehension.

Don't Pull His Head

When loading a horse the first time, don't pull straight forward on his head. This creates resistance because he doesn't want to be dragged into the unknown. He may fight and even bang his head on the trailer.

If you have an assistant, station him at the horse's head. The assistant should encourage and guide the horse, not pull. If a bit of pressure is in order, it should come from the rear, to gently boost the horse in or to encourage him to move forward.

If you are working alone and must give tugs on the rope, pull at an angle rather than straight ahead. Pulling at a 45-degree angle will encourage him to move the nearest foot to balance himself. Release tension when he makes the step.

Patient Loading

Sometimes the horse's first experience with a trailer is when he must be transported somewhere. If you have to combine first experiences in a trailer with hauling, make sure you have extra time to load him, and be very patient and understanding as you go about it. You want no bad experiences to make future loading difficult.

Every horse is naturally somewhat fearful the first time he is loaded. If there is a ramp, your youngster may be afraid of the artificial footing. The enclosed trailer is scary. If it's dark inside, he'll hesitate to enter. When you are acquainting him with the trailer, always place it so that the sun shines into it, giving the horse a better view inside. And don't use a low-roofed trailer for a tall horse.

It helps if you have a well-trained horse for a good example. Load the experienced horse, then the youngster. This works best if the horses are acquainted; it may not work at all if the horse is a stranger and the youngster is timid. The young horse will be more willing to follow a buddy. For two weanlings that are pals, load the bolder one first, even if you have to boost him in; the timid one will be apt to follow.

Your goals now are to gain your horse's trust and to minimize his fears. All too often, the person loading the horse is in a hurry. Frustration can lead to harsh methods, which will convince the hesitant horse that his fears are well-founded. For safe, successful loading, take time and be

If your horse stops as you approach the trailer, let him stop. He should be at ease and comfortable before moving forward.

patient. First, you must convince your horse that you *are* patient, so he will trust you. It is often a good idea to work alone when trailer training, with no other people around to distract him and definitely with no one standing behind him to make him nervous. If he feels boxed in by someone standing behind him, he may want to get farther away from the trailer.

Lead your horse up to the trailer, with a little slack in the rope, the way you taught him to lead, walking freely beside you. As he gets near it he may stop, so you stop, too. Give him time to look it over, and move forward again only after he's relaxed. Take your time, approaching the trailer at his speed; your horse's curiosity will usually get the better of his fear and he'll want to smell it and check it out. If he has confidence in you from earlier leading lessons, you can usually encourage him to step in — especially if it's a large, roomy stock trailer or a four-horse trailer.

Sometimes the youngster is almost willing to enter the trailer but is still hesitant and needs to be persuaded a bit. A helper can gently pick up his front foot and put it in the trailer while you guide in his head. If this does not upset the horse, the other foot can also be gently placed up onto the trailer floor. He may then realize it's not impossible to enter the trailer, and with encouragement may walk right in.

Rump Rope

If your horse is still hesitant, use a rump rope, just as you did when you led him as a foal (see page 88). He'll remember the lessons and realize he must go where you want him to — that he can't avoid the trailer by balking or backing up. If he tries to back up, the rope tightens around his hindquarters. If you are patient but firm, he'll know that he must move forward instead of backward when the rope tightens; then you can lead him step-by-step into the trailer. Success in loading the weanling or yearling the first time often depends on how well halter-trained he is — that is, how well he has progressed with previous leading lessons.

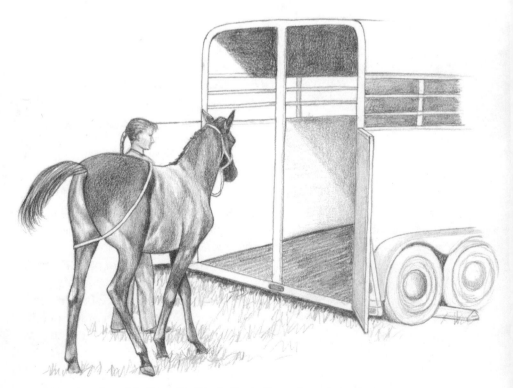

Leading the weanling to the trailer using a rump rope

The Adult Horse

You'll want a few days to acquaint your horse with the trailer — with feed and gradual lessons — because he is too large to boost physically into the trailer as you might be able to do with a foal or weanling. If the first time he sees a trailer is when he must travel, and you have no opportunity for lessons beforehand, patience and tact are your best bet to persuade him it's a good idea. Start as early as possible in the day. Do some ground work in leading to cement his willing responses *before* you approach the trailer.

The Trailer

For lessons, park the trailer on level ground with good footing, preferably inside a pen so the horse can't get clear away if he pulls free. If it's a two-horse trailer and you are working alone, with your aim simply to get him into it as you lend encouragement from the side, leave one door closed so he'll have no confusion about which side he must enter. After he loads on one side, practice on the other side; he needs to learn to load from both sides of the trailer.

Leave all doors closed except the one at the back where he will be entering. Seeing an opening up front may encourage him to get in, but when he can't go on through, he may panic and try to go through the escape door. Keep the escape door and the feed manger door closed so he won't be enticed into thinking there is a forward exit.

Allow Plenty of Time for Lessons

Insufficient training time is the biggest cause of failure in loading lessons. Be prepared to devote several sessions to trailer training. Depending on the horse and your abilities, you may make great progress in one session or it may take several sessions of patient work for him to load successfully. Even after your horse loads, however, plan several more lessons to reinforce his learning so he will load well in the future.

Before you take him to the trailer, polish his leading skills with ground work that requires flexibility and control. Move him forward freely at the walk and the trot, back him, make sure he moves his hindquarters or front legs to the side when asked. Trailer loading is just an advanced form of ground work. Good ground work develops trust, respect, and communication; you can make the proper response easy for the horse and the wrong response more difficult without creating resistance. When leading, your horse should maintain forward impulsion, should not crowd you, and should never pull back on the rope. If he fails at any of these things, he needs more leading lessons before you attempt to load him into a trailer.

TRAINING AIDS

Use a long lead rope (12 feet) so you can direct your horse from a distance when necessary, or give slack when needed, or swing the end of the rope for encouragement. If you are using a training whip as a step-forward stimulus, loop the lead rope over his neck and withers so he won't step on it. Use your left hand on the lead rope, near the halter, with the training whip in your right hand to gently tap his hindquarters when you ask him to step forward. Use the whip *only* if your horse has already been taught to respond to it.

A Step at a Time

Lead your horse up to the trailer in a relaxed manner. If he stops, let him. He mustn't think he's being forced or he may decide he just isn't going any closer. Stepping up into a confined space such as a trailer is an unnatural thing for a horse, and most horses immediately decide that it is too scary and balk. Your job as trainer is to reassure your horse and to dispel his fears. Give him time to look over the trailer. He may decide he doesn't like the looks of it and back up.

As always, be patient. If your horse decides to step back a few steps, let him. Don't create any tension on the halter. Pulling on him or trying to hold him in place if he steps back will make him more afraid and he'll pull back harder. He doesn't want to be forced into something he doesn't trust. If he's on a loose lead, he won't be so fearful and he'll stay more relaxed. Potential danger that he feels he can jump away from is not as bad, in his mind, as potential danger when he is being held tight, unable to back away from it. Every time he stops, let him stand a moment.

If you understand how your horse thinks, you'll be patient and let him step back if he wants to. Move back with him, talking pleasantly, letting him know there's nothing to fear. Rub him and praise him; help him relax. After he is calm, ask for another step toward the trailer. You can tell from his body language when he is ready for another step. Don't rush him, but don't wait so long that you miss the window of opportunity. Sense his receptiveness and make requests when he has an open mind and is willing to expand his comfort zone and level of trust. Whenever he becomes upset, stop and reassure him until his anxiety fades, then ask for another step.

Soon your horse will stand quietly by the trailer, unafraid, looking it over. Let him smell it if he wants to satisfy his curiosity. When you feel he is relaxed again and ready, encourage him to take another step or two toward it, or up the ramp if it has one. One step at a time is sufficient, and it doesn't matter whether he decides to back up again a few times.

With patience and tact, you are changing his mind, convincing him that the trailer is not something to be afraid of, letting him realize that he might be able to enter it. You must be able to sense his change in mood. This indication of changing attitude may be a mere twitch of a muscle or a slight stretching forward of his neck to take a new look inside the trailer — the spark of new interest that shows you he's beginning to reconsider. From that moment forward the lessons are working — as long as you continue to be patient. Don't move ahead too quickly now: Give him a little more time and set his mind at ease.

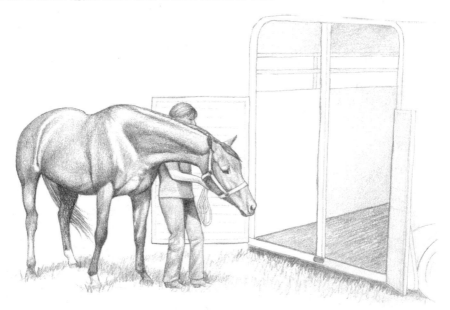

Watch for signs of curiosity or interest, then proceed.

BE PATIENTLY PERSISTENT

This session of patient steps toward and then into the trailer may take only a few minutes with a green horse or several hours with the reluctant spoiled horse, but eventually you will persuade him to enter the trailer. It may take more than one session, but if you are patient and persistent, he will overcome his fear. If you take whatever time is necessary for that particular horse, you will change his mind gradually, without a battle of wills.

Now you can begin again, step-by-step, to encourage your horse into the trailer. You can now use a training whip, just as you used it earlier in teaching him to lead (as an extension of your arm, to touch him on the hindquarters to encourage him to move forward; see pages 134–135). As long as he does not fear a whip, but rather accepts it as your third hand, you can use it to advantage. If you have not used a whip before in his training and he is afraid of it, do *not* use one now. A whip works only if a horse accepts it as a cue to move forward.

At the trailer door, your horse may try to turn away his head, but the only tension on the halter should be to keep his nose pointed at the trailer, not to pull on him. Once he is completely relaxed, standing by the door, encourage him to step into the trailer — even just a step or two or a step up with a front foot. If he decides to back out again, let him. Don't put any tension on his halter to try to keep him in there. He is just trying out this new situation and making sure he can accept it safely.

If he won't step in, tickle his hindquarters with a whip. If he still hesitates, give a light tap. Reward any forward movement — even a shifting of weight in the right direction — with the cessation of tapping and give him praise.

If your horse backs up several steps, don't try to pull him forward or he'll pull back harder. Just keep his head pointed toward the trailer and move back with him, with no tension on his halter, as you continue tapping his hindquarters with the whip. Increase the intensity of tapping until he stops going backward and takes one step forward. Then stop tapping, let him stand, and praise him. Wait until he is relaxed again, then ask him to go forward — a step at a time — until he is once again at the trailer.

If you keep your patience, he'll eventually step in with one foot. It's not necessary for him to go clear in at this point. In fact, it's best if you

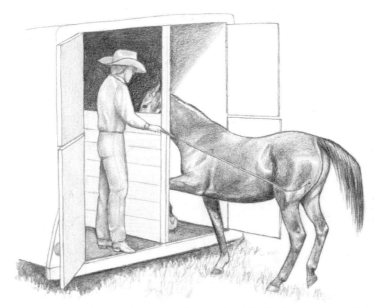

Encourage your horse with a whip by tapping his hindquarters gently to ask for another step. When he complies, stop tapping and reward him with praise.

stop him, praise him, and let him stand there with one or both front feet in the trailer. If you let a horse rush into a trailer, he may panic, not knowing how to get back out. Instead, stop him and let him proceed slowly and gradually, so he is thinking and at ease, not trying to escape into the trailer. Then let him back out (this is part of his training for unloading), and ask him to step in again. He needs to be utterly at ease with getting in and out.

Once your horse goes all the way in, put some grain in the manger (or some hay in a net at the front of a stock trailer), and he will enjoy this reward and relax enough to stay in the trailer without fear. Then close the door or snap the rump chain in place and secure his halter at the front.

Tying in a Trailer

It's usually best to tie a horse when you haul him. In a two-horse trailer, always have the door shut and the rump bar or chain hooked before you tie his head. Otherwise, he may try to back out while tied and have a serious problem. Don't tie too short or he may feel trapped. An adult horse in a two-horse trailer should have enough slack that when he backs up, he

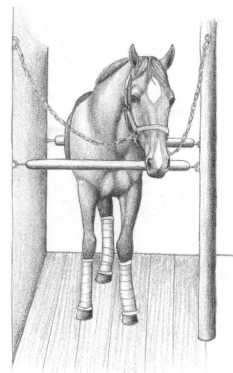

This horse is tied properly in the trailer; note the partition on one side and space bars ahead of and behind him.

feels the rump chain before he feels restrained by the rope. The exception is a small horse that doesn't fill the stall of the trailer. Don't have so much slack in the rope that he could put his head down behind the manger. The rope should be long enough for him to be able to look to the side 90 degrees each way, but not so long that he can turn his head around, or he might think he can turn around in the trailer.

When unloading a horse, always untie his head first before you unhook the rump chain and open the door. Don't open the door until you have the rump chain unhooked, or your horse may start to back out and press so tightly against the chain that it will be difficult for you to unhook.

Another Method for a Two-Horse Trailer

Another way to acquaint the horse with a trailer in the first lesson is to leave the door closed when you lead him up to it and then do some ground work next to the trailer. Stand at the left rear corner, with your back to the trailer, and direct your horse in a semicircle, using the long lead rope as you would a longe line to your right, so he comes to a stop facing the side of the trailer. Let him stand a moment to look at the trailer, then direct him back around you to where he comes right up to the trailer door.

To encourage him to move briskly, swing the end of the lead rope toward his hindquarters or point your training whip — whichever works better for you and your horse. Do this several times until he moves freely and willingly in the small semicircle around you, alternately facing the trailer side and the trailer door. Do this on the other side of the trailer, also.

Once he is calm about this ground work, stand somewhat behind the trailer and direct him between you and the back of the trailer. If he can do this without hesitating, speeding, or trying to push you out of the way, your horse is ready to try loading. Gradually narrow the space between you and the trailer. When he is calm about going back and forth between you and the trailer, get set to open the door.

Open the left door. Stand with your back to it and continue to ask the horse to circle around you. This time, when he comes up to the back of the trailer he'll be facing the open door. Let him put his head in the trailer and give him a few seconds to check it out. While he's still interested and curious — that is, before he decides to back off — direct him away from the trailer. This takes off the pressure and lets him relax. Lead him up to it again and ask him to stick his head farther in the open door. Repeat, building on his curiosity and acceptance of this new idea in small increments, so he never becomes alarmed about the thought of entering.

Take your horse away from the trailer several times before you ask him to put a foot into it. If he does, let him stand a moment and praise and pat him, but don't try to force him to stay there. He needs to know that even attempts to obey you are rewarded. If you are the one who asks him to back up again, before he does it on his own, he will stay in a more willing frame of mind. As he gets bolder, ask him to step in with one foot, then two, but don't let him walk all the way in or he could get scared and lose his trust.

Some horses will put both front feet in during first tries. Others take more time. Tailor your lessons to the horse. Don't ask too much or too little; progress at the rate that seems best, keeping him in a responsive mood. Once your horse is putting both fronts in, back him out only a few steps, then ask him to go right back in. If he will stand there, pat him, then ask him to back out. Each time you bring him to the door, ask him to step in a little farther or to stay longer. If he becomes alarmed, let him stand and relax after he backs out, or do another semicircle and go right back into the trailer.

Eventually your horse must put a hind foot in; if he doesn't offer to do this on his own, ask for a little more briskness as he comes to the trailer. If he stops suddenly before stepping in with a hind foot, tap him on the rump with your whip or the end of the lead rope. If he backs up rather than going forward, do another semicircle and direct him into the trailer again. Usually, once he has reached this level of acceptance and is a little bolder, a bit of impulsion will carry him into the trailer, with one or both hind feet.

OVERCOMING A BALK

If your horse won't go forward after you lead him to the trailer or won't continue in with his hind feet after he's put in his fronts a few times, don't continue tapping his rump; he'll develop a negative association about being there. He simply needs more impulsion. If he won't move his feet into the trailer, get them moving somewhere else first. Direct him into the semicircle to the side of the trailer and back again, and make him move briskly. If he has to hustle, he is more apt to let this impulsion carry him into the trailer rather than walking slowly up to the door and stopping. If you get him more active on his circle, he will put more effort into getting in the trailer.

When your horse gets all the way in, immediately ask him to back out. He has now loaded, a step at a time, and is learning to unload. If you ask him to back out, he'll be more at ease about getting into this confined space. He can back out again before he gets scared. After he backs out, pat and praise him; let him relax or even walk him away from the trailer a moment before asking him to return. If he is calm, load him a few more times, then quit for the day.

In the next day's lesson, you will work with your horse on staying in longer before unloading, teaching him to wait for you to ask him to back out rather than doing it on his own. Before unloading the last time, put some grain in the manger. Using grain this way as a reward is a good idea when doing this type of lesson.

Repetition

Take your horse in and out of the trailer several times before you feed grain, secure him, and go. Several loadings and unloadings before hauling helps ensure good loadings in the future; your horse will know he can do this, that it is not frightening to get in and out of a trailer. A few unstressful lessons will make him willing to load for the rest of his life, as long as he never has a frightening experience.

Your objective during training should focus not on getting your horse loaded, but rather on teaching him *how* to load, to be at ease with getting in and out. Training him in small increments also teaches him to unload — he realizes he *can* back out. Repeat these steps many times, going partway in and back out again, until your horse is comfortable with going in and out and with going completely into the trailer. By then you have accomplished two goals: You have a horse that willingly enters the trailer whenever you ask and one that calmly unloads when you ask.

Unloading

If you haven't practiced during loading lessons, backing out of a two-horse trailer can be as traumatic for the inexperienced horse as is going in the first time. He can't see where he's going and may not want to back up because he doesn't know where the trailer floor ends. He may worry about stepping down. Some horses start to step out and immediately leap back into the trailer. With a young horse — small enough that you can be in the trailer stall with him — you can reassure him and push on his chest to help him back up.

If an older horse refuses to back out, fasten a soft cotton rope to the rear of the trailer, then pass it around his chest and out the rear on his other side. With someone at his head to steady and encourage him to back up, another person pulls back steadily on the rope around his chest, never allowing it enough slack to fall to the floor. Do not rush your horse; the rope is merely to give him support and an incentive to back out. After he has gone in and out a few times, it's not such a scary procedure. A stock trailer makes maneuvering easier, as there is room to turn around and your horse can walk out facing forward.

If a horse rushes backward out of a trailer, spend time going partway in and out. Back him out when he's only partway in, before he has a chance to rush out, until he can back out calmly.

Feeding in the Trailer

If you plan to use feed to entice a horse into the trailer, spend several days letting him get used to eating grain or hay inside (see page 396). Allow him to become accustomed to the trailer at his own speed. He'll soon discover that the trailer will not hurt him, that he is not being forced, and that he can go in and out at will. Some people leave a trailer (well blocked so it can't move) in a horse's pen and feed him in it so he becomes at ease with going in and out.

PRACTICE TRIPS

After you get the horse loaded and are ready to take him some-where, make his first trailer experiences pleasant. If possible, take him on a short practice drive before making a real trip. Give him a little hay or grain to occupy him. Choose a smooth road, and limit your top speed to 40 miles per hour. These first trips help build his confidence if you travel slowly and carefully. Conversely, they can shatter his confidence if he is jostled about.

Using a Well-Trained Companion Horse

For the inexperienced horse, it often helps if you load an experienced stablemate first. If he sees his buddy going right in without hesitation — especially if his buddy starts eating something in there — your horse may realize that the trailer is not a big threat. He may want to hop right in so he can have something good to eat, too. The presence of a calm, experienced horse alongside him on his first few rides in the trailer can also be very reassuring. He'll probably travel much better if you bring along his experienced friend.

Resolving Reasons He Won't Load or Haul Well

Some horses have problems with certain trailers and situations. Many of these problems can be resolved if you can figure out what's bothering the horse. Some horses are afraid to walk up a ramp; it makes noise and has some give, which may alarm them. It's easier for some horses if you use a trailer they can step up into, without a ramp. Often you can situate the trailer so that the step up is only 8 or 10 inches.

> ## RULES FOR SAFE TRAILERING
>
> 1. Tie the horse short so he can't get into an unsafe position.
> 2. Always use a rump strap, chain, or bar behind a horse in a two-horse trailer so he can't back out; don't rely on only the door or tailgate — it could come open.
> 3. Start slowly and smoothly when hauling, and build up speed gradually.
> 4. Always slow down on turns. Right-angle corners should be taken at about 5 miles per hour.
> 5. Be sure the trailer is completely around a corner before you begin to accelerate.
> 6. Take curves slowly. Your horse has to balance and brace himself, and you don't want him to start scrambling.

Use a Bigger Trailer

For obvious reasons, a tall or nervous horse may be uneasy about entering a small or short trailer. Use a tall trailer for a tall horse so he won't hit his head on the roof. Most nervous horses are reluctant to enter a small trailer and will load and travel better in a stock trailer or a four-horse trailer than in a two-horse trailer.

Remove the Partition

If all you have is a two-horse trailer and you are hauling just one horse, he may be more comfortable if you remove the partition so he has more room to move. This may also be the solution for a horse that scrambles as the trailer negotiates corners: He can then spread his legs wider to brace himself on the curves. If you take a corner too fast and your horse tries to brace himself and can't, he must lean against the wall or partition and scramble with his feet on the other side to keep from losing his balance.

This may become a habit, and if the horse feels overly confined, it can become a serious problem. Some horses throw a violent fit as the trailer takes a corner. They may haul well in a truck or stock trailer where there's room to move about and brace themselves but have problems in a two-horse trailer with a partition. The claustrophobic horse may kick and struggle, tremble and sweat. Some horses travel quietly on straight stretches and scramble only on corners.

If you are hauling two horses in a two-horse trailer and cannot remove the divider, remove the lower part or replace it with one that doesn't come clear to the floor. This gives the horse that scrambles more room to spread his feet and brace himself. Some horses like extra foot room to the right, others to the left. Load them on the side that gives them their extra room in the middle, under the partial partition. This will often enable a scrambler to travel comfortably in a two-horse trailer.

Footing

A slippery floor may cause scrambling. Rubber mats can prevent this, but if a horse still has trouble keeping his footing, add some non-dusty bedding for better traction. Make sure the trailer rides level, without a sloping floor. If your hitch is too high or too low, the trailer will slant up or down and put the horse off balance. Tires should be neither over-inflated nor underinflated or the trailer may not ride level. A bent axle, bad wheel bearings, a broken spring, a flat tire, and other problems can also cause a trailer to have a bumpy ride. Routine trailer maintenance is critical to the safety and comfort of your horse.

A scrambling problem is sometimes a tying problem. If your horse doesn't tie well, he shouldn't be tied inside a trailer. If he fights the restraint, he may feel hemmed in and start scrambling. If you plan to tie the horse in the trailer, first make sure he is well trained to tie.

OTHER OPTIONS

Some horses feel more comfortable, and ride better, in a front-load trailer. Here they enter in the front and face the rear. Others prefer a slant-load trailer, in which they are positioned diagonally.

Methods for Loading the Reluctant or Spoiled Horse

A horse that refuses to load may have been in an accident or been handled poorly in earlier training. This horse associates a trailer with discomfort and fear. He may fight strenuously to keep from loading, and it may be difficult to change his mind.

Most trailer loading problems result from fear and resistance. The horse's resistance is reinforced if you are unsuccessful at loading him. If you lack confidence, he knows it. Try to determine whether he is afraid of the trailer or has no respect for your direction, or both.

There are many ways to get a problem horse in a trailer. The success of a loading method depends on the individual horse and what his particular hang-up is and also on how he reacts to the method used. Know your horse, then determine which methods might work and which should be avoided. Work with his personality and problem, and strive to make the loading as nonconfrontational and stress-free as possible.

Advance Preparation: Control His Movements

Techniques for putting an inexperienced horse into a trailer — leading him up to it, taking your time, eventually convincing him it's safe to enter — can work on a horse that's been made afraid of the trailer by poor loading techniques. He may take longer to convince, however; it may take several days of patient handling. He may need lessons in basic leading and with the whip, as when you were teaching him to respond better to voice and whip cues.

With such a horse, it's often a good idea to work alone. He may become more confused and afraid if confronted by several people at once; in this case, it's easier for him to gain confidence if one person handles him. Before you even show him a trailer, give him some lessons in basic handling so he is totally manageable and maneuverable at your command. Acquaint him with the whip and let him know it is nothing to fear. This won't work if he's been abused with a whip unless you have lots of time to work with him very patiently, starting from the beginning to help him overcome his fear. This can be a considerable challenge, though, so perhaps forgo the whip on a previously abused horse.

BE SURE HE'S NOT BLUFFING

Occasionally a horse's refusal to go into a trailer is a bluff. He may have been trailered before but doesn't want to go in now, so he refuses just to see if you're really serious. If you show him a whip or put a rump rope behind him, he'll usually walk right in. Be careful, though: Some horses may become more frantic at the sight of a whip or ropes.

Use a Long Line to Direct His Head

For a reluctant horse that you must get loaded without several lessons, enlist an assistant to help him load: One person leads the horse, the other one offers encouragement from behind. The person leading must *not* pull on the rope; the horse will react by pulling back, rearing, or rushing backward. The rope is merely to keep him facing the trailer. If you are working alone, ahead of time loop a long rope around something solid at the front of the trailer and back again. Snap one end onto the horse's halter, and hold on to your end as you encourage him from behind. Take up slack or give slack, as needed, to keep him facing the trailer. The long rope enables you to keep a forward tension on it, even if you have to be alongside or behind the horse to give him encouragement.

A lead rope of any kind is simply to keep his head straight, not to pull. Most accidents occur when a horse is being pulled by the head — that is, when he rears up or pulls back hard in response to being pulled on. A horse will rarely bang his head on a trailer or rush backward forcefully unless he is being pulled.

Make It Comfortable to Go Forward, Uncomfortable to Back Up

Your horse is rewarded for every forward step by the release of pressure. Refusing to go forward, however, or backing up, is made uncomfortable (not punished!) by an appropriate stimulus, such as someone behind him flicking his hindquarters gently with a whip or poking his rump with a straw broom. (Stand to the side and out of the way in case your horse kicks.) The broom bristles are used on his buttocks, not on any sensitive areas; they're used as an aggravation to move away from rather than as something to cause pain. A plastic bag fastened to the end of a long whip can also be used to "worry" his hindquarters and encourage your horse to move forward, away from it.

Knowing how much to pause — letting your horse think about each step — and how much to encourage is crucial. Be in tune with his attitude and responses. Keep him calm and thinking, rather than pressing too much and triggering panic or resentment. One step forward, a pause where he thinks about going forward rather than back, or the smallest try, should be rewarded with praise and a cessation of the aggravation behind him.

Often a person accidentally punishes positive behavior, by pulling on a horse, for example, or by pestering him from behind when he starts to go forward, inadvertently teaching him that the trailer is an uncomfortable place, to be avoided. The apprehensive horse starts to move forward and the horseman sees progress and tries to encourage it by tapping or spanking from behind — in essence, punishing the horse for his effort. Be mindful of this: Your horse needs to be *rewarded* for positive movement by release of pressure.

Don't worry if he begins to load and then backs out. Continue the aggravation until he begins to walk forward. Give him a clear choice without pulling or pushing. Allow him to enter the trailer because it is his own decision. It may take several minutes of encouragement for him to come to that decision, but if you are patient and consistent, giving rewards when he goes forward and aggravating his hindquarters when he goes backward, eventually he will go in.

Rump Rope

A horse that was trained as a foal with the use of a rump rope may be encouraged to move forward with one. If he protests, tends to kick when a rope tightens around his hindquarters, or tries to swerve away from the trailer as he is led up to it, two helpers — each holding the end of a long rope or longe line — can use the rope against his hindquarters to put pressure behind him. This keeps them out of harm's way if the horse kicks or rushes backward.

The rope behind him encloses the horse and keeps him going straight as he approaches the trailer. Steady tension can resist backward movement. He is not pulled by the rope; its purpose is just to keep him straight and encourage him to go forward.

LET HIM DECIDE

Whatever method you choose to encourage a reluctant horse into a trailer, make it his idea to enter it. Even if you must use firm encouragement from behind, leave it up to him to decide to go into the trailer rather than pulling on him. If it's his decision, he will be more receptive to doing it again; it's a form of training — working with his mind rather than just forcing his body.

If need be, two ropes can be used, one tied to each side of the trailer. Assistants pass behind the horse at a safe distance and cross the ropes behind his hindquarters. Short, firm tugs on the ropes encourage him to move forward. The helpers must be prepared to give if the horse flies backward, so as not to pull his legs out from under him or make him fall over backward. It's better to use a solid "chute" (see next section) to keep the horse from swerving past the trailer, but sometimes this is not an option; the ropes can help convince the horse he must continue to face the trailer.

Give Him Just One Choice

Often the easiest way to load a reluctant horse is to back him up to an enclosure so he can't go past the trailer door; the only route from pen or stall is directly into the trailer. Another way is to make a lane to the trailer: Use a solid fence or wall as one side and create another side with a portable panel.

For safety, when working with a stubbornly reluctant horse, make sure the panel is solid by securing a piece of plywood or even a rug or tarp against it so the horse can't put a foot through. With this "loading chute" behind the trailer, there is nowhere for the horse to go but into the trailer; he knows he cannot dodge past the trailer to escape and therefore does not try. As one person patiently encourages from behind, the horse more readily submits to going in because he knows he can't go any other direction.

Don't Do Battle

It's not worth fighting a horse to get him into a trailer. You'll get him loaded, but the struggle that got him in is one more bad experience; rest assured, he will resist again the next time. On occasion, though, a horse can be forcibly loaded and gets easier to load as time goes on — if he's not hurt in any way and his trailer experience is good — that is, good food to eat, no rough roads, no bad corners.

It's best, however, to begin at the beginning. Use patient methods with a spoiled horse until he gains confidence and goes in on his own. A few days or weeks of nonconfrontational training sessions are better than struggling, which always presents some risk of injury to both the horse and you.

REINFORCE THE LESSON BEFORE YOU HAUL

Though your inclination after loading a difficult horse may be to close the tailgate and go — you're probably running late by then — it's better to give him a few minutes to relax, reward him with praise and feed, and then unload him. Load him several more times until he goes in without resistance. This will make him much easier to load the next time, ending the experience on a positive note.

17

RETRAINING THE SPOILED HORSE

If a horse is handled with forethought and patience from the time he is young, training usually goes smoothly. Each horse has his own personality and quirks, but a good trainer works through any challenges that come along and heads off potential problems and bad habits. If you take time to build trust and confidence, the result is generally a responsive, well-mannered horse.

Sometimes, however, you don't have the option of starting with a young horse. You may need to work with a horse that has had less-than-ideal beginnings or a bad experience along the way. In this case, your challenge is to correct the damage already done. Usually the correction involves going back to basics in the area of training that was left out, rushed, or handled improperly.

There are many ways to deal with problem behavior. This chapter offers some suggestions. Always remember that any training method you choose will have advantages and disadvantages; how you implement a particular training method will dictate your level of success. Even the best techniques may fail if they are done at the wrong time or if they are presented incorrectly. Select a method suitable for your particular horse. Your understanding of the individual horse, and your sense of what is appropriate for you to do with him and when, is most important.

Often a problem is a manifestation of something else, such as a gap in early training. If you go back to elementary lessons and work on basic handling, the problem you are trying to correct may abate, simply because the horse is developing more desirable responses.

> ### SAFETY FIRST
>
> Some bad habits and phobias can be dangerous. A horse that sets back when tied, for example, or one that rears and goes over backward in an attempt to avoid the bit, may be more than you want to deal with. If you feel that a problem is beyond your ability, seek the help of a professional trainer. The best solution may be to get a different horse. *Never* risk your own safety or that of the horse in attempting to correct a problem that could pose serious danger.

The Horse with Bad Ground Manners

A well-trained horse has good manners. He respects humans as the controlling force in his life and does not question or refuse to obey commands. A horse that was properly imprinted as a foal (see pages 80, 82) is receptive to human control, as is any horse that was well handled during early training. A horse allowed to do as he pleases has no respect for humans.

Good manners are crucial to control of your horse and to your safety. A horse that is disrespectful and inattentive can be dangerous, whether he stomps on your toes or takes a parting shot at you with a hind foot as you turn him loose in the paddock.

A horse may have bad manners and an aggressive attitude that was not corrected sufficiently by the person who was handling him. Some pushy horses take advantage of a softhearted or timid person who allows them to do as they please. Bold horses spoil easily if they feel they can have their way in the horse-human relationship. Once bad manners become habit, it takes consistent training to break them and to reform the horse so that he respects his handler.

Some horses are frustrating and difficult to work with on the ground — they fidget when you try to groom and saddle them, step on your toes and bump you, root and tug when led, and drag you along. Showing lack of respect such as walking over you or crowding your space often occurs when a horse is treated as a pet or when training is inconsistent. The horse doesn't have a clear sense of his place in the relationship. To develop a well-mannered horse, you must assume the role of boss, and your horse must fully understand and accept this. Kind but firm handling leaves no doubt in the mind of the horse that his role is follower, not leader.

The Pushy Horse

Establish consistent rules so that your horse understands his limits of behavior and that humans have special status. They give the orders and are to be respected at all times. Humans are not to be touched without permission, never bumped (even accidentally), bitten, kicked, or swatted with a tail. If you spoil a horse by allowing him to nuzzle, rub, play, and push, he learns to consider humans his equals, buddies to roughhouse with. This leads to dangerous consequences, as humans are much more fragile than his herd mates.

Think Like a Horse

Horses are herd animals, accustomed to a social structure in which they must find their niche. Social ranking is part of herd life: who eats first, who leads, who follows. Equine reactions are based on dominance and submission. A horse must figure out how humans fit into that scheme — is he higher in the social order or must he submit to them? If he succeeds at dominating people, he will do as he pleases until he meets someone who can teach him more acceptable behavior. Horses need ground rules that they know they must obey when being handled by people. Then you, not the horse, can make the decisions — a much safer situation.

Don't Allow Pushing

Your horse must learn to keep a respectful distance at all times. This eliminates "accidents," such as stepping on your foot, swinging his rump into you, pushing you into the gate as he crowds through, and smashing your nose when swinging his head to look at something. Be alert to any rule violations. Don't just ignore anything that seems like an accident. Through your own body language, let him know that you have a personal space that cannot be violated or disrespected. Even a gentle horse can be dangerous if he doesn't respect your personal space.

When handling a horse, make sure he knows where his place is and allow him to be comfortable in it. If he pushes into your space, however, make him uncomfortable: Let him run into your hand, lead rope, a stick, or a whip. Create an imaginary box that he is not allowed into. Don't let him rub, hunt for treats, or nibble your hand or hat. Be firm and consistent with young horses that are still learning how to relate to people. Little nippers grow up to be big biters. If you rub on the horse, he will rub back. You may be setting yourself up to be challenged.

If your horse invades your space by stepping on you, bumping you, or rubbing on you, enforce your rule of no uninvited touching by swift and appropriate correction. Firmly but gently push his head or body away, or use a quick, sharp slap or a bump with the butt of a whip. The circumstances and severity of the violation should dictate the degree of reprimand. Do whatever is reasonable to make it clear to the horse that aggressive behavior will not be tolerated.

Be Consistent

Discourage your horse each time he transgresses, insisting that he keep his distance. Move him back or give him a spank (if appropriate), then reward him with a friendly pat as soon as he backs off and stands quietly. Don't overdo the reward; you're trying to instill discipline, not bribe him to correct his faults. If you're consistent, most horses will get the idea and comply with your wishes.

The Body Basher

A horse may step on you and bump or smash you into the stall wall or fence when grooming and saddling, for example. Distinguish between behavior triggered by fear or discomfort and that which is deliberate and aggressive. If your horse is jittery about some aspect of handling, he needs gentle, patient work to help him overcome his fear. If he's trying to bully you, on the other hand, he needs firm and immediate correction.

Be Vigilant

Usually when a horse is trying to get the better of you, he is not showing alarm but instead goes about it purposefully and watches to see what, if anything, you're going to do about it. Any deliberate exhibition of bad manners should be corrected with an immediate reprimand or an appropriate smack — without anger.

After making your point, go about your business so your horse knows that he hasn't won and must do your bidding when you groom or saddle him. A horse that habitually bumps a person must learn to stay back or to move back on command. Tap him on the chest with the butt of a whip to back him up, then reward him with praise when he responds. Anytime he steps into your space, immediately make him get back where he belongs. It's not enough to halt; he must back up. This makes him realize he must submit to you and that you are in control.

CARRY A STOUT STICK

For an aggressive horse that tries to mash you into the wall, carry a short, stout stick, holding it so he runs into the stick instead of into you. After a few times, most horses learn that it's not so pleasant trying to move into your space.

The Hard-to-Catch Horse

Many horses in pens and pastures become evasive for a variety of reasons, but most can be retrained. Fear, resentment, or habit may be at the root of a catching problem. A young or inexperienced horse may avoid people because he's afraid; when he's handled regularly with kindness and consistency, he'll lose his fear. If he was handled improperly as a youngster, however, or was always chased into a corner to be caught, the horse may exhibit elusiveness even after he is no longer afraid.

Sometimes a horse avoids being caught because he doesn't like what's done to him afterward, such as a painful wound treatment or a long ride. Whatever the original reason, some horses become habitually standoffish. Your task is to change your horse's mind so he'll realize that being caught is not something to fear, but rather something to look forward to.

Reward Him

If you consistently give your horse a treat — such as grain to eat while you groom and saddle him — he'll want to come to you. Don't take grain with you to the pen or pasture, however; he may try to get a bite and still keep his freedom. He must learn the rules: If he is caught willingly, he is rewarded with a treat. Give him grain, a carrot, or an apple after you halter him and bring him out of the pasture.

Examine Your Relationship

If your horse becomes hard to catch, carefully consider what you are doing and how you approach him. If he associates you with negative things — long training sessions, endless repetition of boring maneuvers, and long rides day after day — he may want to avoid you and so becomes hard to catch.

THE GRASS IS GREENER

A horse in a paddock that has little grass will welcome a few bites of greenery as a treat. If every time you catch him you let him graze for a few moments in a patch of grass along your yard or driveway, he will look forward to this.

After you catch him, let your horse graze while you groom him in preparation for saddling.

If you are overworking him or overdoing his training, slow down. Give him a break. Sometimes just giving him a treat (or brushing him or cleaning his feet) and turning him loose again will be sufficient reward to keep his interest. Vary the routine, so that some of the time you are not working him, or doctoring him, or doing whatever it is he has come to resent. Make sure pleasant activities are as much a part of his training as is work.

Horses will tolerate hard workouts, but if this is what he expects every time he is caught — especially if he suffers from discomfort and muscle soreness afterward — he'll try to avoid being caught. A horse will work his heart out for you because he respects you and has been trained and conditioned to give you his best, but if you take advantage of this trust, your horse may try to find a way out of hard work.

> ## OUTSMART HIM
>
> You must outsmart your horse and catch him in a way that won't hinder your purpose. If you always run him into a small corral or the barn to catch him, eventually you'll always be able to catch him there, but this will never make him easier to catch in the open. If he ever lives in a large field or gets loose at a show or trail ride, you'll still have a problem. Teach him to be easy to catch under a variety of circumstances.

The Confirmed Avoider

Sometimes you buy a horse, then discover he is hard to catch. A spoiled horse can be a frustrating challenge; to persuade him that being caught brings a reward of treats or some other incentive, you must first catch him. If you have to run him around the pasture every time you want to catch him, you defeat your purpose. Once a horse starts running, he gets in a different frame of mind and may keep running, circling the pasture again and again with no intention of going into a corner or into the corral.

Use a Small Pen

With many problem horses, it helps to keep them in a smaller pasture or pen for a few days or weeks until they get used to being regularly and easily caught. Horses that run from habit rather than fear won't bother to run if there isn't room. Putting a horse by himself can also help if he's been in a group of elusive horses. Horses in a group can be hard to catch if one or two are habitual runners. Keep him by himself and catch him several times a day until he realizes catching is pleasant.

Another method is to put him in a pen with a buddy who is easy to catch. With some horses, this even works in a large pasture. A horse that is a "good influence" makes it easier. Greed and jealousy can be used to advantage. If the hesitant horse sees you catch and feed his buddy, he'll want a treat also and feel left out.

Don't give him the goodies, however, until he allows you to catch him. He may come for his share but not yet be willing to give up his freedom. If he is dominant over the other horse, enlist the aid of a helper so your trainee can't eat the other horse's grain. If you are working alone, take the first caught horse out of the pasture to have his grain, still in close sight of the reluctant one. He'll feel somewhat jealous and be more willing to submit to the halter so he can have some, too.

Another way to teach an evasive horse to be caught is to work him in a round pen, as described in chapter 5 (see pages 126–131). If done properly and with patience, the horse learns it's a lot of work and effort to avoid you and that he'd rather come to you without the fuss.

Walk Him Down

If you don't have a pen, follow your horse patiently around the pasture until he resigns himself to being caught. Allow plenty of time, preferably on a day you don't need to do anything more important with him. If you're in no hurry, you won't be impatient or tense.

Walk toward your horse matter-of-factly, in a casual, nonconfrontational manner. It's better to zigzag than to approach him head on. When he moves away, just follow him calmly and confidently. He may run to the far corner, or incite herd mates into running with him for a while, or stay on the far edge of the group. Ignore the other horses (or you may want to stop and pat the calm ones) if he's in a group, and continue to follow him around patiently.

If he stops, you stop or take a step backward. Read his body language to see if you should walk toward him or wait. His ears and attitude will tell you whether he's receptive or ready to flee again. If he moves off, continue following wherever he goes. Be persistent; you can walk him down, as you don't have to trot or run — just patiently follow. At first he may enjoy this little game, but eventually he'll tire of constantly having to move away from you. Avoiding you takes more effort than he wants to expend. At some point, he'll let you walk up to him. Praise him and pat

---

USE POSITIVE REINFORCEMENT

Try keeping the difficult-to-catch horse in a small pen without feed and water. But catch him several times a day to take him to water or to feed him. This system of catch first, then feed can change his mind fairly quickly. Even if you don't resort to this solution, catching him a few times a day and then immediately turning him loose after a treat or a quick grooming will enable him to associate catching with good things. If you give him some kind of reward every time you catch him, he'll usually reform.

him, but don't halter him immediately. Be casual, then halter him nonchalantly. Turn him loose after a moment and walk him down again or give him a treat and then turn him loose again.

Walk out in his pen or pasture often, to catch and pat him, then turn him loose. After a few times, he'll know you're going to catch him every time you come and that indeed being caught is actually pleasant. Once he realizes this, he will stand when you approach, or at worst just move away a short distance and then let you walk up to him. Soon he'll look forward to your arrival instead of taking off when he sees you coming.

Whip-Break Him

This sounds cruel, but if it is done properly, it is not. The purpose of whip-breaking is to encourage the horse to turn and face you; then he cannot kick you. The whip is used not to punish, but rather as an extension of your arm so you can touch his rump at a safe distance, out of kicking range. This works only in a small pen or a stall, not in a big pasture. The whip can be any type, even willow, as long as it enables you to touch him and stay out of kicking range.

This method works for a spoiled horse that sticks his head in a corner of his stall or pen as you approach, turning his rump to you or even threatening to kick. You can't get to his head without coming up behind him, and it's not safe to approach his rump. Even if he lets you ease past his rump toward his withers, he may turn and run off.

First try to catch the horse quietly without using the whip. If he turns his rump to you, cluck and ask him to move. If his head is away from you, in a corner, whatever way he moves will be toward you and should be rewarded. If he takes a step, stop clucking and praise him. Then cluck again. If he takes another step toward you, praise him and back up a step to reward him. If he doesn't respond or turns away from you, however, tap gently on his rump with the whip. Never tap his legs or hit hard, or he may kick.

Tap gently, as often as needed to make the horse move his rump and turn toward you. Even if it's just a step or he turns his head to look at you, stop tapping. Wait a minute, then cluck again. If you give the verbal cue first, he has the option to turn toward you before you tap. Soon he realizes that after the cluck, you tap him. To make the tapping stop he must turn toward you — and he'll turn as soon as you cluck. He learns that if he turns around to face you, he doesn't get tapped. When he does face you, pat him so he knows he did the right thing.

> ### The Whip Is a Cue
>
> *Never* lose your temper or whip a horse when using this method. If he becomes frightened or excited, he may try to jump the fence or run around the stall and over you. Think of the whip as a cue, not as punishment, and use it accordingly.

Eventually you won't need a whip to catch him; he'll turn to face you when you cluck. This won't make him any easier to catch in a large pasture, but it will teach him not to turn his rump to you in a stall or pen. It will save you the frustration of not being able to get to his head, and it eliminates the risk of being kicked.

The Hard-to-Lead Horse

A led horse should accompany you in a relaxed way. But some horses root, pull, kick, or rear when led. A horse may have become a dragger or a kicker because of the improper way he's been led. Now he resents being pulled on and reacts by lugging into the halter.

Lead Loosely

It takes two for a tug-of-war. If you pull on your horse, you are inviting retaliatory behavior; his reaction is to pull back and move away from pressure. Soon this becomes habit. Pressure exerted to slow him should be intermittent, not continuous. Well-timed short tugs are much more effective than a long, hard pull. Teach your horse to respond to pressure by releasing it when he reacts properly. Start with slight pressure and give him a chance to respond to that before repeating with firmer pressure.

Use a Nose Chain

If a horse drags you faster than you want to go or won't heed commands to halt, you may need a chain. It is used over his nose in the same way that a choke chain is used on a dog. It does not inhibit his movement unless it is engaged either by the horse (going too fast) or by you (asking him to stop). Make sure the chain is correctly attached to the halter; it should not be wrapped around the noseband. See pages 110–111 for more on the use of a nose chain.

Do More Ground Work

Some horses, especially stallions, tend to rear when being led. Because they are impatient, they rear in an attempt to avoid restraint. A horse may rear when he's afraid or if you ask him to do something he doesn't want to do, such as approach a trailer. The solution is more ground work on leading (see pages 109–112), so the horse learns to move forward rather than up on his hind feet when his security is threatened.

For the habitual rearer, put a chain over his nose and be prepared to use it. Give a sharp jerk with it before he actually gets off the ground. If he starts to rear before you have a chance to halt him, allow him to complete the rear. A sharp jerk while he is going up or at the peak of a rear may cause him to throw himself over backward or sideways in his attempt to avoid pain. It could also cause him to strike at you. It's better to halt him just before he gets off the ground or to wait until he's starting back down. Split-second timing is crucial for effective correction and for good control. With proper training, most horses stop this dangerous habit.

POSITION YOURSELF FOR GOOD LEVERAGE

Always stay in the best leverage position — to the side, where you can bend his head toward you — when leading a horse that pulls or rears. *Never* be in front of him and *never* pull down on his head. Try to anticipate a rear. If the horse starts to go up before you can correct him, move back farther on the lead rope, toward his hindquarters, and pull his head around to the side. This makes him move his hindquarters over and makes it more difficult for him to rear.

The Hard-to-Bridle Horse

You may come across a horse that is hard to bridle because of a bad experience, such as a painful bit, pinched ears, clanked teeth — something that left an unpleasant memory. The horse got into the habit of trying to avoid the bridle. The horse has made up his mind that bridling is traumatic. The nervous horse owner, remembering (like the horse!) previous struggles, is worried about the bridling, and it becomes another bad experience each time. This habit is hard to correct — resolving it is more difficult than preventing it would have been in the first place.

Every horse is different, and not all react the same way to the same kind of handling. Some have an individual problem (extra-sensitive ears, an old injury, or some other physical problem that causes them to be evasive) and personality quirks that require a different technique. A problem can start innocently, worsen, then become chronic.

Ordinary bridling methods don't always work for a horse with a problem. He may have been startled or upset when bridled the first time or rushed into it before he was ready. When you encounter a horse that's difficult to bridle, you must figure out why he reacts the way he does, then work around his particular hang-up.

Lure Him

A horse that refuses to open his mouth for the bit can usually be retrained if the bit is coated with something that tastes good. If he likes molasses or honey, smear some on the bit. With patience and sweetening, he may change his mind and willingly take the bit. If he is apprehensive, don't even try to bridle him. Just hold the bit in your hand, without trying to raise the headstall, and let him lick off the molasses. After a few times, he may lose his fear and you can go ahead and bridle him, still with molasses on the bit.

Consider His Ears

Some horses don't like having their ears handled. You may have to use a headstall that can be unbuckled at the side and put it on without going over the ears. If necessary, dispense with brow band and throatlatch and just loop the headstall over his head behind the ears, without touching them at all, and buckle the side after the bit is in his mouth. An ear-shy horse usually is not bit-shy; you can bridle him if you find a way to avoid touching his ears. Patient ear-handling sessions without a bridle can calm the horse and make him more comfortable with gentle handling of the ears. Most ear-shy horses improve over time if bridling causes no pain.

Check for Physical Problems

Check the bit and headstall for proper fit. If they are not causing pain, the extremely head-shy or ear-shy horse may have a physical problem. Consult your veterinarian; she can tranquilize or twitch the horse and examine his mouth and/or ears. The horse may have a wolf tooth irritated by the bit, a tooth problem, a cut tongue, or a sore poll or ears.

Even slight soreness over the poll, from bumping his head, for example, could make him avoid bridling. Ear ticks, warts in his ears, or infection also may make his ears sensitive.

If a horse has a physical problem, it will take time to correct (get rid of ear ticks, treat the warts, fix a tooth), and it may take even more time to overcome his anxiety about bridling. Well after the physical problem has healed, the horse will remember bridling as a painful experience and may still try to avoid it.

Changing his attitude will be a gradual process. Handle his head and mouth frequently without actual bridling. Gradually work toward the sensitive areas until you can gently touch the spots that bothered him before, letting him realize you are no longer causing him pain and that there is no reason to be head-shy. When he finally accepts handling of his ears and placement of fingers in his mouth, try a bridle again — carefully and gently.

Break Bad Habits with Firmness

If a physical problem has been ruled out, the hard-to-bridle horse may be evasive from habit: In some cases, the horse knows better but has learned that being difficult to bridle can delay going to work. He might even get out of work completely if he can bluff the rider. A horse ridden by inexperienced people or small children often develops these tactics. In this case, patience and gentleness are not the answer; the more you pussyfoot around, the more it confirms to the horse that he can keep you from putting on the bridle. The solution for this problem is firmness. Once the horse realizes you *are* going to bridle him, he'll usually give in.

To foil an old, spoiled horse that merely raises his nose out of reach to avoid being bridled, tie his head lower and shorter. Try this technique only on a horse that ties well. Many horses should not be restrained while being bridled. *Never* tie a young, inexperienced horse that is uneasy about bridling, because he will definitely set back. It's better to teach the horse to lower his head on cue, thus making it easier to halter and bridle him (see page 164).

Other horses need a middle-of-the-road approach of firm gentleness. You can't be too forceful or they react adversely, yet you must insist on good bridling manners. For a horse that raises his head and clenches his teeth, grasp the headstall midway (not from the top) so you can use that hand to press on the bridge of his nose as you hold the headstall. This keeps him from raising his head too high as you use the other hand to slip the bit into his mouth. Then use a finger in the corner of his mouth to press on his gum or tongue to make him open his mouth.

You don't want the horse to be successful in avoiding the bit or raising his head, or he may become more sly in the future, gaining boldness from his success. Avoid his devious action with gentle firmness, keeping his head under control. By pressing on his face between nose and eyes, you thwart his head raising and can get him bridled without a struggle.

TYING TIPS

When tying a well-halter-trained horse for bridling, leave a little slack in the rope for freedom of his head and neck so he won't feel claustrophobic but not so much that he can hold up his head out of reach as you try to bridle him.

Get Him Used to Unbridling

A horse should lower his head as you remove the headstall, open his mouth, and let the bit drop out. If he raises his head, the bit will hang up on his teeth. A horse that's had his teeth bumped during unbridling may throw his head in the air in an attempt to avoid pain, and this makes the problem much worse; it's harder to get the bridle off with the horse's head in the air.

One way to get him to relax and lower his head is to rub his forehead. If he has on a halter under the bridle, hold him gently by the halter as you remove the bridle, encouraging him to keep his head low. If he's not wearing a halter, loop the reins over his neck to gently hold his head in a lowered position. If he has a serious phobia about unbridling, take the time to teach him to relax and lower his head on cue (see page 164). You may want to develop a special cue, such as rubbing his neck or his forehead, even if you have to use a treat in conjunction with the cue at first.

The Hard-to-Saddle Horse

Some horses are evasive and grouchy about saddling. There may be a physical reason such as a sore back or pinched skin if the cinch was tightened too quickly. If a horse becomes hard to saddle, check for a physical problem. You may need to change to a saddle that fits the horse better, one that doesn't make his back sore. Always girth up gradually and check for pinched skin.

Go Back to Basics

If your horse is in the habit of fidgeting when saddled and is being evasive from habit rather than pain, go back to fundamentals: Teach him to stand still. More tying lessons, to teach patience, may be helpful, along with additional ground work on *Whoa* and standing still. He needs more basic handling to learn that he must stand quietly when you ask him to.

The Hard-to-Mount Horse

Some horses won't stand for mounting; they run off before you have your foot in the stirrup or fidget and make it difficult for you to mount. First, make sure you're not irritating the horse during mounting. If you notice that you bump his rump with your leg, bump his mouth with the bit, or poke your toe in his side, alter your movements to avoid inadvertent annoyances. If discomfort is not the cause, some trainers use hobbles to retrain a horse to stand still (see the box on page 438), but usually you can correct the problem with patient work.

ELIMINATE THE SOURCE OF THE HORSE'S DISCOMFORT

Sometimes a horse tries to evade mounting because his back hurts. If the saddle shifts as you put weight in the stirrup, for example, it may put pressure on tender areas or on irritated skin. If mounting pulls the saddle to the side, this also can cause discomfort. A rider may create back pain plunking down into the saddle. If your horse's evasiveness is caused by pain, allow his back or girth time to heal if there are tender areas, use a saddle that fits better, or use a mounting block so you can get on without the saddle shifting.

Go Back to Basics

Refresh lessons on *Whoa* — leading or longeing (see pages 133 and 156, respectively) — so your horse knows that the command means to stop and stay stopped. When you apply this to mounting, do some lessons when you don't actually have to get on and ride. Start over as with a green horse (see page 260) and insist that he stand still while you prepare to mount, even if you need a helper to hold him steady at first. Put your foot in the stirrup and insist that he stand. Do a lot of mounting and dismounting without going anywhere. Each time you ride, make a habit of having your horse stand a moment and relax before moving off. He must learn not to move until asked.

If your horse is bad about taking off during mounting, conduct the lessons in a small pen or a box stall, so he realizes he can't go anywhere. If he swings away from you, position him next to a fence or wall. It may take several sessions, but if you have redrilled him on *Whoa* and he isn't being ridden anywhere, he'll resign himself to standing still. A horse that tries to buck as you mount is not just impatient — he's trying to keep you from getting on. In this case, seek out professional help to break this bad habit.

The Horse That Won't Tie

A horse that sets back when tied is dangerous to himself and to you. This bad habit needs diligent attention. Methods described in chapters 4 and 5 — use of a body rope (see page 95), tying to an inner tube (see page 96), or using a long rope looped around a post so you can give and take until the horse loses fear of restraint (see page 140) — will work for most horses when retraining them to stand patiently as they are tied.

Tie High

Try tying your horse to something above his head such as a strong tree branch or a beam in the barn. Secure a sturdy rope to the tree branch or beam, with a solid metal ring tied or braided into the end that hangs down and positioned where you can reach it for ease of tying. Then you can tie the lead rope to the ring. Your horse won't be able to get much leverage to pull back and usually can't hurt himself. He'll stop pulling eventually.

A tall post will work, too — a sturdy post about 8 feet tall is best, so he can be tied higher than his head. For retraining a puller, use a well-set post (buried 4 feet in the ground, set in concrete below ground level) in an open area (no fence to crash into if he fights). A metal pipe also works well for a tie post. Attach a swivel at the top (like a truck axle hub) that moves in a complete circle. Secure to the swivel a stout chain with a sturdy metal ring at the end of it, hanging where you can reach it for tying. Your horse can walk around the pole without getting the rope

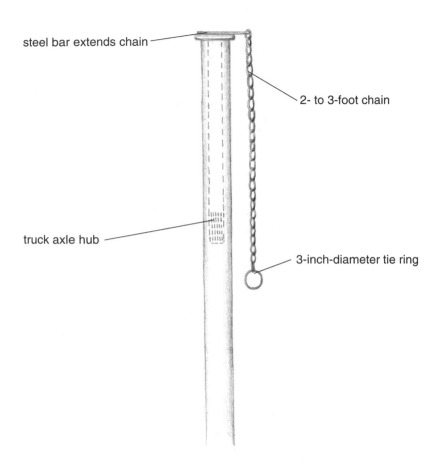

steel bar extends chain

2- to 3-foot chain

truck axle hub

3-inch-diameter tie ring

A sturdy tie post can be made from a tall metal pipe. The pipe is 12 feet long with a diameter of 4 inches. The bottom 4 feet is set in concrete, with the top surface of the concrete well below ground level so it won't hurt the horse's feet.

wrapped around it. Use a sturdy halter and rope, and secure with a quick-release knot.

The Biter

Biting must never be tolerated. A horse that tries to bite as you put on his halter, pick up a foot, or saddle him is showing poor manners and can also be dangerous.

There are several reasons a horse might bite. He may simply do it in

---

FIND THE CAUSE OF DISCOMFORT

If a normally well-mannered horse bites, it's usually a sign that you are doing something wrong. Check equipment to make sure it fits. Establish whether anything you are doing is causing discomfort.

---

playfulness; this is how he interacts with herd members. He may bite if you cause pain; perhaps a saddle pinches him when it is cinched up. Biting may be indicative of a bad attitude or lack of respect. It could also be an expression of masculinity — a stallion's nature is aggressiveness, and he uses his teeth to reprimand other herd members or to establish dominance. Most stallions must be taught, when young, not to bite humans.

Some nervous horses bite when they are frustrated or bored; confinement can lead to irritability, and they take out their frustrations on the people who handle them. A horse that is bored in confinement needs more exercise as an outlet for his energy, more riding and less grain, or a lot more turnout time. A horse may bite because he does not accept the human as boss. Horses also bite when they are upset and when something bothers them. Whatever his reason for being cranky, biting should be quickly halted.

Is He Spoiled?

Most nipping can be halted. A horse's habits and manners are a product of his reaction to people and to the way he's been handled. He'll do whatever he's allowed to do in reaction to a specific human action. He forms habits from these interactions. Horses with an aggressive personality take more training effort, but most can be trained not to bite. Often the worst biters are not outlaws but merely pampered pets that had their own way too often and were spoiled by easygoing handlers. If you are not firm and consistent in handling the headstrong horse, the horse feels no need to respect you.

Discipline Him

A young horse that is nippy must learn more respect. He needs firm and consistent discipline every time he tries to nip — just as he'd get in a herd situation from an older, more dominant horse. Some youngsters are rough and aggressive in their playfulness and are hard to handle if this is allowed to continue. Often the best cure is to put the youngster with an older horse who won't tolerate his pranks. The swift discipline he gets from the older horse makes the youngster understand that he can't be the boss or do as he pleases. A youngster who grows up with other horses is usually much easier to handle than one that has been isolated, with no herd mates to discipline him. (See page 121 for more on herd etiquette.)

Correction for bad actions like biting should be given as the act is happening or about to happen — not afterward, or your horse will think it's a game. He may keep trying to bite in order to see if he can be quicker than you. For best effect, correction should be instant and self-inflicted; that is, his mouth should meet something unpleasant each time he tries, such as your brush handle, a hoof pick, or an elbow instead of your arm. If you stay alert and catch him every time he tries, he'll grow tired of initiating his own discomfort.

Don't swat at a horse that bites. He is faster than you and can nip and jerk his head away quicker than you can hit him. Trying to hit him on the nose will only make him head-shy or, if he thinks it's a game, more sneaky. Meet his nip so he is punished in the act. For the persistent and aggressive biter, conceal a nail or hoof pick in your hand while grooming or saddling; allow him to run into it instead of your hand. Don't jab with it (he can be fast enough to jerk away and you'll miss); just set him up a few times so he runs into it while trying to bite.

Every horse is different. A disciplinary action that causes one horse mental trauma may be ignored by another. Tailor the severity of your method to the individual horse. If he is persistent about biting, make the discipline more sharp until he makes the connection that his own action causes him discomfort. Stallions, especially, may take harsher methods, as biting is part of their normal expression of masculinity.

As you work with the nipper to rehabilitate him, discipline him consistently every time he misbehaves, but don't forget to reward him with kind words and a gentle pat when he does well. Let him know when he pleases you. With proper timing and consistency, the punishment-and-reward system will cure most biters.

The Kicker

Some horses kick at other horses or at people. Horses that kick with intent to hurt need retraining by a professional horseman or should be sold and potential buyers warned about this vice. Horses that kick when startled or upset are more easily retrained. This is part of the fight-or-

---

DON'T TAKE RISKS

Any horse that insists on biting, in spite of harsh and immediate correction, is dangerous. If your horse comes to meet you with teeth when you catch him, you cannot adequately discipline him, and he may make catching or haltering a perilous game. He has a bad attitude or is taking out his frustrations on you. He needs a change of environment (living with a dominant horse to teach him manners) or of handler (a trainer who can deal with this vice) — or he should be sold. An aggressive, biting horse can inflict serious injury.

---

flight response. When a horse is afraid or uncomfortable and can't flee, his response is to kick or fight. Most are nervous or sensitive and can learn to tolerate situations that cause them to kick. A kick at a human is generally an expression of nervous fear or self-defense. A horse that kicks must be reconditioned to stand still and trust his trainer.

Most horses give a warning before they kick. They lay back their ears, roll their eyes, present the rump, clamp the tail, lift a hind leg, or back toward the intruder. Be alert to these warnings. A startled horse may suddenly kick explosively, but usually you can avoid a kick if you are alert to your horse's mood.

Retrain the Sensitive Horse

When rehabilitating the nervous kicker, go slow and easy to gain his trust. Always stay in control of the situation. The first priority is caution: Don't get hurt. The second rule: Don't hurt the horse.

A touchy, kicking horse needs progressive desensitization. He must be adequately restrained so he can't kick or get away from the lessons. Then, gradually expose him to touching, bumping, and brushing his hindquarters until he realizes that these actions don't hurt him. The key is to go slowly and stay beneath his panic level.

Some trainers use hobbles to aid in retraining to keep the horse from kicking, but you may not have to resort to them. Position your horse against a fence or wall so he can't move away from you. Stand at his shoulder. You may need a chain over his nose for control. Once he understands he is restrained, use a folded feed sack or old saddle blanket — something with a familiar smell he won't be afraid of — to desensitize him. Let him smell it, then start rubbing it along his back.

Gradually rub it all over him, under his belly and over his hindquarters, retreating to his back again before he has time to protest. Praise him when he stays calm; scold him if he tries to move or kick, but don't yell at him.

Pole Him

Another way to teach your horse to allow you to touch him on the hind legs (while staying out of kicking range) is to use a small, light "pole" 4 or 5 feet long, such as a broom handle or a whip. (Poling was

HOBBLES

Because the kicker won't like the lessons, he needs to be restrained so he can't hurt you or himself or get away. A hind leg restraint or three-way hobbles can help. Three-way hobbles consist of a regular set of hobbles for the front feet and a foot strap for one hind foot. Hobbles should be soft and strong. The hind foot strap should have a heavy cotton rope attached, which can be tied to the front hobble ring with a quick-release knot. If you haven't had experience with hobbles, get help from someone who has. Don't work with the kicker by yourself. Carry a sharp pocketknife in case the horse gets in trouble and you have to free him quickly.

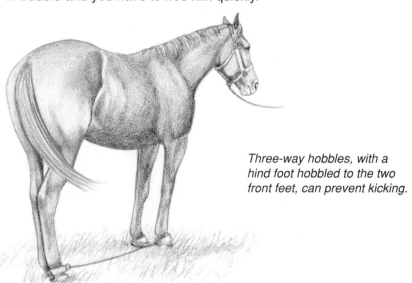

Three-way hobbles, with a hind foot hobbled to the two front feet, can prevent kicking.

used by old-time harness horsemen for getting green horses accustomed to the feel of shafts and harness before hitching them to a cart the first time.) Let the horse smell the broom handle, then rub and touch him with it on his sides, rump, and flanks. Don't raise it or wave it. Let him relax about it, rubbing gently. When he's comfortable with it on "safe" parts of his body, slide the pole down around his hind legs, then between his legs. Use several sessions to convince him that having his hindquarters and legs touched is nothing to worry about.

Desensitize Him

After your horse is used to the pole, and the sack or saddle blanket has been rubbed all over him, it's time for another lesson in desensitizing. For this lesson, have someone hold him. It's best to work with the horse held, rather than tied. If he's tied by the head, he'll feel trapped and is more apt to panic and resist the lesson. Rub the blanket more quickly and let it flop against his body and hind legs. If he flinches or jumps, slow down to a tolerable level of rubbing. You want him to relax, not explode. Gradually try other things like a newspaper and a plastic raincoat. Get him used to these on both sides, under his belly, and around his legs. This may take several sessions with a skittish horse. You can speed the process considerably with the use of a Stableizer; this relaxes the horse (see page 119 for details).

As part of training, move a large-diameter soft cotton rope over your horse's body and legs. Let it dangle and touch his legs. Slide a loop around his legs until he's relaxed about it. This will be helpful if he ever gets tangled in a longe line or driving lines or in a snarl of wire when you are out riding.

Start each lesson with something your horse is at ease with; progress to new experiences only as he is ready for them. Over time, you can gradually desensitize him to something bumping his hind legs.

A horse may kick not from fear but because he's headstrong and doesn't want to cooperate. Your horse needs to learn that the human is boss. He must respect your status and learn never to kick you, just as he would never kick a more dominant herd member. The first step is to establish authority so he doesn't test you. Once he respects you, proceed to lessons to halt kicking. Stand at his head, controlling him with halter and lead, and give him touching lessons with a whip. If he tries to kick, tap him below the hock. Don't whip him; just tap smartly. You don't want to create fear or enter into a battle.

The Turnout Terror

Some horses have bad manners when they are taken out of their stalls, put out at pasture, or turned loose after work, and a trainer may foster a horse's bad habits by letting him have his way. Often a person is unaware of this negative influence, which could result from inconsistency, timidity, lack of discipline, or inappropriate discipline. If a horse runs off bucking and kicking the instant he's unhaltered, the handler may try to quickly unsnap the lead while hurrying through the gate, hoping the horse will continue on through and go a few steps before starting his obnoxious — and dangerous — behavior. But a timid, undisciplined approach leads to less control of the horse and more trouble. If your horse becomes a problem, analyze your body language. Are you nervous? Are you timid? Are your methods of correction ineffective? You must create a clear distinction between acceptable and unacceptable behavior.

Stay in Control

To prevent misbehavior, think ahead and stay in control. This means using a halter and lead shank when taking a horse to and from pasture or paddock, no matter how gentle or well trained he is or how short the distance. Lead only one horse at a time. Pay attention, and have a calm and confident manner. If you always keep him under control while leading, he won't get into the habit of impatience when you take him "home."

To prevent charging at turnout time, have your horse stop and relax before you let him go (see page 16). Loop the rope around his neck before taking off his halter; in this way, he can't rush off when he thinks he's free. When he's calm and relaxed, quietly walk away. With persistence, you can retrain a horse that charges off at a gallop.

Bad Manners Under Saddle

Some horses handle very well on the ground but exhibit problems while being ridden. Most problems stem from inadequate or inappropriate training or a phobia that developed after a bad experience. Often, the retraining required to eliminate the problem is more challenging and takes longer than training the horse correctly in the beginning, so if your horse has a problem plan to invest sufficient time to correct it.

Barn Sour

Some horses don't like to leave the security and comfort of the barn, pen, or pasture. They don't have enough trust in the rider and don't feel confident away from home and their buddies. At home, they are normal, reasonable individuals, but if you try to ride them, they become panic-stricken, upset or angry, unwilling to listen to you. The barn-sour horse may be afraid because he remembers bad experiences out there or he may just want to stay in a familiar, secure environment. He will not relax until the two of you return home.

A horse can become barn sour if he is ridden infrequently or taken from the barn only to work. A rider may unwittingly contribute to the problem by routinely rewarding him for returning by unsaddling and turning him back out with pasture buddies or into his stall. It's important to break this pattern of being comfortable at home and having to work when taken away. Take your horse on an outing just for the fun of it. This way, he'll have pleasure outside his stall in addition to comfort within.

Is He Emotionally Insecure?

A horse may have a serious phobia about leaving home and do everything he can to avoid it. He may rear, run backward, whirl, or buck. His fear is stronger than normal self-preservation instincts; he may not care what he crashes into in his efforts to stay in the barn or pen. If a rider becomes apprehensive about this and rides the horse less frequently, it makes the situation worse. The horse resents even more these sporadic intrusions into his comfort and security and becomes more determined not to budge.

If a rider in frustration resorts to force and punishment for the bad behavior, the problem is compounded. Your actions confirm that a ride is a bad experience, to be avoided at all costs. Do not give your horse more reason to fear leaving home.

Rehabilitate Him

If your horse has a serious problem, the best way to overcome it is to plan a retraining program. Allow within the training many days of patient work. Start slowly and progress gradually. Bring your horse out of his stall or paddock several times a day, just to groom him, lead him, let him graze a little grass as you brush him, or do some easy ground work. Make these sessions varied, pleasant, and nonconfrontational. As your horse gains confidence and is more tolerant of what you are doing with him, gradually increase his time and distance from "home."

If his phobia flares up only when he is ridden away from home, give him daily or even twice-a-day short rides near the barn or pen. Keep a calm attitude; you'll impart your relaxation to your horse. Convey calm through your reins and body cues.

In your early attempts to rehabilitate your barn-sour horse, you may want to ride with a friend on a calm horse when you want to take him farther from home. Insecure horses will go with another horse more readily than by themselves. Eventually, though, retraining must include solo work; you don't want your horse dependent on a companion.

KEEP HIM GUESSING

Each time you take your horse back to the barn, immediately leave with him in another direction. Keep him guessing, so he'll never know whether he'll be put away or ridden more. Instead of riding a long way and having to struggle with him until you return, make many short trips. Thus, you can take a several-mile ride without getting more than a quarter of a mile from home. Head down a road, lane, trail, or pasture and come back before your horse starts to get upset (the actual distance depends on the horse), then go out another way until you've ridden back and forth in all possible directions. Once he can leave and return confidently at that distance, start increasing it. Ride in a different sequence each time or return another way. Now your horse won't be able to anticipate what he'll be doing next.

Ride Him Farther from Home

Ride in slow easy circles, gradually moving farther from his stall or pen. If your horse starts to get nervous or upset, don't use force. Just make the circles smaller again, closer to home, until he can relax. Once he is comfortable with circles, start making short rides away, stopping him at the first signs of nervousness. Don't go so far that he explodes in temper or panic. Stop him and relax. Take his mind off his fears with a reward, if necessary. Ride him a short distance from the barn, dismount, and let him eat grass. After a few sessions like this, your horse will view leaving the barn as something pleasant.

If he is fractious about being mounted again that far from the barn, lead him back. Soon he'll enjoy being ridden away from the barn to get his treat of grass, and you can ride him farther to earn this reward. If your destination is a comfortable place where you let him rest, relax, and graze — or give him grain you brought along — the idea of a ride will be more appealing to him.

Once he loses his fear about leaving home, you won't have to continue the grazing or grain reward. Keep taking short rides. If your horse ever blows up or gets overeager on his return, halt him and make him stand until he is calm again, even if you must ride circles in place to take his mind off his anxiety, frustration, or anger. Never let him move toward home in an uncontrolled frame of mind. Wait until he is calm. If he's not listening to you, make sure he's working against himself and not against you (see The Horse That Rushes Home, page 445).

If your horse balks on the trail and tries to turn to go home, put him to work doing circles and serpentines, sidepasses, and other maneuvers. When he needs a rest, ride forward beyond the spot where he balked and let him relax. If he discovers that every time he balks he gets put to work, he'll soon realize that going forward on the trail wherever you ask him to go is to his advantage.

Ride a little farther each session. Talk or sing to your horse to keep him relaxed. Stop now and then, both going and coming, to let him stand quietly. Always go past home a short distance before returning to put him away. Ride back and forth past the barn or pen so he'll know he isn't always going home to end the ride. "Home" should not be synonymous with the end of work. Many horses that rush or try to bolt home acquired that habit because they expected to be put away as soon as they arrived back at the stall or pasture. Working your horse hard at home and going away to let him relax will make him rethink his bad habit.

Vary the Routine

Ride on by the barn many times, dismount in a different place every day, tie your horse for a while before putting him away, do a schooling session in the stable area before you end a ride. Keep him guessing. You must be the one to decide when the ride is over; your horse must learn to accept your decisions. With consistent, patient, and regular sessions, during which you strive to get him to relax, the barn-sour horse learns that leaving home is not traumatic and he doesn't need to be in such a hurry to return.

Prevent Barn Sour

Some horses become barn sour because of a frightening experience away from home. This can happen when a green horse is exposed to new situations prematurely. Give the youngster training sessions before you ride. Safely acquaint him with the big wide world before you try riding away from the barn. Spend time leading him, with an assistant if necessary to follow and encourage, and accustom him to new sights and sounds until he is comfortable with new situations and obstacles. Walk around the neighborhood to get him used to what he will encounter.

If the horse is quite insecure or wants to rush home during his first rides out, have a helper walk along with you, prepared to snap a lead line onto your horse if needed (see page 267).

Dealing with the Spoiled Bluffer

Some horses discover that if they create a scene when leaving home, a timid rider may decide not to continue. The horse gets rewarded for bad behavior by not having to work. If your horse won't leave the barn area, don't force him. Just put him to work right there in the barnyard, doing circles, gait changes, figure eights, and backups. Set up an obstacle course with cones or buckets and make him weave in and out among them.

Give him more work than he'd have on an easy ride away from home. Don't punish him; just work him. When he seems ready for a rest, work some more, then ask him again to leave the barnyard. If he does, ride a short way, then let him stop and relax. Ride back to the barn and head out again in a different direction. If he resists, again make him work hard in the barnyard. Keep this up until your horse realizes that the place where he gets a rest is actually away from the barn.

THE GATE-SOUR HORSE

The gate-sour horse is like a barn-sour horse: When working in an arena, he always wants to return to the gate because that's how he gets back to the barn. If he hurries going toward the gate and you have to urge him when going away from it, work on this just as you would with a barn-sour horse. Make him work hard, in small circles and other vigorous movements, when he's near the gate and let him relax and rest at the other end of the arena. After a while, he won't be so eager to go to the gate.

The Horse That Rushes Home

Sometimes a horse leaves home willingly but is a handful when you turn around to start back. He's so impatient to return home that he doesn't pay much attention to the rider. In such a case, use diversion. Head in the other direction for a while, make circles, or do figure eights or serpentines so he must concentrate on something else. He'll soon realize that you're not yet going home. A rider may react to the hores's impatient behavior by pulling or jerking on the bit, but this makes the horse more upset. He may become angry and frustrated and react by bucking. Don't try to hold back a horse. The more you pull, the more he'll pull against you, and you won't win this battle. If he starts to trot, turn him around in a tight circle — this will slow him from a trot to a walk — and continue on with a loose rein. If he breaks into a trot again, spin him in the other direction. Stay relaxed; let him have a loose rein again. Rather than fighting, you are asking him to make a nice spin and continue on at a walk.

Diversionary tactics are always more effective than punishment. Punishment creates an adversarial relationship, which aggravates the problem. Avoid a fight; any argument defeats your purpose in training.

If a horse tries to hurry home, turn him in a tight circle to slow him to a walk again, then continue on a loose rein.

Conflict will make your horse more insecure or may fuel his anger into fighting harder, bucking, or rearing. When a horse is trying to rush home — especially one that might buck if he isn't allowed to bolt — circle him when he becomes unmanageable and repeat the process as many times as is necessary to keep him under control. Making him trot in circles gets his mind off bolting home.

If you're on a narrow trail where you can't circle him, bend his head and neck a little to one side, without force, to slow him. Use reins for the neck bend and your inside leg to keep him going straight. When he drops to a walk, immediately cease rein and leg pressure. If he starts trotting again, bend his neck the other way. This works better than pulling equally on both reins, which would give him more to brace against, and makes it uncomfortable for him to continue rushing forward. If each time he slows to a walk you release the rein, he'll soon get the picture.

HELP HIM EXPEND SOME ENERGY

Trot to settle down your horse. If he can't walk home on a loose rein and starts rushing, turn him around and head away from home at a fast trot. When he settles down, let him go in the direction of home. As long as he walks on a loose rein, he can continue toward home. This works for some horses, but others won't walk until they've had a lot of riding.

Another version of this method is to anticipate his impatience and alternate trotting with walking. If you know your horse wants to trot, ask him to trot before he breaks. In this way, you avoid having to correct him *after* he breaks, and thus can maintain his good attitude toward you. Trot a short way, then slow to a walk. Walk a short distance, then ask him to trot. If you alternate gaits and keep him under control, he'll relax and walk more. If you stay a jump ahead by asking him to trot before he breaks, you stay better in control.

Stay Relaxed

The calmer you are, the longer your horse will stay relaxed before worrying about going home quicker. Talk to him or sing, keeping his mind interested, focused on you, and receptive to your cues. Once he tunes you out in eagerness or panic to get home, you've lost the game. You must do some circles or other diversionary tactics to settle him down again.

Enlist the Help of Another Horse

A horse is usually most determined to get home quickly when he hasn't been ridden regularly. During first rides after a winter vacation or layoff, even a secure horse may be more difficult to handle. After several rides, though, he'll settle down and become more willing to listen to you. To reduce frustration on those first rides and to make a potentially dangerous ride a pleasant one, invite along another rider on a calm horse. Another horse to keep yours company will ease his mind.

With a companion horse, he is less apt to work himself into a tantrum or try to buck. Daily rides can take the edge off his explosive homecoming exuberance, and he will be more content to pay attention. After a few rides with another horse and by gradually getting back to your regular work routine, a horse that was a handful will become manageable.

The Herd-Bound Horse

A horse may not want to leave the security of his herd mates or riding companions and will throw a fit if you expect him to leave the other horses. The horse is calm and content as long as he's with buddies, but doesn't like getting behind and doesn't want to leave them. If you split off from the group, he goes into a panic, whinnying and trying to get back to them. Even in a riding arena, he won't pay attention to you when doing solo work; he's always trying to move closer to his pals.

Some herd-bound horses are merely insecure and apprehensive when separated from their buddies; this isn't as difficult to deal with as the individual who rears, bucks, and otherwise throws a temper tantrum when asked to separate from the group. The insecure, panicky horse can usually be safely ridden by a firm, patient rider who can keep him steady and heading in the proper direction. The angry horse is more of a challenge. You must keep him under control and obedient without getting bucked off or having him go over backward during his fit of temper.

Distraction

If you get into a dangerous situation, distract the horse. He can focus on only one thing at a time. If he's angry, give him something else to consider so he can't think about bucking or rearing.

Spin your horse in a circle. Use your direct rein firmly to pull him around and your outside leg to make him turn. Keep spinning him until he gets his mind off his anger and starts paying attention to you. This should defuse a potentially dangerous outburst and keep your horse under control. If he has to make a tight circle, he can't buck and he can't rear.

If your horse needs this kind of handling, use a snaffle so you can direct-rein him (see page 187). If he's trained in a curb or is inclined to buck when he becomes angry, use a Pelham (see page 194). You can keep his head up and keep him from bucking with the action of the curb, yet still be able to direct-rein him with snaffle reins to spin him. If he is well trained to neck-rein and also responds to leg pressure, you can spin him adequately without the snaffle.

Be Patient and Work Him

Punishing your horse for screaming and whinnying when you leave other horses is counterproductive. Horses that need the security of other horses are highly emotional and will continue to scream in spite of punishment; it just makes them jumpy or head-shy. Allay his fears with patient work and miles of riding. Long daily rides are usually the best therapy for the horse that is insecure without companionship.

Once a horse has been ridden regularly enough to take the edge off his emotional insecurity — that is, he now displays resigned acceptance when leaving other horses or working apart from a companion — his reactions to other horses he hears or sees in the distance will be more predictable and he will also scream less. Be patient; eventually you'll notice a change in his attitude.

The best cure for the herd-bound horse is lots of work — daily rides to impress on him that he can function and listen to you without other horses

---

CONDITION HIM GRADUALLY

One way to overcome the problem of the horse who loves company is to part ways with other riders during some planned lessons. Go a short distance from them but travel alongside them, gradually riding farther apart each day. After several sessions, you can split up to meet again at a designated area. A horse may never cease looking and watching for the other horse, but he'll become better behaved and calmer while he is alone. He may be upset until you are clear out of sight of the others, then settle down and pay attention to you — until, of course, he meets the other horses again.

around. Cattle work is ideal, as the horse must focus his mind on following or checking cattle instead of on other horses or on rushing home.

Give Him Special Lessons

Take a trip in open country or along a jeep road. Have two riders go with you so you can continually change positions with them. Take turns trotting ahead, slowing to a walk, and letting the others pass. Continue this walk-trot leapfrog of riders, adjusting the distances to the comfort level of your horse and increasing distances as he becomes more manageable.

If your horse wants to trot to catch up with one ahead, don't hold him back. Ask him to walk, and release pressure when he does. Maintain your body rhythm at a walk, using all your cues to keep your horse relaxed and walking. Whenever he breaks gait, ask for a walk. If he won't walk, have your friends make the distance gap smaller until he can. You can do this leapfrogging lesson for miles, and gradually your horse will be at ease even though his buddies are farther away.

Use His Buddy as Motivation

This is an arena exercise to calm a herd-bound horse. Station a friend on a buddy horse at one end of the arena, leaving room between the horse and the fence, while you ride back and forth in the arena. When you go to the far end, halt your horse so that he is facing the buddy horse. If he wants to go to his buddy, let him, but when he gets there, make him work — trotting or cantering circles around the buddy horse — until he decides he'd rather rest.

Ride him to the far end again, then ask him to turn and stand with a loose rein. Allow him to rest and pat him. If he wants to go back to his buddy, let him, but make him do circles again. When he realizes he must work when he's with his buddy and can rest at the other end of the arena — with you for security — he'll start to rethink things. Once he calms down and can relax a moment at the far end, ride him quietly back to his pal, let him stand there a moment, then ride back to the far end. You want him to learn that he can be comfortable away from his buddy but that you aren't trying to keep them apart all the time; he must be able to come and go as you direct.

With patient work, his level of insecurity will change. He'll find he can tolerate and even be content working apart from the other horse, especially if he starts to trust you as his working partner. You become his security. This is why patience is always better than punishment: Your horse must come to trust you and find his security in you.

The Head Tosser

Some horses toss their heads because of a physical problem — a sore mouth aggravated by bit pressure, for example, or a poorly fitting bit that pinches, or a bit that bumps wolf teeth. If head tossing begins or continues in spite of good horsemanship, check your horse's mouth. Make sure the bit fits and is adjusted to rest comfortably on the bars. A bit too narrow or wide that is sliding around, or a headstall that is rubbing the corners of his mouth because it's too tight or is clanking against his teeth because it's too loose, could be the cause of his fretting. In other instances, the horse tosses his head because of impatience or annoyance.

It's frustrating to ride a head tosser who's always pulling and rooting at the bit. An overeager horse may root in reaction to a rider trying to hold him back. Another frets and tosses his head when following another horse because he is afraid of being left behind. Still another tries to get more slack in the reins in order to keep his mouth from being bumped by a rough-handed rider. Once a horse has the habit of tossing his head, it takes time and patience to correct it.

Three types of horses are particularly prone to head tossing.

- ◆ **The overeager horse.** When a horse tries to go too fast, the rider's tendency is to hold him tighter, to stop the prancing and head tossing. Unfortunately, this usually makes the horse fight the constant restraint and toss his head even more. Under some riders, a nervous or overeager horse is never able to relax. The rider is upset and frustrated, and these feelings are transmitted to the horse through tight reins and the rider's tension.
- ◆ **The horse that frets at being left behind.** Many nervous horses don't like to follow another horse, especially if they get a little bit behind. They root at the bit, trying to go faster and catch up, tossing the head. If the rider gets frustrated and tries to make the horse stop prancing, the horse tosses his head even more.
- ◆ **The annoyed horse.** Sometimes a horse is well trained but his rider doesn't know how to handle the reins — perhaps bumping the mouth with the bit or keeping the reins too tight — so the horse gets into the head-tossing habit. Though the horse is always under control and

always at the right gait, he tosses his head or grabs the bit. He is protesting the manner in which his mouth is being handled.

Improve Your Signals

To rehabilitate your horse, you need better communication and must fine-tune your cues. Communication is through hands, bit, leg pressure, and weight shifts. Calm hands can relax your horse and keep from irritating a sensitive mouth. If a horse performs poorly, it's usually the rider's fault. If you can't get him to do what you want or to settle down or he starts tossing his head, don't automatically blame your horse. Look at your own horsemanship to find the communication gap. Some riders resort to a tie-down or a martingale, but with good hands you can prevent or halt head tossing without such crutches.

Frustration is catching, but so is relaxation. If you are relaxed, your horse will feel it. Try to communicate relaxation through your fingers and your whole body (see pages 298–301 for more on this technique). If you stop pulling and use give-and-take actions on the reins, he will stop rooting.

Most head tossers pull at the bit and throw their heads up as a reaction to being pulled at or bumped in the mouth. If you don't pull with the bit or bump his mouth, your horse won't toss his head. If you aren't pulling, he won't. It's as simple as that.

Too much restraint creates more head tossing. You and your horse must develop better rapport so you won't need much restraint with the bit. This takes time and patience. Eventually, you'll reach the point where the horse can relax, knowing his mouth won't be pulled. If he starts to toss his head again, give him more slack, leaving his mouth alone. He should settle back into a relaxed gait, having learned confidence in your light hands. If he trusts you, he will be content to stay at the gait you want. He won't toss his head unless you bump his mouth or hold him up too tightly.

MAKE WALKING THE EASY CHOICE

A horse can be talked out of prancing and jigging by asking him to do other maneuvers besides just going down the trail. If you request turns and changes of direction, he will eventually figure out that these require a lot of effort on his part, and that it would be easier just to settle down and walk.

Horses That Resist and Avoid the Bit

Most horses start resisting or avoiding bit pressure because the bit has been overused (see The Jaw on page 331). The best way to prevent or to overcome this problem is to use the reins as lightly as possible. Never rely on reins alone as signals, even just for slowing and stopping your horse, and never use reins to recover your balance.

Ride well balanced. Give cues through your legs, your seat, and weight shifts to augment or reinforce prompts given with the bit. This combination helps to ensure that you won't exert excessive bit pressure. Perfect your riding so you can go through all gaits and back to a walk and halt on a loose rein, signaling more with your seat and legs than with the bit. With light hands, you can cue with slight checking actions while still on a loose rein. It may take time to retrain a horse that resists the bit, but once he realizes that he is not abused by it, he will get softer in his responses and learn to relax his jaw.

The Bolter or Hard-Mouthed Horse

Communication is thwarted if a horse doesn't respond at all to the bit. He keeps doing what he's doing, rooting into the bit or tightening his jaw and neck muscles, so a pull on the mouth is ineffective. A horse that lugs or pulls on the bit becomes dangerous when he defies control if he becomes a bolter.

Lack of responsiveness to bit pressure is usually a sign of insufficient training or improper training and can result in a spoiled, headstrong horse. He has been allowed to get away with bad actions and now does them out of habit.

Some horses become hard-mouthed because a rider was careless. A green horse, or any horse that hasn't learned how to collect when ridden, travels heavy in front and pulls against the bit; the rider must be careful not to use too much bit pressure. Harsh action with a bit causes pain and can damage the nerves in the bars of the mouth (the toothless space where the bit rests), deadening the feel and creating an insensitive mouth. These are the most difficult cases to retrain.

Other horses travel heavy in front due to poor conformation of the neck and back; they cannot flex very well. They seem insensitive to a bit because it is difficult for them to collect and respond properly. These horses often become hard-mouthed.

An insensitive mouth or a bit-grabbing horse is frustrating to ride. There can be no precision or fine-tuned control of his actions if you can't communicate at all with his mouth. It is almost impossible to make him supple and responsive.

Use Caution

The hard-mouthed horse can be dangerous, because he won't respond properly when you ask him to slow or stop. He may discover he's stronger than you are and become a bolter or a runaway. If a horse bolts just once, because of a frightening stimulus, he should be forgiven, not punished. Halt his flight and try to determine the cause of his fear. If it is something he may have to encounter again, see if you can acquaint him with the worrisome stimulus under calmer circumstances. Go back to the spooky rock or the clothesline flapping and halt. Let him stand, relaxed, at a safe distance until he wants to check it out.

He needs to realize his flight was unnecessary. Deal with the situation in such a way that his run-first, think-later reaction won't become a habit. A horse that bolts out of fear needs reassurance and patient training. But if your horse charges off whenever he becomes frightened or upset, he needs extra work on control so that *you* are the one in charge of the situation. He must become disciplined and trusting enough to accept your direction and control.

DON'T GIVE UP

Most hard-mouthed horses or bolters can be retrained to be more responsive to the rider's hands. Remember, though, that each horse is different. The habitual bolter may need a more forceful remedy than the honest horse with an insensitive mouth. It's up to you to figure out the correct retraining procedure.

Dispense with the Bit

For the truly hard-mouthed horse that lugs into the bit when asked to stop or clamps his teeth on it so that you have no control, either change bits or use a side-pull for a while. You need some means of control during the retraining process. You don't want your horse bolting or pulling at the bit while you are reschooling him; you must change tactics and start at the beginning.

Another solution is the mechanical hackamore (see page 205). With this aid, you can control your horse without using his mouth at all. It has a headstall like a bridle but lacks a bit. Pressure points are his nose, chin, and poll instead of his mouth. The mechanical hackamore allows you to leave his mouth alone while you teach your horse to respond more to your body and leg aids.

Reschool Him

Your goal here is to get your horse to respond to all cues — to your whole body and its signals — and not just to a pull on the reins. If you learn to control him with subtle cues, you can depend less on bit pressure and thereby circumvent his problem. He'll also become more responsive and less inclined to try to avoid bit signals.

Your horse must pay attention to what you are telling him — to speed up or slow down, to collect and extend, to come to a stop — just by the feel of your body language, weight shifts, and subtle movements, which should always accompany the bit signals. The way you ride gives him clues about what you are asking him to do. Only the inexperienced rider relies mainly on a bit. Cues given with a bit should be refined and varied, not just a yank or a steady pull. You can go back to a bit again after your horse responds well — as long as you continue to control the horse with your whole body.

Start Over

If the horse has an unresponsive mouth because he has never been trained properly and thus ignores bit pressure, go back to the basics. Teach him to respond to leg pressure and how to flex, collect, and extend. Soon he'll find that the bit is just part of the total communication package. In the process, he will learn how to respond to a broadened application of the aids and also to the bit.

When giving signals with the bit, be gentle. This horse is accustomed to being pulled at and pulls hard in response. If you use a hard, steady pull to slow or stop, he may clamp his jaw. He'll pay more attention to a softer touch, especially if the bit moves around or vibrates in his mouth. A soft give-and-take is more effective than a strong steady pull on the reins. Pull and slack, and he'll pay attention. When he responds, reward him with relaxation of pressure. He'll learn that obeying your signal to slow or halt brings instant relief and he'll start to relax his mouth more.

It may take a lot of time and patience, and your horse may revert to pulling into the bridle or trying to bolt if he becomes startled or upset. If he misbehaves in the company of other horses and pays more attention to them than to you, do a lot of quiet training with just the two of you until he demonstrates good patterns of response.

Tire Him Out

One way to counteract bolting is to spend a lot of time riding miles at the walk, day after day, until your horse becomes more relaxed. He may still bolt a few times until you get further along in his rehabilitation, so be prepared to keep him under control. The confirmed bolter may need a firm hand to break him of taking the bit in his teeth and doing whatever he pleases.

But remember, your horse is stronger than you are. If he gets up speed and you can't stop him, the worst thing to do is to pull straight back on his mouth; he'll just lean into the bit and keep going. A racehorse runs leaning into the bit, so bit pressure is *not* what you need to stop a horse.

The best way to control a runaway is to pull his head around to the side. This can be accomplished with quick give-and-take actions on one

To stop a bolting horse, brace one hand against his neck and take a short, firm hold on one rein with the other hand. Then, with a quick pull, take the horse's head away, bringing him around in a circle so he has to stop.

rein to bring his head around before he has a chance to set his jaw and neck muscles against your pull. Use your legs and body weight at the same time to make your horse turn and circle. You don't want to keep him running straight and blindly with his head pulled around to the side. Once he's circling, make the circles smaller and smaller until you can stop him.

Use a snaffle bit so you can direct-rein him to pull his head around (see page 55 or 187) or try a Pelham with four reins, which has the action of the snaffle as well as the curb (see page 193). For a bad runaway with the bit in his teeth, you may have to take one rein in both hands and pull him around. Reach well forward on the rein and, holding it low, give a series of quick, strong pulls with both hands to bring him around.

If your horse has made a habit of headstrong flight, make it unpleasant for him so he won't want to bolt. If he gets up speed and doesn't want to stop, pull him around and make him keep going in a controlled circle until his desire to run has worn off and he is quite willing to stop at your signal. A few sessions of making the bolting into an enforced work session will do much to persuade your horse that there's no future in running away. In the meantime, keep up the reschooling sessions to develop better communication and responsiveness.

Improve Your Riding

To prevent or correct a hard mouth, always make sure you do not bump or jerk the mouth unintentionally. Learn to follow the horse's head and neck movements without losing contact with his mouth or jerking it. Move your hand(s) in response to the horse's balancing movements; his head bobs at each stride at the walk and at the canter and gallop. Your own balance must be excellent so you never "catch him in the mouth" accidentally.

Better horsemanship can help you retrain a hard-mouthed horse, developing his ability to respond to *all* your cues — legs, voice, weight shifts — as well as to subtle touch on the bit. A good mouth is the result of careful training, encouraging the horse to accept the bit willingly with no fear of being hurt.

The well-trained horse accepts bit contact with a relaxed and yielding jaw because he has confidence in your hands. To retrain the hard-mouthed horse (and thwart the bolter), you must start over and teach him the basic responses, encouraging him to become more pliable and maneuverable. When he learns or relearns how to respond to your whole body, you won't have to rely so much on the bit. As you develop more subtle communication, this may improve your own horsemanship, eliminating the conflict and confrontational situations that earlier resulted in a tug-of-war stalemate between you and your horse.

The Balker

Some horses refuse to go past, over, or through an obstacle because they're afraid to deal with anything beyond their range of experience. Others balk at things they associate with earlier anxiety. They won't go near a creek because they were yelled at and whipped in an attempt to make them go into water. They balk at a jump because of an earlier failure or insecurity.

You need to determine whether your horse, in trying to avoid the task at hand, is acting from fear, inexperience, apprehension, bad memories, lack of confidence, or overwork. Is he truly afraid or merely reluctant? An inexperienced horse needs a different training approach from that of a confirmed balker. The frightened horse needs reassurance. A recalcitrant horse may need some different incentives.

KNOW YOUR HORSE

Know your horse and have an idea why he is balking so you can approach the problem from the proper direction. Force may work for a stubborn animal, but it won't be successful with the truly fearful one. Tactics that might work for one balker may not work for another. Never use force in a contest that you might lose, or the horse will realize he is in control and he will balk again.

Patience and Persuasion

A young or inexperienced horse may be so unsure of a situation such as crossing water or stepping over a log that he won't even try it. Whenever there's a problem, always first try patience; this tactic can persuade most horses. Occasionally, however, you encounter one that reacts stubbornly to anything he perceives as difficult or impossible. With this horse, you may have to use enough coercion to prompt him into the water or over the log so he can learn that what you asked was *not* the terrible, impossible request he perceived it to be. He discovers that he can get his feet wet and that he can step over the log without it biting him.

Coercion

If you must resort to force, it is essential to use it wisely. Use only the amount necessary to persuade your horse, then immediately reward him when he complies. Talk gently to him and pat him as he nervously stands in the water, praise his accomplishment after he has stepped over the log. Repeat the action. Go over the log or through the water again until your horse gains confidence in his ability and loses his fear or apprehension. Repeating the activity a few times will reinforce his change of attitude.

With the horse that balks because of insecurity, incentive (persuasion) and reward (praise and letting him relax) after he performs the required action will usually solve the problem. He'll see that the situation is not as terrible as he thought and will find that the coercion ceases as soon as he complies.

Work Gradually

Often the best way to overcome the phobia of a balker is to acquaint him gradually with the obstacle so he can lose his fear. If the situation is something he had a bad experience with, however, gradual rehabilitation may be in order. Start with a similar but smaller obstacle. When he learns to deal with it, then build up his experience and confidence. Go back to ground work if necessary. This system works both for training the jumper and for retraining the jumper who balks. Lead him across poles on the ground (see page 147) and build step-by-step from there.

Sometimes it helps to have someone follow your horse as you lead him over a pole or log, to gently encourage with a whip if he balks. Once the obstacle loses its scariness, go back to mounted lessons. If your horse still has a hang-up, resort to a previous aid or cue, such as the touch of a whip or a following helper, that your horse associates with an earlier success. He learns that he can do it without confrontation.

Now, gradually work back up to the original obstacle that caused the problem — the puddle, perhaps, or a log or jump. If you take rehabilitation one small step at a time, building on previous successes, your horse's confidence level will increase until he can tackle what was once a frightening obstacle. At this point, he knows he can do it.

Be Creative

If your horse's problem is noise at a show, have friends make applauding sounds. If he balks at certain jumps, create obstacles that are really easy to hurdle. Success breeds success. Make the jumps so low that your horse can do them easily, no matter what position he tries to get into. After he jumps, praise him, even if he jumps imperfectly or runs off afterward. Make the jump a positive experience in his mind, even if you have to lower the obstacles. To ensure success, put the pole clear down on the ground for a while.

At first, slow to a trot so your horse can get a better look at a jump if he wants to, but once he understands he must go over it no matter what, work at normal speed. If he tries to avoid the jump, use the reins and your legs to keep him facing it. Sidepass back and forth or back him up, but keep him facing it. Don't turn in a circle to make a new approach. Your horse must learn he can't have a second chance to look at the jump; he has to take his best shot the first time. He can't escape that responsibility by balking or running out. He'll soon find that evasive action only makes it harder because he'll have to jump from a standstill.

Don't force him at jumps too much, however; you are trying to build confidence. You don't want his fear to override your commands. Try to stay in tune with your horse's attitude so you can encourage him but not push too hard. Always praise his accomplishments. Your goal is to train him to want to jump rather than avoid it by balking; he must trust you and gain confidence in his ability.

Reasonable Expectations

Don't expect more from your horse than he can perform. Don't destroy his confidence by pushing him into something he's not ready for. If he balks, whether out on the trail, over jumps, or some other aspect of training or work, it's up to you to find the cause and correct it. Respect his fears and phobias, and together you can work through them. Patience is key. Don't lose your temper when things go wrong or the horse will remember the conflict and be reluctant to try the obstacle again. A balker is often created by the rider.

The Rearer

Rearing can be more dangerous than bucking. Usually when a rider is thrown, the result is a few bruises or a broken bone or a concussion, injuries that are serious enough. But if a rearing horse goes over backward, the rider may be crippled or crushed unless he is agile or lucky enough to jump free.

To reform a rearer, you must determine the cause. Horses rear in an attempt to avoid something. Always eliminate the possibility of a physical problem before looking any further. If it's a physical problem, your horse is trying to avoid pain. If a horse starts rearing for no apparent reason, have your veterinarian check his mouth and teeth for injury or soreness. If he's ridden in a curb bit, check his lower jaw to see if the curb chain or strap is too tight or rubbing him. Examine the corners of his mouth. If he has a mouth injury or the bit is causing pain, don't ride him in a bit until it heals.

HE MAY BE IN PAIN

A horse may start rearing when he experiences pain that is aggravated by forward movement, from ill-fitting tack perhaps, or a sore back. He rears to avoid going forward. An old back injury or some other problem may cause pain if a saddle puts pressure on it.

Fear, Frustration, Emotion

Rearing may be caused by fear of something in front of the horse or along the trail; he's trying to avoid going closer. Rearing can also be caused by frustration at conflicting demands — for example, the rider asks him to go forward with strong leg aids but restricts forward movement with the reins.

Rearing may be an emotional response to something the horse doesn't like; he refuses to perform a specific maneuver because he has been drilled too much or has had a bad experience with it. Occasionally a horse will rear due to fatigue and overwork; it's a way of protesting and telling the rider he's not going to go forward anymore. If the horse finds out he can avoid a situation or maneuver by rearing, it may become a bad habit.

Another cause of rearing is the horse learning to get behind the bit in order to avoid control. Most horses that learn this trick have been overbitted and rushed in training. They may have been put in a curb bit before they had mastered the fundamentals in a snaffle.

Some rearers result when trainers work too hard at setting the horse's head instead of trying to achieve true collection. They think they can collect a horse by working his head, collecting him from front to rear instead of the other way around. They pull on the bit instead of pushing him forward onto the bit. A horse that is behind the bit has not learned how to respond to the rider's legs or reach for the bit, because the trainer has taught him only to back off the bit (see page 330). Some horses follow that to the ultimate degree and start rearing.

Improve Your Riding

The most common cause of rearing is a heavy-handed rider who tries to make up for her own insecurity by using a severe bit. A timid rider, confronted with a high-spirited horse, may be afraid to let the animal have its head for fear of being bucked off or having a runaway, so instead keeps a tight rein. The rider may even switch to a more severe bit if she has trouble controlling the horse. In self-defense, the horse starts rearing to escape abuse of his mouth. If the timid rider is frightened by the rearing and doesn't know how to regain control (basically letting the horse take control) or dismounts hastily, the horse soon learns he can avoid an unpleasant situation by rearing. If this happens more than once, it could become habit.

The Barn-Sour Horse

Other horses rear when upset and insecure. The barn-sour horse doesn't want to leave the security of stall or paddock, and when a rider tries to force him to leave his familiar place, he may start rearing. Or he may leave reluctantly but be in a hurry as soon as the rider turns toward home. The horse may be so frantic to get home that if a rider tries to hold him back, he starts to rear.

Other horses are not happy unless they are with their buddies or in front of the pack; these horses get upset if the rider wants to stand still, especially if his pals are moving away. Because the horse wants to go with them or to be out in front, he may throw a tantrum and even resort to rearing.

NIP IT IN THE BUD

It's easier to thwart rearing the first time a horse tries it than to stop it once he has begun using it as an evasion tactic. After it becomes a habit, the horse tends to do it automatically whenever he is confronted by a situation that causes him frustration.

Prevent Rearing

To prevent rearing in the first place, use a mild bit. Don't rush the horse in training. Build on the basics and don't progress to more difficult work until he is ready. Many rearers are the result of trainers being in a hurry; the horse starts rearing when he is confused or resisting the rider because he is incapable of performing what is being demanded.

Never "reward" a horse for rearing by getting off and ending the ride, or he will use this tactic to avoid work. If a green horse rears because of fear and insecurity, have another rider on an experienced horse go with you to "baby-sit" the young one out on the trail a few times until he gains confidence in new situations.

If he rears because he doesn't want to leave the barn, ride around the barn or in the barnyard awhile — continue the ride so you can end on a positive note and with you in control. Then figure out a way to prevent a repeat performance on your next ride.

Never ignore a rearing episode, even if it's the first time a horse tries it. Determine the cause so you can prevent recurrence and head off this evasive action before it becomes a pattern. Was he reacting from pain, stubbornness, fear, anxiety, frustration, or because he was confronted with something he didn't understand and didn't know how to cope with? Knowing the reason for rearing gives you a starting point for how to deal with it and how to prevent its happening again.

Alternate Methods

When working with a habitual rearer, remember that a horse "on the aids" (see page 55) cannot rear unless you make him do it. Being "on the aids" begins by being on the bit; this should be your first goal, especially with a horse that has learned to get behind the bit or one that has been abused with a bit.

Work with Him on the Ground

A safe way to start retraining the rearer is to longe him in a bitting rig (see pages 153 and 213, respectively). Adjust the side reins just snug enough for your horse to feel the snaffle bit while doing a medium trot but not tight enough to make him flex at the poll. Make him trot freely. If he balks or tries to rear, use a longeing whip to make him move briskly forward against the light pressure of the snaffle. Give him some breaks, walking with the side reins loosened, but don't let him stop.

Continue these lessons in making him go forward against the bit until your horse accepts light bit pressure quite well. The next step is to ground-drive him in the bitting rig. Gradually shorten the side reins and ask for a little more flexion and a firmer hold on the bit. If he's reluctant to go forward, encourage him with the whip. Ground-drive your horse in circles and spirals — this keeps his neck slightly flexed toward center. In this position, his hind feet make a slightly larger circle than his fronts, which puts more weight on his forehand and makes it more difficult for him to back off the bit.

Ground-drive him this way for several days, until he's relaxed and no longer afraid of bit pressure. Then drive him from behind and ask him to halt. Don't demand an abrupt stop and don't pull on the lines. Just give the verbal *Whoa,* fix your hands (don't pull), and gradually slow him down as you slow down yourself, and let him feel the bit. Each time he slows, give slack as a reward, then fix your hands again. Once your horse has stopped and stands quietly, praise him. If he throws his head up or tries to back off the bit, you've been too heavy-handed. Go back to driving him in circles and spirals until you both relax.

Work with Him Under Saddle

Go through these movements mounted, using a mild snaffle. Ride with fixed and yielding hands (no pulling), and develop your horse's responsiveness to leg aids. He can rear only if he is standing still (even if only momentarily) with his legs well under him or if he is moving backward. If he's responsive to leg cues and you can urge him forward, he can't rear. If you handle him with consistent firmness and no abuse, he'll forget about trying to rear.

Substitute an Alternative Action

Correcting the rearer can also be accomplished by substituting an acceptable movement for rearing; that is, you train him to do another action in place of it. A good alternative is the turn on the forehand (see pages 178 and 367), which puts his weight forward and on his front legs, a position from which he cannot rear.

Teach your horse this movement from the ground, cuing him to move his hindquarters away from pressure. Then do the lesson mounted, teaching him instant obedience to the leg cue. Ask him to keep his front legs in place, using firm rein contact, and use your leg — the one you want him to move away from — to ask him to move his hindquarters over. Then ask him immediately to move forward again. Ask for the turn on the forehand whenever you feel your horse shift his weight backward to rear.

By not confronting the horse and using a trained maneuver to defuse a potential rear, you can win the battle without a fight. The key to reforming the horse is to develop trust. It's possible that by simply inspiring trust and a feeling of security, you can reduce or alleviate the instances in which your horse feels threatened enough to rear.

The Bucker

Another dangerous habit is bucking. Some horses buck when they are startled or annoyed. When you first start a horse under saddle, for example, he may try to buck off the saddle or the rider. Most horses discontinue this tactic once they are accustomed to the new experience. A few, however, learn they can get rid of a rider by bucking and will keep doing it. A confirmed bucker may need professional training or should be sold, but a horse that bucks only occasionally can be retrained by a good rider.

Be Prepared

If you are always alert and in control, you can generally keep your horse from bucking or at least from bucking hard. He must be able to get his head down at least to knee level to perform a powerful buck; if you have contact with his mouth and control of his head, you'll be able to ward off a buck.

The best way to keep a horse from bucking is to pull his head quickly around to the side so he can't get a strong downward pull on the reins. A horse can't buck while circling. If you're unsure about your ability to keep his head up, ride with his halter rope tied to the saddle horn. Have just enough slack in the rope to give him freedom of head for traveling normally but not enough to get his head down far enough to buck. Usually, after a horse is continually thwarted in his attempts to buck, he'll stop trying, but some must always be ridden with extra vigilance so that the rider is never caught off guard.

EPILOGUE

The fundamentals of training a horse are simple: Train him right the first time. Build step-by-step on previous lessons. The good trainer starts slowly and acquaints the horse with new experiences little by little, moving ahead as the young horse gains trust and confidence.

Creating a good foundation for future training is like building a house. You want it solid and strong. If you rush through the basics, the result will be shaky and won't hold up under pressure.

Most of the problems with horses — the spoiled horses, the ones with bad attitudes and bad habits — hark back to their early experiences. If you take the time to start a young horse properly, you'll encounter far less trouble later on.

When you develop a good partnership with your horse, together you can handle any unexpected situations that might come along. Teamwork enables both of you to overcome any challenges with a strong, secure relationship and greater trust and confidence in one another. And that, in essence, is the hallmark of a well-trained horse.

APPENDIXES

A. Equine Anatomy

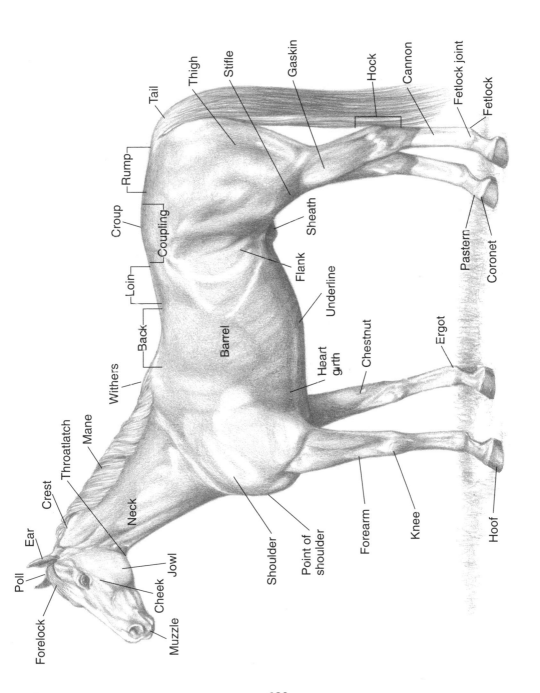

B. Gait Sequences

The walk. *The legs come to the ground in sequence for a four-beat gait — right hind (1), right front (2), left hind (3), left front (4). As a hind comes forward, the front foot on that side prepares to leave the ground and takes off a split second before that hind foot lands.*

The trot. *This is a two-beat gait, with the horse's legs moving in diagonal pairs — right front and left hind (1) moving in unison, and left front and right hind (2) hitting the ground together. During the fast trot, there is a moment of suspension in which all four feet are in the air, but during the slow trot one diagonal pair is striking the ground as the other diagonal pair is pushing off.*

The canter (left lead). *The canter is a three-beat gait. After the brief moment of suspension when all four feet are off the ground, the first leg to land will be the right hind (1), then the left hind (2) will come to the ground, leading, and landing at the same time as the right front (2), the second beat of the cadence. Then the left front (3) will come to the ground, leading. When it leaves the ground again, all four feet will be briefly in the air. At a slow canter, the horse has one or three feet on the ground at once, except for the moment of suspension.*

The canter (right lead). *If a horse is cantering on the right lead, his left hind leg lands first (1), then his right hind comes to the ground farther forward and loads (2), landing simultaneously with the left front (2) for the second beat of cadence. Then the right front reaches farther forward and comes to the ground for the third beat (3), leading. When the right front is lifted again, there is a brief moment of suspension.*

The gallop (left lead). *The fast gallop is a four-beat gait. There are two feet on the ground at once (both hinds, then the leading hind and nonleading front, then both fronts), but each foot hits the ground separately. In the left lead, the right hind (1) comes to the ground first, then the left hind (2), followed by the right front (3) and then the left front (4), followed by a brief moment of suspension — then the sequence repeats.*

C. RESOURCES

Magazines

If you are interested in training, driving, or a certain style of riding, you may want to subscribe to a magazine that specializes in a particular aspect of horsemanship. There are also many good articles that deal with various methods and aspects of training. Here are a few of the many magazines that occasionally publish training articles.

The American Quarter Horse Journal
1600 Quarter Horse Dr.
Amarillo, TX 79104
www.aqha.com
Mainly Western riding and performance sports

America's Horse/AQHA
P.O. Box 200
Amarillo, TX 79168
Western riding and performance

Arabian Horse Times
P.O. Box 1736
Waseca, MN 56093
www.ahtimes.com
Arabians; breeding and training

The Chronicle of the Horse
P.O. Box 46
Middleburg, VA 20118
www.chronofhorse.com
Hunting, jumping, eventing, dressage

Dressage Today
656 Quince Orchard Road #600
Gaithersburg, MD 20878
www.equisearch.com/dressagetoday
Dressage competition

Driving Digest
P.O. Box 110
New London, OH 44851
www.drivingdigest.com
Driving in harness

Driving West Magazine
P.O. Box 395
Jamul, CA 91935
Driving in harness

Equine Journal
103 Roxbury Street
Keene, NH 03431
800-742-9171
www.equinejournal.com
Mostly English riding

The Equine Journal
P.O. Box 5299
Laguna Beach, CA 92651
949-376-4900
Hunter, jumper, dressage, eventing

Equus
656 Quince Orchard Road #600
Gaithersburg, MD 20878
www.equisearch.com/equus
Horse health care and handling

Flying Changes
19502 NE 134th Place
Battle Ground, WA 98604
360-687-0203
www.flyingchanges.com
Sport horses

Horse & Pony
P.O. Box 2050
Seffner, FL 33584
General-interest horse subjects

Horse & Rider
P.O. Box 4101
Golden, CO 80401
Western riding

Horse Illustrated
P.O. Box 6050
Mission Viejo, CA 92690
Horse care, horse subjects

The Horsemen's Voice
5805 Warren SW
Albuquerque, NM 87105
505-873-0150
www.horsemensvoice.com
Horse care, events

Horsemen's Yankee Pedlar
83 Leicester Street
North Oxford, MA 01537
508-987-5886
www.pedlar.com
Mostly English riding

John Lyons Perfect Horse
P.O. Box 5656
Norwalk, CT 06854
800-424-7887
www.perfecthorse.com
Horse handling and training

Performance Horse
P.O. Box 9768
Ft. Worth, TX 76147
www.performancehorse.com
Cutting and reining horses

Practical Horseman
P.O. Box 420235
Palm Coast, FL 32142-0235
English riding

Western Horseman
P.O. Box 7980
Colorado Springs, CO 80933
www.westernhorseman.com
Horse care, Western riding

Books

You'll find books on training horses in most libraries, or you can order from catalogs. Books are also advertised in horse magazines. Following is a list of just a few good titles.

Brainard, Jack. *Western Training.* Colorado Springs, Colo.: Western Horseman, 1989.

Camarillo, Sharon. *Barrel Racing.* Colorado Springs, Colo.: Western Horseman, 2000.

Dunning, Al. *Reining.* Colorado Springs, Colo.: Western Horseman, 1996.

Fielder, Paul. *All About Long Reining.* North Pomfret, Vt.: Trafalgar, 2000.

———. *All About Lungeing.* North Pomfret, Vt.: Trafalgar, 2000.

Harris, Susan. *The UPSC Guide to Longeing and Ground Training.* Indianapolis: Howell, 1997.

Herbermann, Erik. *The Dressage Formula.* North Pomfret, Vt.: Trafalgar, 2000.

Hill, Cherry. *Arena Pocket Guides* (English and Western series). North Adams, Mass.: Storey Publishing, 1998.

———. *101 Arena Exercises.* North Adams, Mass.: Storey Publishing, 1995.

———. *101 Horsemanship & Equitation Patterns.* North Adams, Mass.: Storey Publishing, 1999.

———. *Horsekeeping on a Small Acreage,* Second Edition. North Adams, Mass.: Storey Publishing, 1990.

———. *Horse Care for Kids.* North Adams, Mass.: Storey Publishing, 2002.

Kevil, Mike. *Starting Colts.* Colorado Springs, Colo.: Western Horseman, 1990.

Kidd, Jane. *Dressage Essentials.* Indianapolis: Howell, 1999.

———. *A Young Person's Guide to Dressage.* North Pomfret, Vt.: Trafalgar, 1999.

Loriston-Clarke, Jennie. *Lungeing and Long-Lining.* Boonsboro, Md.: Half Halt Press, 1994.

———. *The Young Horse.* North Pomfret, Vt.: Trafalgar, 1999.

Lyons, John. *The Making of a Perfect Horse.* Greenwich, Conn.: Belvoir, 1998.

———. *Private Lessons: John Lyons Answers Questions about Care and*

Training. Greenwich, Conn.: Belvoir, 2000.

Martin, Marty. *Problem Solving*. Colorado Springs, Colo.: Western Horseman, 1998.

Miller, Robert M. *Imprint Training*. Colorado Springs, Colo.: Western Horseman, 1991.

———. *Understanding the Ancient Secrets of the Horse's Mind*. Neenah, Wisc.: R. Meerdink, 1999.

O'Connor, Sally. *Commonsense Dressage*. Boonsboro, Md.: Half Halt Press, 1990.

Parelli, Pat, and Kathy Kadash. *Natural Horse-man-ship*. Colorado Springs, Colo.: Western Horseman, 1999.

Podhajsky, Alois. *The Complete Training of Horse and Rider*. Manhattan Beach, Calif.: Wilshire, 1982.

———. *My Horses, My Teachers*. North Pomfret, Vt.: Trafalgar, 1997.

Richter, Judy. *Riding for Kids*. North Adams, Mass.: Storey Publishing, 2003.

Savoie, Jane. *Cross-train Your Horse*. North Pomfret, Vt.: Trafalgar, 1998.

Shrake, Richard. *Western Horsemanship*. Colorado Springs, Colo.: Western Horseman, 1987.

Strickland, Charlene. *Western Practice Lessons*. North Adams, Mass.: Storey Publishing, 2000.

———. *Western Riding*. North Adams, Mass.: Storey Publishing, 1995.

Mail-Order Suppliers

There are many, many training videos available from numerous sources, including mail-order catalogs. Most of these companies will provide catalogs on request.

American Livestock Supply
613 Atlas Ave.
Madison, WI 53714
800-356-0700
www.americanlivestock.com
Horse products catalog

Bit of Britain Saddlery
141 Union School Rd.
Oxford, PA 19363
800-972-7985
www.bitofbritain.com
Horse tack

Dover Saddlery
P.O. Box 1100
Littleton, MA 01460
800-406-8204
www.doversaddlery.com
Horse tack and supplies

Freedom Rider
P.O. Box 4187
Manchester, NH 03108
888-253-8811
www.freedomrider.com
Horse tack and supplies (including Western safety helmet)

Libertyville Saddle Shop
306 Peterson Rd., Hwy. 137
Libertyville, IL 60048
800-872-3353
www.saddleshop.com
Horse tack and supplies

Miller's
P.O. Box 406
Westford, MA 01886
800-784-5831
www.millersharness.com
Horse tack and supplies

Nasco
4825 Stoddard Road
Modesto, CA 95356-9318
800-558-9595
www.enasco.com
Horse supplies

Professional's Choice
2709 Via Orange Way
Spring Valley, CA 91978
800-331-9421
www.profchoice.com
Horse tack

State Line Tack
P.O. Box 910
Brockport, NY 14420-0935

800-228-9208
www.statelinetack.com
Horse tack and supplies

Supracor Equestrian
2050 Corporate Court
San Jose, CA 95131
888-924-6773
www.supracor.com
Saddle pads

United Vet Equine
14101 West 62nd Street
Eden Prairie, MN 55346
800-328-6652
www.unitedvetequine.com
Horse equipment, medication, and supplies

Valley Vet Supply
1118 Pony Express Hwy
Marysville, KS 66508
800-419-9524
www.valleyvet.com
Horse equipment, medication, and supplies

Western Ranch Supply
P.O. Box 1497
Billings, MT 59103
800-548-7270
www.westernranchsupply.com
Horse equipment, medication, and supplies

Training Clinics

Clinics have been popular since the 1980s. Clinics are training demonstrations given by professionals in various fields of horsemanship to help educate both horse and rider. Some are lectures and demonstrations; some are hands-on training sessions in which the participants bring their own horses. If someone whose style of training you admire puts on a training clinic in your area, take advantage of it. There is always more to be learned about training horses.

Associations

If you are interested in a particular sport, the parent association can often direct you to local clubs and groups that can be helpful in furthering your own and your horse's education in that sport. Following is a partial listing of associations. (Additional listings can be found in the *American Horse Council Horse Industry Directory*, which you can obtain from the American Horse Council, 1700 K Street, NW, Suite 300, Washington, DC 20006-3805; 202-296-4031.)

American Driving Society
2324 Clark Road
Lapeer, MI 48446
810-664-8666
www.americandrivingsociety.com

American Endurance Ride Conference
P.O. Box 6027
Auburn, CA 95604
530-823-2260
www.aerc.com

American Hunter and Jumper Foundation
P.O. Box 369
West Boylston, MA 01583
508-835-8813
www.ahjf.com

National Cutting Horse Association
260 Bailey Avenue
Fort Worth, TX 76107
817-244-6188
www.nchacutting.com

National Reining Horse Association
3000 NW 10th Street
Oklahoma City, OK 73107-5302
405-946-7400
www.nrha.com

United States Dressage Federation
220 Lexington Green Circle, Ste. #510
Lexington, KY 40503
859-971-2277
www.usdf.org

United States Eventing Association, Inc.
525 Old Waterford Road NW
Leesburg, VA 20176-2050
703-779-0440
www.eventingusa.com

GLOSSARY

aid The rider's voice, reins, legs, and weight, used as cues. "On the aids" refers to a horse being "on" the bit and perfectly controlled by bit and leg pressure — able instantly to respond.

animated Lively; moving with energy and greater action.

balk To refuse to go forward.

barn sour Reluctant to leave the barn or pen.

bars Interdental space in the mouth where there are no teeth; the bare gums, consisting of skin over the bone, where the bit rests.

"behind the bit" Term used to describe when the horse avoids bit pressure by tucking his chin excessively.

bellyband Surcingle; part of the harness that goes around the girth.

billet Leather piece that attaches the girth to the saddle.

bit Metal mouthpiece of a bridle.

bitting Teaching the horse to carry a bit, yield his jaw, and be responsive.

bitting harness Harness used for getting the horse accustomed to the bit.

blinkers Also called *blinders* or *winkers*, these protrusions from the driving bridle block the horse's side and backward vision so he can look only straight ahead.

body language The way a rider or horseman moves or stands, conveying attitude and intent.

body rope Rope around the horse's girth; the free end comes through the halter ring for tying.

bolt To run away; also implying hard to stop.

bosal Braided rawhide noseband on a hackamore.

breast collar Strap around the lower part of the neck in front of the shoulder and attached to the saddle to hold it in place.

breastplate Leather strap around the lower part of the neck in front of the shoulder; part of a driving harness.

breeching Part of the harness that goes around the hindquarters.

bridle Headgear for controlling the horse, consisting of headstall, bit, and reins.

bridoon Snaffle bit with small rings.

broke Trained and reliable.

brow band Leather strap of bridle that goes around the horse's forehead.

cadence Rhythm of hoofbeats.

canter Three-beat gait; collected gallop.

cantle The back of the saddle seat.

capriole Classical dressage movement in which the horse springs into the air and thrusts out his hind legs. Originally a military maneuver used against foot soldiers.

carriage (head) How the horse carries his head (e.g., high, low, nose down or up).

cart Two-wheeled vehicle pulled by a horse.

cavalletti Series of poles on the ground, followed by a jump standard, for teaching a horse how to space his strides.

cavesson Noseband attached to a headstall. A cavesson used for longeing has rings for the longe line or long reins and a jaw strap to keep the cheek pieces from getting in the horse's eyes.

change of lead To change from one leading set of legs (at the canter or gallop) to the other.

charging Rushing forward.

check To contact the mouth with the bit; to cue the horse to slow or stop or pay attention.

checkrein The strap to the bit on a harness or bitting harness that holds the horse's head in a certain position.

cheek Side piece of a headstall or side piece of a bit.

chin strap *See* **curb strap.**

chute Narrow alleyway for confining an animal.

cinch ring Metal ring on the end of a cinch that fastens to the saddle billet or latigo.

closed bridle Driving bridle with blinkers.

clucking Making a kissing or clicking sound to encourage the horse to move or move faster.

coldbacked Resentful when saddled; the horse humps his back and protests, especially when the cinch is tightened.

collar Neck piece on a harness that rests on the shoulders; the horse leans into it for pulling.

collected Moving with animation, impulsion, and elevated action; shortened frame.

conformation How the parts of a horse's body are put together.

contact Bit contact; constant communication via the reins.

courbette Classical dressage movement in which the horse rises on his hind legs, folds his front legs, and hops on his hind legs.

crop Short riding whip.

cross-canter Cantering on one lead in front and the opposite lead behind; *see also* disunited.

croup Highest point of the rump.

crownpiece Top part of the headstall of a bridle.

crupper The part of a harness that runs from the back pad under the dock of the tail.

curb Bit with shanks. Puts pressure on the mouth, chin groove, and top of the head when reins are pulled due to leverage action.

curb chain, curb strap Fastened to the top part of a curb bit, this puts pressure on the chin groove when the bit is tipped by a pull on the reins.

cutting Sport competition in which horses are judged for their ability to handle and outmaneuver a cow — sorting her out of the herd and keeping her from rejoining it.

D-ring Metal ring on the saddle to which the leather is attached for holding a cinch.

dally   A loop of rope around a post; to loop a rope around a post, pole, or saddle horn.

dam   Mother of a horse.

diagonals   At the trot, the horse's feet move in diagonal pairs; the rider posts on the left or right diagonal.

direct-rein   To turn a horse with a pull on one rein (to the side), pulling his head around to that side.

disunited   Cantering on one lead in front and the opposite lead behind; cross-cantering.

double   To bend the horse sharply and pull him around so that he is facing in the opposite direction.

draw reins   Reins that run through the bit and attach back to the saddle, giving extreme leverage action.

dressage   Highest form of training and horsemanship. Teaches the horse to be totally responsive, via subtle cues, to the rider.

driving   Controlling the horse with a bit and long lines (as when he is pulling a vehicle) rather than riding him.

driving in long lines   Ground training in which the horse is controlled by a bit and long lines while walking behind him.

extended   Moving with a lengthened frame; uncollected.

flex   To bend at the poll and yield to bit contact; to bend the body to the curve of a circle.

flying lead change   To change leads at the canter, lope, or gallop without slowing to a trot.

forehand   Front end of the horse; shoulders and front legs.

gallop   Fastest gait of most horses.

Gee   Verbal cue to turn right.

gentle   "Broke" or well trained; to train a horse.

girth   Strap under the belly to hold an English saddle in place. Also, the body area of the horse where this goes.

good hands   Light contact on the mouth at all times, never bumping the horse in the mouth.

green horse   Inexperienced, untrained horse.

groom   To brush and clean a horse.

hackamore   Bitless bridle that uses a bosal for control.

hair coat The horse's body hair.

half-halt Temporary pressure on the bit (quick pull and release) to signal him to slow down or pay attention.

half-pass Lateral movement in which the horse travels sideways as well as forward, with his body curved in the direction he is going.

halter-break To teach the horse to lead and tie.

hame The frame that holds the driving collar in place; used in conjunction with a driving collar.

haunches Hindquarters.

Haw Verbal command to turn left.

head a cow Horse and rider move into position to turn her in the proper direction.

head shy Insecure about having anything approach the head; not wanting the head touched.

headstall Headpiece that holds a bit in place.

heat Estrus; period in mare's estrous cycle when she is, or is becoming, receptive to the stallion.

heavy hands No flexibility; pulling or bumping the mouth.

herd-bound Reluctant to leave the security and companionship of other horses.

hobbles Leather, rope, or cloth loops used on legs for restraint.

hock Large joint on the hind leg, above the cannon bone.

horse psychology Understanding the mind, emotions, and social nature of horses.

horse sense Understanding how a horse thinks; being able to predict what he might do in certain situations.

imprint training Handling a foal at birth to get him accustomed to touch and to manipulation of all body parts so he will later be submissive under human handling.

impulsion Forward movement and energy.

incisors Front teeth.

indirect rein Rein opposite the direction the horse is turning. When neck-reining, the indirect rein is used, pressed against the horse's neck.

interdental space The area between the incisors (front) and molars (back) where there are no teeth.

jigging Prancing and fretting, refusing to settle down and walk.

jog Slow and unanimated trot.

joining up Bonding with the trainer; becoming willing to respond to the trainer.

lariat Small-diameter hard-twist rope used for roping cattle.

lash The long, flexible end of a whip.

lateral Sideways movement.

latigo Leather strap that fastens the cinch to the saddle.

leads At the canter or gallop the horse leads with the legs of one side or the other, the legs on that side going farther forward.

legged up Conditioned, improved muscle fitness developed by gradually increasing the horse's work.

leg yield Maneuver in which the horse travels sideways and forward, with his head and neck curved slightly away from the direction of travel.

levade Classical dressage movement in which the horse crouches on his hind legs and rears slowly.

light hands Maintaining constant light contact with the bit; *see also* good hands.

light-mouthed horse A horse with a sensitive mouth; very responsive to the mildest bit pressure. A horse that is very responsive to a slight signal with the bit.

locking up "Freezing"; unresponsive to cues. Also, locking the hind legs for a sliding stop.

longe (also spelled **lunge**) To exercise or train a horse in a large circle, controlled by a long line held by the trainer standing at the circle's center.

long line Long rein for ground-driving or for longeing a horse.

lope Western term for the canter or slow gallop.

lug To pull against the bit.

martingale Combination of straps attached to the reins (*running martingale*) or noseband (*standing martingale*) at one end and the saddle girth at the other, to keep the horse from tossing his head or carrying it too high.

mecate Braided horsehair reins tied to a bosal on a hackamore.

mullen mouth Straight mouthpiece; no port, no joint.

"muscle memory" Having done a maneuver once, it becomes easier for the horse not only mentally but also physically.

navicular Small bone inside the hoof; also a term for a disease of this bone, often caused by too much concussion on hard surfaces.

near-side Left side of the horse.

neck-rein To turn the horse by use of the indirect rein, moving it across his neck.

noseband The strap of a halter or bridle that goes across the horse's nose.

off-side Right side of the horse.

"on the bit" Willingly responding to constant but gentle bit pressure, with relaxed and yielding jaw; constant bit contact.

open bridle Bridle without blinkers.

overcheck Cord running from the bit to the top of the headstall, then back to the harness (or saddle), to keep the horse from putting his head down.

Paso A breed of gaited horses from Central and South America.

passage Dressage movement; slow to medium trot, highly collected.

Pelham Bit with both curb and snaffle action (curb mouthpiece with shanks and snaffle rings).

piaffe Dressage movement; prancing in place.

pirouette Dressage movement; galloping in a small circle with hind legs practically stationary and front legs moving around them.

poll Top of the horse's head; junction of head and neck.

pommel Front of the saddle.

pony horse The horse being ridden when leading another horse alongside.

ponying To lead one horse from another.

posting Rising to one beat of the trot and staying out of the saddle for the next; rising (or sitting) as the same front foot comes to the ground.

quarters Hindquarters

rack One of the gaits of a five-gaited horse; a swift four-beat gait (also called a *singlefoot*), faster than a walk, and with more up-and-down motion of the legs.

rail   Fence line; also term for a pole.

rein back   To back the horse by first pushing him into the bit with leg pressure, holding the bit steady with fixed hands rather than pulling on it. Leg pressure tells him to move, and because he cannot move forward, he moves backward.

reining   Sport competition in which horses are judged by their ability to do various maneuvers at speed.

reins   Lines, usually leather, attached to a bit, used to give signals to the horse.

rigging   Straps connecting the cinch to the saddle.

rollback   Maneuver in which a horse stops quickly from a gallop, pivots on the hind legs, and again takes off at a gallop in a new direction.

rowel   Small wheel (often with points) attached to spur shank.

rubberneck   To turn the head in response to rein pressure without turning the body.

sacking out   Getting the horse accustomed to various things touching all parts of his body.

serpentine   Winding pattern or loops or half-circles.

set back   To pull back strongly on the halter rope when the horse is tied.

shafts   The small-diameter poles attached to a cart, between which the horse is positioned for pulling.

shank   The portion of a spur that sticks out past the rider's heel.

shanks   The lower portion of the cheek pieces of a curb bit or Pelham.

shoulder-in   Dressage movement. Horse travels sideways as well as forward, his whole body flexed away from the direction of travel.

shy   To leap sideways or spin away from a frightening object.

sidepass   To move sideways, crossing one leg over the other.

side piece   Cheek piece of a halter, bridle, or a bit.

singletree   Spacer bar attached to a cart or wagon, behind the horse, to which the harness traces are attached.

snaffle   Bit with rings attached at the mouthpiece. Rein pressure is directly to the back or side, putting pressure on the tongue, bars, and mouth corners.

splint Bony enlargement on the cannon bone due to injury, often caused by concussion on hard surfaces or striking it with a foot.

spook To shy or jump away from something frightening.

stifles Large joints at the top of the hind legs.

stirrup bows Support for the rider's feet, in a Western saddle.

stirrup irons Support for the rider's feet, in an English saddle.

stirrup leather Straps attaching stirrup irons or bows to the saddle.

supple Flexible.

surcingle A broad girth that goes around the horse; *see also* bellyband.

tack Equipment used on the horse.

tapadero Leather hood over the front of a stirrup bow.

throatlatch Strap of the bridle that goes under the throat to keep the horse from rubbing off the headstall. Also, this area of the horse's head.

tie-down Strap connecting the noseband to the cinch or breast collar, to keep a horse from getting his head too high.

traces The portion of the harness (leather straps or chains) that attach to the singletree to pull the vehicle.

tree Inside foundation of a saddle, traditionally made of wood but now sometimes fiberglass or other materials, upon which the leather is attached.

trot Two-beat, diagonal gait.

tug Portion of the harness at the rear of the horse that attaches to the vehicle.

two track To travel sideways as well as forward.

unanimated Moving with little energy or extra motion.

walk Slowest gait of the horse, a four-beat gait.

war bridle Cord running through the mouth and over the poll; when tightened, it is a means of restraint.

"way of going" How the horse handles himself, or how he moves himself, at various gaits.

Whoa Command used to tell the horse to stop.

INDEX

Note: Numbers in *italics* indicate illustrations.

OTHER STOREY TITLES YOU WILL ENJOY

The Horse Doctor Is In by Brent Kelley. Combining solid veterinary advice with enlightening stories from his Kentucky equine practice, Dr. Kelley informs readers on all aspects of horse health care, from fertility to fractures to foot care. 416 pages. Paperback. ISBN 1-58017-460-4.

Horse Handling and Grooming by Cherry Hill. This user-friendly guide to essential skills includes feeding, haltering, tying, grooming, clipping, bathing, braiding, and blanketing. The wealth of practical advice offered is thorough enough for beginners, yet useful for experienced riders improving or expanding their horse skills. 160 pages. Paperback. ISBN 0-88266-956-7.

Horse Health Care by Cherry Hill. Explains bandaging, giving shots, examining teeth, deworming, and preventive care. Exercising and cooling down, hoof care, and tending wounds are depicted, along with taking a horse's temperature and determining pulse and respiration rates. 160 pages. Paperback. ISBN 0-88266-955-9.

Horsekeeping on a Small Acreage by Cherry Hill. The essentials for designing safe and functional facilities, whether you have one acre or one hundred. 320 pages. Paperback. ISBN 1-58017-535-X. Hardcover. ISBN 1-58017-603-8.

Horse Sense by John J. Mettler Jr., DVM. Invaluable advice on preventive health care for horses. Explains how common and serious equine diseases, including colic, choke, and founder, can be prevented. 160 pages. Paperback. ISBN 0-88266-545-5.

101 Arena Exercises by Cherry Hill. Suitable for both English and Western Riders, this classic best-selling equestrian workbook presents both traditional and original exercises, patterns, and maneuvers. Can be hung on the wall or on an arena fence for easy reference. 224 pages. Paperback with comb binding. ISBN 0-88266-316-X.

101 Horsemanship & Equitation Patterns by Cherry Hill. This sequel to *101 Arena Exercises* is a compendium of the essential patterns for Western Horsemanship and English Equitation competition. 256 pages. Paperback with comb binding. ISBN 1-58017-159-1.

101 Jumping Exercises for Horse & Rider by Linda L. Allen with Dianna R. Dennis. This must-have workbook, the newest addition to Storey's successful "read-and-ride" series, provides a logical and consistent series of exercises with clear maps and straightforward instructions. 240 pages. Paperback with comb binding. ISBN 1-58017-465-5.

Storey's Guide to Raising Horses by Heather Smith Thomas. A comprehensive guide to facilities, feeding and nutrition, daily health care, disease prevention, foot care, dental care, selecting breeding stock, foaling, and care of the young horse. 512 pages. Paperback. ISBN 1-58017-127-3.

Storey's Horse Lover's Encyclopedia edited by Deborah Burns. This hefty, fully illustrated, comprehensive A-to-Z compendium is an indispensable answer book addressing every question a reader may have about horses and horse care. 480 pages. Paperback. ISBN 1-58017-317-9.

These and other books from Storey Publishing are available wherever quality books are sold or by calling 1-800-441-5700. Visit us at www.storey.com.